ATOMS AND INFORMATION THEORY

ATOMS
AND
INFORMATION THEORY

AN INTRODUCTION TO STATISTICAL MECHANICS

Ralph Baierlein
WESLEYAN UNIVERSITY

W. H. FREEMAN AND COMPANY
San Francisco

Printed in the United States of America
Library of Congress Catalog Card Number: 71–116369
International Standard Book Number: 0–7167–0332–7

1 2 3 4 5 6 7 8 9 10

CONTENTS

PREFACE

Two modest objectives led me to write this book. My primary objective was to provide a specific alternative to the conventional developments of statistical mechanics, an alternative based on Richard Cox's contribution to probability theory, Claude Shannon's development of information theory, and Edwin Jaynes' view of statistical mechanics.

To outline that alternative, I must start with the meaning adopted here for the notion of probability: a probability expresses numerically the degree of belief to which an inference is rationally entitled on the available evidence, that evidence being (typically) sufficient only to render the inference "likely" to a greater or lesser extent. This is surely the sense in which one talks about the probability of there being life on Mars, and it seems eminently suited to statistical mechanics, where the severe limitations on the information available compel one merely to estimate the behavior of the physical system under consideration. The adopted meaning is, to be sure, a minority view in contemporary probability theory, but it is certainly not a radical innovation, for it is very much the view held by Laplace, John Maynard Keynes, Erwin Schrödinger, and Sir Harold Jeffreys. Richard Cox has made that general view, plus two intuitively reasonable axioms, the foundation for a compelling development of probability theory as a theory of reasoning from inconclusive evidence.

It is desirable, however, to go beyond Professor Cox's little book, *The Algebra of Probable Inference*, and to choose general criteria for the assignment of probabilities. Shannon's information measure, that remarkably rich function of the probabilities, $-K \Sigma P_j \ln P_j$, enters at this point. It is used to construct the second of two simple criteria for the assignment of probabilities, a decision cogently advocated by Edwin Jaynes. One has then established a kind of estimation theory— a general procedure for making assessments from data too meager to permit exact prediction, even in principle—as the framework for the development of statistical mechanics proper.

That development leads, in Chapter 4, to the canonical probability distribution, and most of the book is devoted to examining its implications: in the quantum and classical forms, with and without the neglect of interparticle forces, for identical fermions and bosons, and so on. Thus the actual results in statistical mechanics are the same as those derived by the more conventional approaches, and not surprisingly so, for the conventional "fundamental postulates of statistical mechanics," such as Richard Tolman's, are special cases of the criteria chosen in Chapter 3. What merits, then, might I claim for this alternative?

The merits most readily pointed out are the naturalness of the probability concept and its appropriateness for statistical mechanics, and then, when confronting the perennial difficulty of assigning probabilities, the explicitness of the approach and the generality of the solution. More subtly, some of the mystery surrounding statistical mechanics is dispelled when one starts, as here, with the candid recognition that one is merely estimating the behavior of the physical systems under consideration. The ideas developed in Chapters 2 and 3 do, moreover, carry far beyond their use in statistical mechanics.

My secondary objective in writing the book was to develop statistical mechanics, and to apply the theory, without an admixture of thermodynamic arguments. Statistical mechanics can stand by itself, and the structure of the theory is most clearly seen when it does stand alone. I certainly concede that a mixture of statistical mechanics and thermodynamics often provides the easiest approach to the solution of a difficult problem. Landau was a master of the technique, but what is a powerful tool in the hands of a master may be a mystery for the apprentice. The conceptual bases of the two disciplines are vastly different—thermodynamics is inherently macroscopic and deterministic, whereas statistical mechanics is inherently microscopic and probabilistic—and so a premature mixing may be the source of much confusion and grief. If one first sees each by itself, then the distinctions will be as evident and as meaning-ful as the similarities. The connection between statistical mechanics and thermodynamics is made in an extensive appendix (and there Shannon's information measure helps us to understand the behavior of thermodynamic entropy, another merit of this alternative).

A few words should be said about the background in physics that the reader is assumed to have. The book was written with a one-semester course for college seniors and first-year graduate students in mind. It is expected that the reader knows a bit of kinetic theory, the perfect gas law, and elementary quantum mechanics. The amount of the last needed is indeed modest: (a) a familiarity with the Heisenberg uncertainty principle, (b) the idea that quantum mechanics describes physical systems in terms of wave functions or quantum states, and (c) a nodding acquaintance with the Schrödinger equation. The states and energies of a particle in a cubical box are used ex-tensively, but they are presented in detail and are discussed. Wave functions for two or more identical particles are constructed when needed. Although the development of statistical mechanics is here explicitly based on quantum mechanics, with the classical version being casually derived by a limiting argument, nothing more is required than the modest background outlined here. (The last chapter is an exception, but even there the outline of the needed quantum-mechanical principles in the first section should make the chapter accessible to a reader with only the background described here.)

To the many students in my classes who read early versions of the manu-script and gave me their criticisms and suggestions, I extend warm thanks. John Fox and Irvin Winer must be singled out for special appreciation. To my colleague Richard Lindquist go thanks for help in clarifying and tightening the presentation. And lastly I thank my wife, Jean, for her words of good sense and for cheerfully enduring.

Oxford, 1970

RALPH BAIERLEIN

ATOMS AND INFORMATION THEORY

1

THE AIM OF
STATISTICAL MECHANICS

[Statistical mechanics provides] the methods that must be employed when
we wish to predict the behaviour of a mechanical system on the basis
of less knowledge as to its actual state than would in principle be
allowable or possible. Such partial knowledge of state is
in reality all that we ever do have, and the discipline of
statistical mechanics must always remain necessary.

Richard C. Tolman
The Principles of Statistical Mechanics

We can best appreciate what statistical mechanics is about by confronting the
need in physics that it satisfies. So let us turn at once to a problem, homely
and perhaps overworked, but here familiarity will be an asset.

Suppose we are presented with a box filled with a dilute gas—helium, say—
and are told a few things about the container and the gas. We are informed of
the total mass of the gas and the mass of a single helium atom, the behavior
of the interatomic forces, and the volume of the cubical box. The gas has
remained undisturbed for a considerable period of time. Finally, we are told
that the total energy of the gas is somewhere close to a particular value $\tilde{E}$; to
be more precise, let us suppose the energy is known to be within 1 per cent of
that value. All the preceding is reasonable enough, but the question we are now
asked is perhaps not so reasonable: given this information at a time t_1, what
is the pressure exerted by the gas on the box walls at a specified later time t_2?

1

1.1. THE CLASSICAL TREATMENT

If we approach the stated problem in terms of classical physics—we will discuss the quantum-mechanical approach later—we visualize a host of small spheres, say 10^{20}, in some sort of very complicated motion. To answer the question directly, we would have to calculate which atoms actually strike the box walls at (or around) the specified time t_2. If we knew the velocity of each atom just before it hit the wall and assumed that each bounced off without any change in its energy, that is, assumed that the sole effect of the collision with the wall was to reverse the momentum component perpendicular to the wall, then we would be very close to an answer. (We might have to insist that we be entitled to time-average over the very small time duration of a collision.) In order to calculate an answer in this manner, by using Newton's laws of motion, we would need to know the position and velocity of all 10^{20} atoms at the initial time t_1. Quite obviously such knowledge has not been given us. The very few pieces of information we have been granted are totally inadequate for such a direct analysis. If we want at least some kind of answer to the question, we must adopt a different approach. To make up for the paucity of the information we have been given, we must engage in some reasonable guessing, in an attempt to arrive at an answer that ought to be close to the truth.

For the sake of simplicity, let us momentarily neglect the interatomic forces. Then the atoms move freely except for their collisions with the walls. The symmetry of the cubical box suggests that we may then imagine that, in effect, one-third of the atoms travel straight back and forth between a particular wall and the opposite one. This is analytically convenient, plausible, and certainly not ruled out by the information given to us. The remaining two-thirds of the atoms travel between the other walls, and we can forget about them for the moment. Now we pick an atom from our chosen one-third and denote its speed by v. If the length of an edge of the cube is L, then in a time interval Δt, the atom makes $v\Delta t/2L$ round trips. At each collision the atom imparts to the particular wall a momentum $2mv$, the factor of 2 arising because its momentum is first reduced to zero and then increased to mv in the opposite direction. Hence the momentum per unit time that the chosen atom imparts to the particular wall is

$$(2mv) \times \left(\frac{v\,\Delta t}{2L}\right) \times \frac{1}{\Delta t} = (\tfrac{1}{2}mv^2) \times \left(\frac{2}{L}\right).$$

This is the time average—over many round trips—of the force that the atom exerts. We should now sum a result of this form over our chosen $\frac{1}{3}N$ of the total number N of atoms, and then divide by the wall area L^2 to get the pressure.

By now we are pretty close to some kind of reply to the original question. We need only relate the kinetic energy, $\frac{1}{2}mv^2$, summed over $\frac{1}{3}N$ of the atoms, to the given total energy of the gas. Our information provides us with no definite relation, but it also suggests no reason why we should not use a symmetry argument and choose $\frac{1}{3}\tilde{E}$ as the most reasonable answer. Our full result for the estimated pressure p_{est} on the particular wall is then

$$p_{\text{est}} = (\tfrac{1}{3}\tilde{E}) \times \left(\frac{2}{L}\right) \times \left(\frac{1}{L^2}\right) = \frac{2}{3}\frac{\tilde{E}}{V},$$

with $V = L^3$ being the volume. Since there is no reason to believe that this one particular wall is in any way preferred, this result is also our estimate for any wall.

When we take into account the specified uncertainty in the total energy, we can write our estimate of the pressure as

$$p_{\text{est}} = \frac{2}{3}\frac{\tilde{E}}{V} \pm 0.01\left(\frac{2}{3}\frac{\tilde{E}}{V}\right). \tag{1.1.1}$$

This is only a guess for what the actual pressure on a particular wall at the specified time will be. But, given nothing more than the meager information provided us, this is about the best guess we can make. A more sophisticated distribution of atomic velocities, such as all directions being equally probable, leads, after more algebra, to the same result. And that computation, too, is only a guess, though we are apt to ascribe to it a higher probability of being close to the unknown true situation. Only if we begin to use our knowledge about the interatomic forces can we produce any significant corrections to the preceding estimate. We need not, however, bother with that here, for we can now make the point this little exercise has been leading up to.

We have arrived at an answer to the original question by a process of reasoned guessing. The answer is almost certain not to be exactly correct, but it is also unlikely to be very far from the truth. The essence of the situation is that we are *compelled* to merely estimate the answer. We are compelled, *not* because the direct approach with Newton's laws of motion is too complex, but because we do not possess the information, the atoms' initial positions and velocities, that we need in order to use it. Both for later times and for the initial time we lack the data needed for an exact prediction of the pressure.

Although complexity is not here the crucial element, it is often a formidable obstacle to the solution of a problem that could be solved exactly. In such a case, we might well be willing to sacrifice accuracy in the answer for ease in the computation. For instance, suppose we had been given the initial positions and velocities of the particles, all $2 \times 3 \times 10^{20}$ pieces of data. If we could feel confident of getting a reasonably accurate answer by the statistical arguments we used above, then using them would certainly be preferable to applying Newton's laws to 10^{20} particles. We would be neglecting much of the information provided us, but we would immeasurably simplify the calculation.

The extent of the simplification is both staggering and amusing. Let us estimate the height of the stack of paper that would be needed to write down the initial positions and momenta for a modest laboratory number, $N = 10^{20}$, of helium atoms. Six-figure accuracy for each variable, typed on standard $8\frac{1}{2}'' \times 11''$ paper, means about one particle (3 position and 3 momentum variables) per line and so at most 100 particles per page. There are then 10^{18} sheets in the stack. Since a ream of paper has 500 sheets and is about 5 cm thick, we get 100 sheets per centimeter. This gives a stack 10^{16} cm high. For comparison, the distance from the sun to the outermost planet, Pluto, is about 6×10^{14} cm. The stack would extend well beyond the limits of the solar system, though it would fall short of reaching the nearest star.

So we begin to see the need in physics for a systematic procedure by which we can *estimate* the properties of a physical system about which we have (or choose to use) *only limited information*. To provide such a procedure is the aim of statistical mechanics.

1.2. THE QUANTUM-MECHANICAL TREATMENT

We have yet to look at the quantum-mechanical approach to our problem. We should really describe the system of N interacting helium atoms with a single wave function ψ for the entire system. The wave function will depend on $3N$ variables, the three position variables for each particle, say, and on the time. The quantum-mechanical analog of Newton's laws of motion is the time-dependent Schrödinger equation. It prescribes the evolution of the wave function in time. With the interatomic forces given, we may assume that the associated interatomic potential energies are known, and may proceed to set up the Schrödinger equation for the time-development of the single wave function. If we designate the energy operator by $\mathcal{H}$, we may write the time-dependent Schrödinger equation as

$$i\hbar \frac{\partial \psi}{\partial t} = \mathcal{H}\psi,$$

where $\hbar$ is Planck's constant h divided by 2π.

Though quantum mechanics has many intrinsically probabilistic aspects, the Schrödinger equation is a deterministic equation. Given the wave function at one time and given the energy operator, we can unambiguously calculate the wave function at a later time. (An outline of a proof of this contention will be given later, in Section 11.1.) Only in drawing conclusions from the wave function for a system does the probabilistic aspect appear.

Suppose for now that we know how to calculate the pressure at any specified time if we are given the wave function at that time. Then the question is whether we can, by using the Schrödinger equation, determine the wave function at the time t_2. If we knew the wave function at time t_1, the Schrödinger equation would be exceedingly complicated, but we could in principle solve it for the wave function at time t_2. But we certainly do not know the wave function at time t_1. The little information we are given does not begin to specify a function of $3N$ variables. So we cannot compute, even in principle, the wave function at time t_2.

The quantum-mechanical situation is strikingly analogous to the classical one. The crux of the problem is not complexity, though that is certainly present. Rather, it is that we do not have enough information to be able to perform an exact calculation, even in principle. Again we are forced to a process of estimating.

Let us see how we might form an estimate by quantum-mechanical reasoning. Since we will take this up in quite careful detail later, we need not pause to justify fully each step now. If the steps seem plausible, that will be sufficient, for the aim is to sketch the approach in familiar terms.

First we must decide on the quantum-mechanical states or wave functions that we might use to describe the system. The information given us in the problem provides us with a description of the atomic constituents, their interactions, and the environment in which they are located. This description suffices to determine the quantum-mechanical energy operator. We are also told that the energy of the system is within 1 per cent of a value $\tilde{E}$. Since the system is isolated and has already been undisturbed for some time, our knowledge about the system really does not change with time. The estimate we make for the pressure at any one time is appropriate for any other time. At the computational level this is a great asset, for we need not explicitly concern ourselves with the time-development of wave functions. Still, we must face up to picking one or more quantum-mechanical states for the description of the system at some one time. Since we are given rather specific information about the energy of the system, let us try using the energy eigenstates of the system. This is a reasonable and convenient choice; later, in Chapter 4, we will discuss why consistency with the information we possess implies that this is the only reasonable choice.

The ensuing algebra will be immeasurably simpler if we suppose there is only one helium atom in the box. Everything we say and do can be directly translated into the more realistic case of N particles: we need only replace the one-particle wave function by the wave function for the entire N-particle system, augment the energy operator by adding the parts for the $N - 1$ other particles and by including the interatomic potential energies, and so on. The statistical approach is independent of whether we deal with a one-particle or a many-particle system.

So we look for the energy eigenstates of a single helium atom confined to a cubical box. Fortunately, those states are not difficult to determine. In Cartesian coordinates, the energy operator for our atom of mass m is given by

$$\mathcal{H} = -\frac{\hbar^2}{2m}\left(\frac{\partial^2}{\partial x^2} + \frac{\partial^2}{\partial y^2} + \frac{\partial^2}{\partial z}\right) + U(x, y, z). \qquad (1.2.1)$$

The potential energy $U(x, y, z)$ is due solely to the forces exerted by the walls of the box. They are responsible for restricting the particle to the prescribed volume. For a cubical box of volume L^3, with one corner at the origin, we may write the potential energy explicitly as

$$U(x, y, z) = \begin{cases} 0 & \text{if } 0 < x < L, 0 < y < L, 0 < z < L, \\ \infty & \text{otherwise.} \end{cases} \qquad (1.2.2)$$

The (infinitely) steep rise in potential energy at the edge of the box represents the force that the walls exert to contain the particle.

The probability of finding the particle outside the box is to be zero; hence the wave function must be zero outside. Mathematical continuity implies that the wave function is zero at the walls (as well as outside). In short, an acceptable wave function vanishes at the walls and outside the box, thus describing a particle confined to the box. The infinite step in the potential energy serves to enforce this behavior at the mathematical level.

If we label each distinct energy eigenstate by a letter j, as in ψ_j, and write the corresponding energy eigenvalue as E_j, the equation we must solve is:

$$\mathcal{H}\psi_j = E_j\psi_j. \qquad (1.2.3)$$

The properly normalized solutions are

$$\psi_j(x, y, z) = \begin{cases} \left(\frac{8}{V}\right)^{1/2} \sin\left(\frac{\pi n_x x}{L}\right)\sin\left(\frac{\pi n_y y}{L}\right)\sin\left(\frac{\pi n_z z}{L}\right), \\ 0 \quad \text{outside the box,} \end{cases} \qquad (1.2.4a)$$

with

$$E_j = \frac{\pi^2 \hbar^2}{2m V^{2/3}} (n_x^2 + n_y^2 + n_z^2).$$
(1.2.4b)

The numbers n_x, n_y, n_z, are positive integers, and to each j corresponds a specific set of them. For instance, to $j = 13$ might correspond the set $n_x = 3$, $n_y = 5$, $n_z = 2$. Replacing a positive integer by its negative merely changes the sign of the wave function and does not lead to a new state. There are an infinite number of energy eigenstates, and it is worth noting that several distinct states can have the same energy, for example, the three states with the sets of integers (2, 1, 1), (1, 2, 1), and (1, 1, 2). The states are distinct in that the spatial properties (behavior as a function of x, y, z) of the wave functions are different, even though the energies are not. This phenomenon—called *degeneracy*—may be familiar from the hydrogen atom, where a number of states with different values of angular momentum possess the same energy. In our problem an energy level is typically six-fold degenerate; that is, there are typically six distinct energy eigenstates with the same numerical value for the energy. The index j designates a specific quantum-mechanical state, not merely a specific energy level.

We now have a set of quantum-mechanical states to use in describing our one-particle system. Since we do not have the information needed to select a single wave function, we will have to consider many possible wave functions. Here we must engage in some further reasonable guessing. The question is the appropriateness (for our physical system) of the description provided by each wave function. We must consider each state in our set of possible states and decide how good a description it provides. This amounts to associating a probability with each state in the set; the probability expresses the extent to which we believe that the state ψ_j provides the appropriate description of the system.

The question of the best way to assign those probabilities—as well as the precise meaning of the term "probability"—is one that we will deal with intensively in the following chapters. For now let us simply adopt a reasonable method. The information specifies that the system has, to within 1 per cent, an energy $\tilde{E}$; so we may reasonably exclude all states whose energy falls outside the limits $\tilde{E} \pm 0.01\tilde{E}$. (More technically, we associate zero probability with each such state.) For those states whose energy lies within the limits, we have no reason for preferring one more than any other, and it is therefore natural to treat them equally. (More technically, we associate the same probability with each.) So our limited information and a little reasonable guessing leads to our saying that the physical system can (so far as we know) satisfactorily be described by any energy eigenstate ψ_j for which the corresponding energy E_j satisfies the relation

$$\tilde{E} - 0.01\tilde{E} \leqslant E_j \leqslant \tilde{E} + 0.01\tilde{E},$$

and that these states are to be given equal weight in the statistical description.

Now we face the last preliminary task, that of determining the pressure associated with a specific energy eigenstate. Doing so is not difficult if we think a bit about what pressure means. If we imagine uniformly expanding the box a little, the pressure does some work. After all, pressure is simply force per unit area, so [pressure] times [surface area] times [displacement] is the work done on the walls. For an infinitesimal change dV in volume, which is effectively [surface area] times [infinitesimal displacement], the amount of work done by the gas is $p\, dV$. By the principle of energy conservation, this must lead to a diminution of the energy within the box. If we look at the change in energy of a specific state ψ_j when we change the volume slightly, we will be able to determine the corresponding pressure p_j. Remembering a minus sign because of the diminution in energy, we find

$$
\begin{aligned}
p_j &= -\frac{\partial E_j}{\partial V} \\[2mm]
&= -\frac{\partial}{\partial V}\left[V^{-2/3}\,\frac{\pi^2\hbar^2}{2m}\,(n_x^2 + n_y^2 + n_z^2) \right] \\[2mm]
&= +\frac{2}{3}\frac{E_j}{V}.
\end{aligned}
\qquad (1.2.5)
$$

Finally, we are in a position to form a quantum-mechanical estimate of the pressure exerted by the system. The only energy eigenstates that are admissible candidates for the description of the system are those states with energy in the range $\tilde{E} \pm 0.01\tilde{E}$. In view of the relation between pressure and energy just derived, any one of those states will give a pressure within the limits set by $(2/3)(\tilde{E} \pm 0.01\tilde{E})/V$. So our estimate for the pressure is within these limits and might be written as

$$p_{\text{est}} = \frac{2}{3}\frac{\tilde{E}}{V} \pm 0.01\left(\frac{2}{3}\frac{\tilde{E}}{V}\right). \qquad (1.2.6)$$

That this agrees so well with our estimate based on a classical analysis is partially happy coincidence. We should not always expect such good agreement.

All that we said by way of general inference from the classical analysis is confirmed by the quantum-mechanical analysis. We are rarely, if ever, presented with enough information about a physical system to enable us to specify its initial quantum-mechanical state. It is this aspect, rather than the complexity of the problems, which *forces* us to a statistical approach.

To be sure, even if we could specify a unique initial state for a complicated N-particle system, we would probably prefer a scheme simpler than that of integrating the Schrödinger equation or of computing $3N$-fold integrals with N-particle wave functions. If we could not only compute more simply an approximate answer to a question, but also estimate the uncertainty and confirm its relative smallness, we would almost certainly choose the less precise but easier approach. We would throw away some of our initial information, but the loss in accuracy could be offset by a tremendous gain in calculational ease. We should bear in mind, however, that here it would be a matter of practical choice, not of absolute necessity.

The quantum-mechanical analysis, as well as the classical analysis, exhibits the need that statistical mechanics is designed to satisfy. It might be well to say again that the aim of statistical mechanics is the development and application of a systematic procedure by which we can *estimate* the properties and behavior of a physical system about which we have (or opt to use) *only limited information*. In the specific context of quantum mechanics, "limited information" means information insufficient to enable us to specify a unique quantum-mechanical state for the system.

In the next few chapters we will develop the foundations for a theory of statistical mechanics. The example we have worked out has, hopefully, suggested the general pattern of the approach we must take. But we must make some notions more precise, and we must systematize the procedure. As the first step, we must reach a clear understanding of what the term "probability" means, at least as we will use it. Then we must establish the criteria to be used in assigning probabilities. With that accomplished, we will have a statistical framework within which we can pose and answer questions about physical systems.

PROBLEMS

1.1 The zero value of the wave functions (of Section 1.2) at the walls of the box may seem peculiar. Suppose the walls are not infinitely hard, so that the potential for a "one-dimensional box" has the shape shown in figure P1.1. The potential still goes to infinity, but not infinitely rapidly: the wall is impenetrable but soft. The shape of the new ground-state wave function is also shown. Write down arguments in support of the given qualitative form of that new wave function, and then consider how it would be modified as the potential becomes more steep. What does the wave function look like in the limiting situation of an infinitely sharp rise in potential?

FIGURE P.1.1.

1.2 (a) Verify that the energy eigenstates and eigenvalues for a single particle in a cubical box are correctly given by the equations in Section 1.2.

(b) Show explicitly that those wave functions have the property

$$\int_{\text{all space}} \psi_k^* \psi_j \, dx \, dy \, dz = \begin{cases} 1 & \text{if } k = j, \\ 0 & \text{if } k \neq j; \end{cases}$$

that is, they are mutually orthogonal (and correctly normalized). The asterisk indicates that ψ_k^* is the complex conjugate of ψ_k, which is irrelevant here because the wave functions are not complex.

1.3 Construct an argument, based on the Heisenberg uncertainty principle, and on the relation

$$\Delta x \, \Delta p_x \gtrsim \hbar,$$

to show that if a particle is confined to a cubical box of volume L^3, then its kinetic energy is likely to be of order $\hbar^2/2mL^2$ or greater. Compare this with the energy of the ground state in the exact calculation.

2

THE MEANING OF
THE TERM "PROBABILITY"

It is difficult to find an intelligible account of
the meaning of "probability," or of how we are ever to
determine the probability of any particular proposition; and yet
treatises on the subject profess to arrive at complicated results of
the greatest precision and the most profound practical importance.

John Maynard Keynes
A Treatise on Probability

To ascribe a definite meaning to the term "probability" is no easy task. Still, we must not only make the attempt, but also reach an acceptable conclusion. A theory of statistical mechanics depends, at least for its conceptual content, on its use and interpretation of the notion of probability. It behooves one to adopt an interpretation both fruitful and intellectually satisfying.

In the first section of this chapter we will consider the meaning to be ascribed to the term "probability," at least as we will understand it in our development of statistical mechanics. Following that, we will set up the rules for the combination and decomposition of probabilities—an algebra for reasoning with probabilities. That enterprise will form the bulk of the chapter. Then we will construct a procedure for estimating the value of a numerical function when we know merely the probabilities that the function's argument takes on specific values. The procedure provides an essential link between a theory of probability and its application to physical problems. The chapter will close with the analysis of two special topics, one of them relevant to the foundations of probability theory.

2.1. PROBABILITY AND COMMON SENSE

Before we ascribe a definite meaning to the term "probability," let us look at two situations in which the term is colloquially used. These will give us an indication of how we might want to specify the meaning.

For the first situation, suppose we are given a die and contemplate the result of rolling it once. We specify that, to all appearances, the die is symmetric and the faces properly marked. On just this information, we are likely to agree that the probability of rolling a two is the same as the probability of rolling a one. We might even venture to say that the probability of rolling a two is one in six or perhaps express it as $\frac{1}{6}$. Almost certainly we would say that the probability of rolling an even number is greater than the probability of rolling specifically a two. All this seems quite natural, but from a formal point of view we have been assigning probabilities and stating relations among them.

For the second situation, suppose we observe a group of students emerging from a mathematics examination. We might hear one dejected student say, "I probably passed, but ..." Another student might comment, "I might get a B, but it'll probably be a C." This, too, seems natural.

The situations are quite dissimilar, yet the idea of probability arises in both. After pondering over such use of the notion of "probability," we might tentatively say that in both situations the notion is associated with a degree of rational belief in an outcome or inference, that degree of belief being based on some knowledge of the context. Certainly some discussion of this tentative conclusion is in order. Let us return to the illustrative situations.

If a single die is to be rolled once, we are likely to feel that neither the appearance of a two nor that of a one is certain, though each is possible. If someone (with no more information than we possess) were to say, "A two will appear," we would not fully believe him. Nevertheless, we could not assert that he is necessarily wrong (though we might wonder why he is so brash). After all, a two might appear, and then we would have to concede that he was correct. More reasonably, though perhaps begrudgingly, we would accord his statement some degree of belief. If we agree that the probability of rolling a two is the same as that of rolling a one, we might look on this as meaning that we have the same degree of belief in the assertion that a two will appear as in the assertion that a one will appear. To each assertion we accord the same degree of belief, that itself being intermediate between full acceptance and utter rejection.

If the preceding appears reasonable, we may move on and interpret statements to the effect that the probability of rolling a two is one in six or is $\frac{1}{6}$ as statements in which we make quantitative assertions about the degree of belief. The same probability may be put into numerical form in a variety of equivalent ways, but in any such procedure one is representing a degree of belief by some numerical expression. The crucial point is the possibility of making "degree of rational belief" a quantitative notion.

To conclude the discussion of the die case, we need to look at the proposition that the probability of rolling an even number is greater than that of rolling specifically a two. Neither event is certain, neither impossible. Given that a properly marked die is rolled once, each outcome is naturally entitled to some degree of belief. We are willing to reason, however, that since any appearance of a two is the appearance of an even number, and since some other even number may appear if a two does not, the belief in the appearance of an even number is stronger than the belief in the appearance of specifically a two. We can recognize different degrees of belief and order them in a sequence of greater and lesser.

Now for the mathematics examination. The student who says he has "probably passed" is surely expressing a degree of belief in his having passed the exam. We can imagine him contrasting two contradictory possibilities in his mind, having passed and having not passed, and giving more credence to the former than the latter. If based on past examinations, time studied, and the like, this can be a rational judgment. He may find it difficult to give quantitative expression to the degree of his belief in having passed, but he has at least asserted a definite inequality between the strengths of two contradictory beliefs.

The second student has apparently not granted even a shred of belief to the possible grades A, D, and F. He feels that the possibility of a B warrants consideration and some measure of belief. But we may reasonably interpret his comment as implying a greater degree of belief in the statement that he will receive a C than in the statement that he will receive a B. He, too, has implicitly compared degrees of belief and found an ordering possible.

The analysis of these two situations supports the tentative conclusion. Indeed, the supplementary idea emerges that "degrees of rational belief" can be compared, ordered in meaningful sequences of greater and lesser, and represented in numerical form.

Yet the common property of the two situations—the association of the notion of probability with a degree of rational belief in an outcome or inference, given specific information—should not be allowed to mask a salient difference. In the die-rolling situation, the probability assignments and relations are, in a limited sense, "testable." By rolling the same die many times and noting the frequencies with which a two or a one or an even number appears, we could, in a certain sense, test the correctness of the probabilities and relations that we have asserted for a single roll. We could, if we wished, interpret the probability statements as statements about anticipated frequencies of occurrence. The probability of rolling a two could here be given the meaning of the anticipated ratio of the number of twos to the total number of rolls, in the limit as the total number of rolls becomes infinite. One could even go so far as to say that the probability of rolling a two has no meaning until the die has been rolled many (perhaps infinitely many) times and the actual frequency of occurrence of a two determined.

For the mathematics examination, any such approach is impossible. There is only one examination to be considered. A probability in this case—for example, the probability that the first student has passed—could not meaningfully be tested by many repetitions, nor could it be interpreted as a frequency, anticipated or actual. (The mere thought of taking the same examination many times, perhaps infinitely many times, is sufficiently appalling to dispose of the possibility.) Yet a student certainly feels justified in talking about the probability of having passed or failed, though whether he has passed or failed is already determined once the examination is over. Here a probability can be interpreted only as a degree of rational belief. The same conclusion holds for the second student.

These two examples may seem now to have created more problems than they have solved. Indeed, the meaning to be ascribed to the term "probability," and the domain to which a mathematical theory of probability is applicable, are still the subject of debate among statisticians, physicists, and philosophers of science. In particular, is the notion of probability to be restricted to situations in which a frequency interpretation is possible? To insist on a frequency interpretation would be to limit severely the situations in which a theory of probability could be used. In discussing the box filled with helium, we spoke of the probability that, given our data, a specific state ψ_j provided the appropriate description of the gas at a specific time. There is no notion of frequency here, but there is certainly an idea of a degree of reasonable belief that the state ψ_j is the appropriate state (or that it is not), based on the information granted us. If we want a theory of probability that we may directly apply to the typical estimation problems in statistical mechanics, we would do well to associate "probability" with "degree of rational belief."

In this book we will adopt the broad view and will take "probability" to be *a quantitative relation*, between a hypothesis and an inference, *corresponding to the degree of rational belief in the correctness of the inference, given the hypothesis*. The hypothesis is the information we possess, or assume for the sake of argument. The inference is a statement that, to a greater or lesser extent, is justified by the hypothesis. Thus "the probability" of an inference, given a hypothesis, is the degree of rational belief in the correctness of the inference, given the hypothesis.

In any specific case the degree of rational belief in the correctness of the inference, given the hypothesis, may be large or small, but it is quite naturally limited by the extreme of certainty, and by that of impossibility, of the inference's being correct. In slightly different words, the extremes are those of complete acceptance and of complete rejection of the inference on the basis of the hypothesis. Needless to say, in most situations the probability is somewhere between those extremes.

Probability theory, for us, is not so much a part of mathematics as a part of logic, inductive logic, really. It provides a consistent framework for reasoning about statements whose correctness or incorrectness cannot be deduced from the hypothesis. The information available is sufficient only to make the inferences "plausible" to a greater or lesser extent. Such a theory is precisely what we need for coping with the estimation problems of statistical mechanics.

A remark of some importance: a probability acquires its objectivity because it refers to a degree of *rational* belief. Probabilities are not to be assigned or manipulated capriciously. The process must, as part of the notion of rational, be founded on principles and must be consistent.

In taking "probability" to be a *quantitative* relation, one means that the probabilities themselves—the degrees of rational belief—can be compared, ordered in meaningful sequences of greater and lesser, and represented numerically. These properties, in turn, mean that mathematical methods may be applied in the development of a theory of probability. In the mathematical treatment, we will represent the degree of rational belief by a (real) number chosen from some range, with the extremes of the numerical range corresponding to certainty and impossibility. Though the degree of rational belief and the number that represents it are not identical, we will, for convenience of expression, refer to both as "the probability" of the inference, given the hypothesis.

Much of the preceding has a very formal and abstract ring to it, but it is really just a moderately precise statement of what we mean by probability when we use the idea colloquially. The view we have adopted encompasses both those cases where probability could be given a frequency interpretation and those where such an interpretation is manifestly impossible. We will find the concept of " probability " adopted here a fruitful one. I can only hope that the reader will, as I do, find it intellectually satisfying as well.

Since we often consider more than a single inference, and since we can entertain a variety of related hypothesis, there are two basic problems in any theory of probability:

1. Given a particular hypothesis and a particular inference, how does one determine the degree of rational belief in the correctness of the inference and assign to it, the probability, a numerical value?

2. Given certain probabilities, how is one to establish relations among them, and to compute others, in a logical fashion?

Remarkably, the second problem is easier to solve than the first, and, to ensure consistency in the solution of the first, must actually take precedence in the development of a theory. To this we now turn.

2.2.   AXIOMS FOR THE RELATIONS AMONG PROBABILITIES

In a slim volume, *The Algebra of Probable Inference*, Richard T. Cox has provided a general solution to the second problem, that of establishing relations among probabilities. Before we consider his axioms and the resulting theorems, we should develop notation appropriate for a mathematical description of relations among probabilities.

Since we talk about hypotheses and inferences, we are really dealing initially with statements. For example, in the case of the helium gas, the hypothesis might be taken to be the statement of our given information (total mass, total energy and associated limits, and so on), plus the statement that quantum mechanics is applicable. An inference might be the statement that a specific quantum-mechanical state provides the appropriate description of our physical system. We have already dealt, in a colloquial way, with the probability that, given the hypothesis, the inference is correct.

The statements that may be considered to be a hypothesis or an inference are of a special kind (in a logical sense). One must be able to say that the statement is either true or false (although one may not know which of these alternatives is correct). With a hypothesis, the need for this restriction arises because we must know, or assume, that the hypothesis is correct in considering the probability of an inference from that hypothesis. When we turn our attention to the inference, we find the same need. We are concerned with the degree of rational belief in the correctness of the inference, given the hypothesis, so the inference must have the property of being either correct or incorrect. This restriction excludes (as inadmissible) a sentence such as "Red houses are wiser than blue houses." That sentence may be regarded as neither true nor false; it may be poetic, or it may be merely nonsensical, but in any event it is outside the domain of our concern. In the future we will take the term "statement" to mean a sentence that is either true or false, though we may not know which alternative is correct. A clear example of a "statement" is provided by the sentence, "The flower has five petals." This sentence is either true or false—either the flower does or does not have five petals—though only additional knowledge, of whether the flower in question is a rose or a periwinkle, could tell us which alternative is correct.

In the mathematical description we will denote statements by capital letters: $A, B, C, \ldots$ or $A_1, A_2, A_3, \ldots$. For the probability of the inference A, given the hypothesis B, we will write

$$P(A|B).$$

The vertical bar serves to separate the inference (on the left) from the hypothesis (on the right). One may regard the bar as having the meaning "given that." Since a probability expresses a quantitative relation, here between statements A and B, the probability is (represented by) a number. The numerical range over which the number $P(A|B)$ can vary will depend on our choice of the scale for the numerical representation of degrees of rational belief. We may defer that choice until later.

For every statement there is a contradictory statement. For example, if A denotes the statement, "The flower has five petals," the contradictory statement, "not A," is the statement, "It is not true that the flower has five petals." One might also write the contradictory statement more simply as "The flower does not have five petals." A concise notation for the contradictory statement will be convenient. Let us denote the statement expressing the contradiction (or denial) of a statement A by the standard convention: $\sim A$. The negation sign ($\sim$) may be read as "It is not true that" and has precisely that meaning.

Next on the agenda is a symbolic expression for the sum of two statements, for the term "and." For the sake of illustration, let us modify the flower example so that the statement, "The flower has five petals," is denoted now by the symbol A_1. Another statement, "The flower is red," we denote by A_2. The statement, "The flower has five petals and the flower is red," is the sum of the two individual statements. We could write this sum as "A_1 and A_2". Let us, however, use the more concise conventional notation, and denote by $A_1 \cdot A_2$ the statement that is the sum of statements A_1 and A_2. The centered dot represents "and." The statement $A_1 \cdot A_2$ is called the *conjunction* of the statements A_1 and A_2.

The preceding suffices to enable us to express symbolically Cox's axioms for a theory of probability in which "the probability" is the degree of rational belief, represented by a real number, in the correctness of an inference, given a hypothesis. That some axioms, as in Euclidean geometry, are necessary is to be expected. Still, one would like the axioms to be simple, few, and reasonable, to agree, if possible, with common sense. The axioms that Cox suggests do appear to fulfill these desires.

The two axioms will be presented in the following manner: first we will consider an example of reasoning that might lead to the axiom; then the axiom will be set out as a quotation from Cox's book; and finally we will express the axiom in symbolic form.

Suppose for the moment that we are devotees of the track. We have a favorite horse, and, on the basis of his past performance and present condition, we want to assess the probability of his winning the race. (This might be a matter of some small financial concern.) Somehow we decide on the initial probability of his winning, and the race begins. Our horse is off to a poor start. Mentally we begin to reduce the probability of his winning and, concurrently and sadly, increase the probability of his not winning. But then our horse forges into the lead. We reverse the trend of the probability of his winning and of its contradiction, the statement that he will not win. As the race goes on, and our horse maintains his lead, we confidently increase the probability that he will win and virtually neglect the small probability of its contradictory.

The point of this little episode is that we quite naturally feel that the probability of an inference and the probability of the contradictory inference are intimately connected, even deterministically connected. As one goes up, the other goes down. This is sufficient preface for Cox's first axiom:

"The probability of an inference on given evidence [the hypothesis] determines the probability of its contradictory [the contradictory inference] on the same evidence."

Now we put this into symbolic form. If we let A represent the inference and B represent the hypothesis, the axiom asserts that the probability $P(A|B)$ somehow determines the probability $P(\sim A|B)$. Since probabilities are (represented by) numbers, we may introduce a presently unknown function

f_1 and write

AXIOM I:
$$P(\sim A|B) = f_1[P(A|B)].$$

This is the content of the first axiom in symbolic form.

For an introduction to the second axiom, we return to the start of our horse race. Suppose we are not so sanguine about the prospects of our favorite's winning. Second place might be good enough, but we still have an interest in how well our horse will do. We might want to assess the probability that our horse will come in second *and* that he will be no more than three meters behind the winner. How would the reasoning go? We might try to assess the probability directly, but more likely we would find it easier to split up the problem. We might first assess the probability that our horse will come in second, and then, on the additional assumption that he does place second, assess the probability that he is actually no more than three meters behind the winner. Some combination of these two probabilities ought to provide us with the desired final probability. (To supplement the general hypothesis in the second step is natural, for unless we assume that our horse places second, there is no point in going on with the calculation. Moreover, we have taken account of our uncertainty about second place when assessing the first probability.) We might also try the other tack: assess the probability that he finishes no more than three meters behind the winner, and then, on the additional assumption that this is true, assess the probability that he is second rather than third, fourth, and so on. Each of these three approaches seems reasonable. We are likely to feel that either of the two split-up-the-problem approaches, if done "correctly," will lead to the same final probability as the direct approach. Nothing more than this is the point of our second episode at the track.

This provides a preface for Cox's second axiom:

"The probability on given evidence that both of two inferences are true is determined by their separate probabilities, one on the given evidence, the other on this evidence with the additional assumption that the first inference is true."

Again we can codify the axiom neatly in symbolic form. We let A_1 and A_2 represent the two inferences and let B represent the hypothesis. The axiom asserts that the probability $P(A_1 \cdot A_2 | B)$ is determined, in some way, by two related probabilities: the probability of A_1 given B, $P(A_1|B)$; and the probability of A_2 given both B and A_1, $P(A_2 | B \cdot A_1)$. To put this axiom into the form of an equation, we introduce a second, presently unknown function f_2 and write

AXIOM II:
$$P(A_1 \cdot A_2 \mid B) = f_2[P(A_1 \mid B), P(A_2 \mid B \cdot A_1)].$$

This expresses the content of the second axiom in symbolic form. We could repeat the equation with A_1 and A_2 interchanged in the role of "first inference," as that role is described in the verbal statement of the axiom, but doing so is actually unnecessary.

At first glance, it would not appear that much, if anything, could be deduced from such general axioms, for they contain two unknown functions, f_1 and f_2. Indeed, something more is needed, something to determine the form of the functions, but that something is remarkably modest and natural.

Quite independently of any connection with probability theory, relations among *statements* conform to the rules of common logic. Two examples may help to clarify the meaning of this assertion.

The first relation is the equality

$$\sim(\sim A) = A,$$

whatever the meaning or truth of the statement A. The equality sign signifies that the two statements, $\sim(\sim A)$ and A, though different in form, are the same in meaning. For example, the lefthand side might read, "It is not true that 'it is not true that the coin is a dime'", whereas the righthand side would read simply, "The coin is a dime." Though different sets of words are used, the two statements are identical in meaning.

A second relevant example of relations among statements is the equality

$$A_1 \cdot A_2 = A_2 \cdot A_1.$$

The two sides differ in the order in which the statements A_1 and A_2 are asserted, but the meaning of the expressions is the same, whatever the statements A_1 and A_2 may be. The conjunction produces a symmetric combination of statements.

The first addition to the two axioms is the demand that relations among probabilities be *consistent* with valid relations among statements. The consistency requirement is the major supplement to the bare axioms, and implies severe mathematical constraints on what will be acceptable as functions f_1 and f_2. An example is in order.

Let us see what consistency with the relation $\sim(\sim A) = A$ implies about the function f_1 in Axiom I. Since A in that axiom is any inference, the axiom holds if A is everywhere replaced by $\sim A$:

$$P(\sim(\sim A) \mid B) = f_1[P(\sim A \mid B)]$$
$$= f_1[f_1[P(A \mid B)]].$$

The step to the second line follows from Axiom I as originally presented. Now we may impose the consistency requirement. Since $\sim(\sim A) = A$, consistency with that relation among statements implies that the function f_1 must obey the following equation:

$$f_1[f_1[P(A \mid B)] = P(A \mid B).$$

Since this equation must hold for any probability $P(A \mid B)$, the demand for consistency has provided a mathematical constraint on the function f_1.

The second addition supplementing the axioms is concerned with the mathematical tractability of the two unknown functions. It is the assumption that the functions f_1 and f_2 are twice-differentiable functions of their arguments and that those second derivatives are continuous. This does impose a condition on the otherwise unspecified functions, but it is rather mild. Most functions met in physics are at least as well-behaved, mathematically speaking, and only a purist would bother to list this as an assumption.

Finally we must say something about the scale to be used in the numerical representation of degrees of rational belief. The length of a piece of lumber is still the same length whether it is measured in centimeters, meters, or light-years; yet if one wants to represent the length numerically, one must choose some one scale. A comparable situation exists for degrees of rational belief or probabilities. We must, by a set of conventions, specify the scale to be used for them.

First we adopt a convention about one endpoint of the numerical range. Let us choose the extreme of certainty and agree that, if the inference A is certain to be correct, given the hypothesis B, then we give to the probability the numerical value one. We are always free to choose the numerical value of at least one endpoint of a scale, so this is not an assumption but a convenient convention.*

The development of the theory, as given by Cox, indicates that this single convention does not totally fix the scale. There is still some freedom in how one numerically represents "the degree of rational belief." The much more

---

* An example of a different convention is provided by that commonly used in weather forecasts, as in "The probability of showers tomorrow is 3 out of 10." There certainty is represented by 10. Many conventions are possible and in use; ours is the most convenient analytically.

limited theory of probability based on a frequency interpretation has its own natural scale for the representation of what it calls a probability. (It is the scale one commonly uses in discussing dice problems and other games of chance.) Since any choice of scale in our theory is merely a matter of convention, it makes sense to adopt the scale already in use in the frequency theory, and so let us choose this option. Unfortunately, the exact manner in which this decision on scale is to be implemented cannot easily be specified before the derivation of theorems from the axioms is well under way; so let us halt the discussion at this point.

The preceding is all that is needed to establish a consistent theory of probability, with "probability" understood in the broad sense we have adopted. Great pains have been taken to describe the two axioms in precise language and to display the subsidiary requirements. Now we relax and examine—without the proofs—the two theorems that follow by deduction from this axiomatic framework. Though the proof itself is often instructive, the essence of a theorem is what is assumed and what is proved; the detailed proofs are all in Cox's little volume.

The subsidiary requirements determine the functions f_1 and f_2, so that, in effect, each of the two axioms gives rise to a theorem.

THEOREM I:

$$P(A \mid B) + P(\sim A \mid B) = 1. \tag{2.2.1}$$

THEOREM II:

$$P(A_1 \cdot A_2 \mid B) = P(A_1 \mid B) \, P(A_2 \mid B \cdot A_1). \tag{2.2.2}$$

The first theorem states explicitly the manner in which the probability of the contradiction of A is determined by the probability of A. The second theorem provides the answer to how one computes the probability $P(A_1 \cdot A_2 \mid B)$ when one wishes to split up the problem. All further relations among probabilities follow deductively from these two theorems by manipulation of logical relations among statements.

Could we have written down the two relations without all the preceding labor and the appeal to Cox's derivation? Not if we want to avoid arbitrary decisions on essential points, and not if we want to be sure of having consistency. The deep significance of Cox's work lies in his demonstration of the *consistency* and the *uniqueness* of the quoted relations. If one imposes the two subsidiary requirements—that relations among probabilities be consistent with relations among statements, and that, loosely, the unknown functions be mathematically well-behaved—and if one adopts the usual conventions about the numerical scale, then there is only one solution for each of the two unknown functions appearing in the axioms. In our context, in which a probability is understood as a degree of rational belief and in which Cox's two axioms are adopted, the two theorems provide a consistent theory of probability, and, in that context, they are unique in doing so.

2.3. SOME IMMEDIATE CONSEQUENCES OF THE THEOREMS

As the first use of the theorems, let us confirm that to the extreme of impossibility (or complete rejection) corresponds the numerical value zero. The principle step toward this end consists of taking the general relation given by Theorem I and specializing it. One replaces the hypothesis B by the conjunction of B and A:

$$P(A \mid B \cdot A) + P(\sim A \mid B \cdot A) = 1.$$

Given the hypothesis $B \cdot A$, the inference A is certain to be correct, regardless of B, for the correctness of A is included in the hypothesis. Given the same hypothesis, it is impossible for the contradiction of A, $\sim A$, to be true. Since to certainty corresponds (by our convention) a probability of one, the equation becomes

$$1 + P(\sim A \mid B \cdot A) = 1.$$

From this we may conclude that the probability of an impossible (or completely rejected) inference is indeed assigned the value zero. Thus the numerical value of a general probability $P(A \mid B)$ is restricted to the range

$$0 \leqslant P(A \mid B) \leqslant 1. \tag{2.3.1}$$

In Theorem II, giving the decomposition of the joint probability $P(A_1 \cdot A_2 \mid B)$, the statements A_1 and A_2 appear asymmetrically on the righthand side. Some of this asymmetry is necessary, but the theorem is still true if we interchange A_1 and A_2 on the righthand side, thereby changing radically the two individual probabilities in the product. To prove this, we start by interchanging A_1 and A_2 on *both* sides. This is just like changing names and must be permissible:

$$P(A_2 \cdot A_1 \mid B) = P(A_2 \mid B) \, P(A_1 \mid B \cdot A_2).$$

Now we need only invoke the symmetry on the lefthand side. The relation among statements, $A_2 \cdot A_1 = A_1 \cdot A_2$, implies that we may write

$$P(A_2 \cdot A_1 \mid B) = P(A_1 \cdot A_2 \mid B).$$

A comparison of this equation with the preceding equation gives us the desired result:

$$P(A_1 \cdot A_2 \mid B) = P(A_2 \mid B)\, P(A_1 \mid B \cdot A_2). \qquad (2.3.2)$$

At this point it would be well to interrupt the development of rather abstract theorems with an example. As yet we have not said much about the methods for assigning numerical values to degrees of rational belief, but since for rolling a die we are not likely to disagree about the numerical values, we may reasonably use the rolling of a symmetric, "true" die in an example.

Suppose our hypothesis R is that we roll a standard die once. For inferences we take the statements, "An even number will appear" (D_1), and "A two will appear" (D_2). Since there are six faces on a die, three with even numbers and a single face with a two, we are likely to agree on the following numerical probability assignments:

$$P(D_1 \mid R) = \tfrac{3}{6} = \tfrac{1}{2},$$

$$P(D_2 \mid R) = \tfrac{1}{6}.$$

Now we wish to determine, on the same hypothesis R, the probability of the joint inference $D_1 \cdot D_2$, that is, of the statement, "An even number will appear *and* a two will appear." As in the introduction to Axiom II, we may imagine splitting up the problem. First we determine the probability of an even number, given one roll, $P(D_1 \mid R)$. Then we assess the probability of a two, given a single roll *and* given that an even number has appeared, $P(D_2 \mid R \cdot D_1)$. These two probabilities are then to be combined, according to Theorem II, to give an unambiguous numerical result for the probability $P(D_1 \cdot D_2 \mid R)$.

Since a two is one of the three possible even numbers, we can agree on the probability assignment

$$P(D_2 \mid R \cdot D_1) = \tfrac{1}{3}.$$

If there is no dispute about the probabilities so far assigned, then the desired probability follows from Theorem II as

$$P(D_1 \cdot D_2 \mid R) = P(D_1 \mid R)\, P(D_2 \mid R \cdot D_1)$$

$$= (\tfrac{1}{2}) \times (\tfrac{1}{3}) = \tfrac{1}{6},$$

a not unreasonable result.

As a check on consistency, we can invert the order in the split-up-the-problem approach and see whether the same probability for the joint inference emerges. This means using equation (2.3.2), and so we must first determine the probability $P(D_1 | R \cdot D_2)$. Since the hypothesis $R \cdot D_2$ contains the statement (D_2) that a two appears, the inference D_1, that an even number will appear, is certain to be correct. Hence

$$P(D_1 | R \cdot D_2) = 1.$$

When we insert this and the agreed-upon value of the probability $P(D_2 | R)$ into equation (2.3.2), we find

$$P(D_1 \cdot D_2 | R) = P(D_2 | R) \, P(D_1 | R \cdot D_2)$$
$$= (\tfrac{1}{6}) \times (1) = \tfrac{1}{6},$$

which checks. But note that the individual probabilities appearing as factors in the two products are radically different. The origin of this difference is not in the inferences, for they are the same: D_1 and D_2. Rather, the hypotheses are different in the two schemes:

$$R \cdot D_1 \neq R \cdot D_2.$$

Since a probability is the degree of rational belief in an inference, *given a hypothesis*, it depends crucially on *both* inference *and* hypothesis.

The present simple example makes evident the logical fact that we cannot consider the probability of an inference in isolation. That has no meaning. Remembering that "probability" expresses a *relation* between an inference and a hypothesis can spare one much grief.

2.4. AN ESSENTIAL THEOREM

Thus far our ability to write out relations among statements and probabilities in symbolic form has been a decided asset. To continue this mode of expression we need to introduce one more symbol, that for the notion that *at least one of* the statements A_1, A_2, is correct. To return to our flower, we need a symbol that will enable us to write concisely the statement "Either the flower has five petals (A_1) or the flower is red (A_2) or both ($A_1 \cdot A_2$)." For this kind of statement, we use a conventional symbol and write the statement as $A_1 \vee A_2$. The statement $A_1 \vee A_2$ is called the *disjunction* of the statements A_1 and A_2, and the symbol $\vee$ is called a disjunction sign.

The symbol for "at least one of" considerably increases our symbolic vocabulary and permits us to frame an essential theorem. For purposes of illustration, let us continue with the flower and assume some constant hypothesis B—perhaps, "The gardener likes small, cheerful flowers." Now we pose a question: how is the probability $P(A_1 \vee A_2 \mid B)$ of the inference that "At least one of the statements A_1, A_2, is true" related to the probabilities $P(A_1 \mid B)$ and $P(A_2 \mid B)$? A little reflection should indicate that $P(A_1 \vee A_2 \mid B)$ cannot be less than either $P(A_1 \mid B)$ or $P(A_2 \mid B)$, but is it equal to the larger, to their sum, or to what? The unique answer follows from the two basic theorems plus some logical manipulations with statements.

Let us permit the symbols A_1, A_2, and B to revert to the status of representing general statements. The question raised in the preceding paragraph is answered by the theorem

$$P(A_1 \vee A_2 \mid B) = P(A_1 \mid B) + P(A_2 \mid B) - P(A_1 \cdot A_2 \mid B). \qquad (2.4.1)$$

The derivation is not obvious; we can, however, reduce it to a tidy sequence of five steps if we first establish some preliminary results.

The first preliminary is an expression for the disjunction in terms of conjunction and contradiction:

$$A_1 \vee A_2 = \sim(\sim A_1 \cdot \sim A_2). \qquad (2.4.2)$$

The righthand side says, in effect, "It is not true that 'A_1 is not true and A_2 is not true.'" That statement is logically equal to the statement on the left, that "At least one of the statements A_1, A_2, is true." Thus we have the disjunction written in terms of the logical notions—conjunction and contradiction—that appear in the two basic theorems of our probability theory. Reexpression of the probability of the disjunction has become possible.

For the next preliminary results, we must do some manipulating. We write down Theorem II as it stands and then again with A_2 replaced by $\sim A_2$:

$$P(A_1 \cdot A_2 \mid B) = P(A_1 \mid B) \, P(A_2 \mid B \cdot A_1),$$

$$P(A_1 \cdot \sim A_2 \mid B) = P(A_1 \mid B) \, P(\sim A_2 \mid B \cdot A_1).$$

Addition of these two equations gives

$$P(A_1 \cdot A_2 \mid B) + P(A_1 \cdot \sim A_2 \mid B) = P(A_1 \mid B)[P(A_2 \mid B \cdot A_1) + P(\sim A_2 \mid B \cdot A_1)]$$

$$= P(A_1 \mid B).$$

The second step follows with the aid of Theorem I.

This relation is a noteworthy general result in itself. We will need two specific instances of it: once with A_1 replaced by $\sim A_1$,

$$P(\sim A_1 \cdot A_2 \mid B) + P(\sim A_1 \cdot \sim A_2 \mid B) = P(\sim A_1 \mid B), \qquad (2.4.3a)$$

and again with A_1 and A_2 interchanged,

$$P(A_2 \cdot A_1 \mid B) + P(A_2 \cdot \sim A_1 \mid B) = P(A_2 \mid B). \qquad (2.4.3b)$$

Now for the derivation. We begin by using the expression for the disjunction, given in equation (2.4.2), to write

$$P(A_1 \vee A_2 \mid B) = P(\sim(\sim A_1 \cdot \sim A_2) \mid B).$$

Theorem I permits us to reexpress the righthand side as

$$P(A_1 \vee A_2 \mid B) = 1 - P(\sim A_1 \cdot \sim A_2 \mid B).$$

The preliminary result (2.4.3a) allows us to reexpress the far right term as

$$P(A_1 \vee A_2 \mid B) = 1 - [P(\sim A_1 \mid B) - P(\sim A_1 \cdot A_2 \mid B)].$$

We apply Theorem I to the first two terms on the right, with the result

$$P(A_1 \vee A_2 \mid B) = P(A_1 \mid B) + P(\sim A_1 \cdot A_2 \mid B).$$

The first term on the right is now one of the desired terms. To reexpress the term on the far right, we use the last of our preliminary results, equation (2.4.3b), remembering that the conjunction is a symmetric relation. With this move, we have finished:

$$P(A_1 \vee A_2 \mid B) = P(A_1 \mid B) + P(A_2 \mid B) - P(A_1 \cdot A_2 \mid B). \qquad \text{Q.E.D.}$$

It is unfortunate that the derivation is so far from being transparent, but the derivation does have a compensating virtue: it shows the power of symbolic reasoning.

The appearance of $P(A_1 \cdot A_2 \mid B)$, and with a minus sign, is perhaps surprising, but the term compensates for a possible overestimation of the probability $P(A_1 \vee A_2 \mid B)$ when we try to set that equal to the sum of the individual probabilities.

2.5. MUTUALLY EXCLUSIVE
AND EXHAUSTIVE INFERENCES

Often one meets a situation in which, given the hypothesis and a specific pair of inferences, it is impossible for the conjunction of the two inferences to be correct: for example, the appearance of an even number and the appearance of a three when a proper die is rolled once. Indeed, such situations are frequent enough and significant enough to merit special consideration and the introduction of a definition.

If, given the hypothesis B, it is impossible for the inference $A_1 \cdot A_2$ (formed by addition from the inferences A_1 and A_2) to be correct, then

$$P(A_1 \cdot A_2 \mid B) = 0,$$

and we call such a pair A_1, A_2, of inferences *mutually exclusive on the hypothesis B*. Quite simply, given the hypothesis B, the inferences A_1 and A_2 cannot simultaneously be correct. Neither of the two probabilities, $P(A_1 \mid B)$ and $P(A_2 \mid B)$, need be zero; the example in the preceding paragraph illustrates this.

In approaching a statistical problem, we will often consider several inferences, $A_1, A_2, \ldots, A_n$, and a single hypothesis B. Typically it will be the case that, on the hypothesis, none of the inferences is certain to be correct; that is, no inference has unit probability. Yet the hypothesis may enable us to assert that, necessarily, one or more of the inferences in the set is correct, though we may not know which inference. (An example follows shortly.) In that case, the inference that "at least one of the inferences $A_1, A_2, \ldots, A_n$, is correct," written as $A_1 \vee A_2 \vee \ldots \vee A_n$, is necessarily correct on the hypothesis B. The corresponding probability has the value unity:

$$P(A_1 \vee A_2 \vee \ldots \vee A_n \mid B) = 1.$$

The inferences $A_1, A_2, \ldots, A_n$, in such a set, for which the hypothesis B implies that, necessarily, one or more is correct, are called *exhaustive on the hypothesis B*.

For example, for the single rolling of a standard die, the inferences that we have labeled D_1 ("An even number will appear") and D_2 ("A two will appear") are not exhaustive on the hypothesis. It is not true that, necessarily, one or both of them is correct, for an odd number might appear. If, however, we include a third inference, "An odd number will appear" (D_3), then the inference $D_1 \vee D_2 \vee D_3$ is necessarily correct, given the hypothesis. So the augmented set of inferences, D_1, D_2, and D_3, does form an exhaustive set on the hypothesis R that a standard die is rolled once.

If, in addition to being exhaustive on a hypothesis B, the inferences $A_1, A_2, \ldots, A_n$, in the set are also mutually exclusive, then repeated use of the theorem in equation (2.4.1) leads to the relation

$$\sum_{j=1}^{n} P(A_j | B) = 1. \tag{2.5.1}$$

This is a crucial result; so it would be well to follow the line of derivation.

Since the inferences are specified to be exhaustive on the hypothesis B, we may assert as a starting point the equality

$$P(A_1 \vee A_2 \vee \ldots \vee A_n | B) = 1. \tag{2.5.2}$$

The derivation consists of showing that the probability appearing here on the lefthand side can be reduced, with the aid of equation (2.4.1) and the specified mutual-exclusion property, to the proposed summation.

We proceed by rearranging the statements, by inserting parentheses:

$$A_1 \vee A_2 \vee \ldots \vee A_n = A_1 \vee (A_2 \vee \ldots \vee A_n).$$

The combination of statements in the parentheses we regard momentarily as a single statement. Then the theorem in equation (2.4.1) permits us to write

$$P(A_1 \vee A_2 \vee \ldots \vee A_n | B) = P(A_1 \vee (A_2 \vee \ldots \vee A_n) | B)$$
$$= P(A_1 | B) + P(A_2 \vee \ldots \vee A_n | B) - P(A_1 \cdot (A_2 \vee \ldots \vee A_n) | B).$$

Since A_1 is mutually exclusive with all other A_j on the hypothesis B, it is mutually exclusive with their disjunction, and so the term on the far right is zero. We have separated out the first term in the desired series, and need contend only with a probability involving a combination of $n - 1$ statements:

$$P(A_1 \vee A_2 \vee \ldots \vee A_n | B) = P(A_1 | B) + P(A_2 \vee A_3 \vee \ldots \vee A_n | B).$$

The probability on the far right may now be treated in the same manner to extract $P(A_2 | B)$, yielding

$$P(A_1 \vee A_2 \vee \ldots \vee A_n | B) = P(A_1 | B) + P(A_2 | B) + P(A_3 \vee \ldots \vee A_n | B).$$

This process, when repeated $n - 1$ times, will lead to

$$P(A_1 \vee A_2 \vee \ldots \vee A_n | B) = \sum_{j=1}^{n} P(A_j | B).$$

Now we may combine the partial results to provide the desired proof. Equation (2.5.2), a consequence of the exhaustive property alone, and equation (2.5.3), a consequence of the mutual-exclusion property alone, together imply the relation

$$\sum_{j=1}^{n} P(A_j \mid B) = 1 \tag{2.5.4}$$

for a set of n inferences A_j that, on the hypothesis B, are exhaustive and mutually exclusive.

We can gain some understanding of the result if we remember that, given the hypothesis B, the exhaustive property implies that *at least one of* the inferences A_j must be true, and the mutual-exclusion property implies that *no more than one* of the inferences may be true. Given the hypothesis, one but only one of the inferences is true, though we may not know which. That the sum of the probabilities adds up to unity, the number used to represent certainty, should not be too surprising. In the future we will refer to the result in equation (2.5.1) as the *normalization condition* for the probabilities of a set of inferences which, on the given hypothesis, are exhaustive and mutually exclusive.

Once again we should note well the relational aspect of probabilities. One can assert that the inferences in a given set are exhaustive, or mutually exclusive, or both, only after the hypothesis has been explicitly specified. The inferences D_1, D_2, and D_3 (the appearance of an even number, a two, and an odd number, respectively) of a preceding paragraph are exhaustive given the hypothesis that a standard die is rolled once. They do not possess the exhaustive property if the hypothesis is that a single card is drawn from a standard deck, for in the latter situation a face card might appear. Even the inferences D_1 and D_3 are not necessarily mutually exclusive. On the hypothesis of the roll of a single die they are, but not if the hypothesis is that two dice are rolled simultaneously. The inferences in a particular set can be exhaustive, or mutually exclusive, or both, only *relative* to a given hypothesis.

2.6. BAYES' THEOREM

A complete theory of probability should enable one to answer the question, how does the degree of rational belief in the correctness of an inference change when the initial hypothesis is altered by the addition of further relevant information? An answer is provided by Bayes' theorem. Although the theorem itself is elegant and the proof amazingly short, the establishment of the context is a rather long-winded affair.

BAYES' THEOREM :

Let A represent the inference and B the initial hypothesis. The original probability is then expressed as $P(A \mid B)$. Suppose the initial hypothesis is augmented by the assertion of the statement B'. Then the new hypothesis is represented by the conjunction $B \cdot B'$. Our concern is with the probability $P(A \mid B \cdot B')$ and its relation to $P(A \mid B)$. Provided that $P(B' \mid B)$ is not zero, the relation is

$$P(A \mid B \cdot B') = P(A \mid B) \times \left[\frac{P(B' \mid B \cdot A)}{P(B' \mid B)} \right].$$

The proof is, for a change, as simple as the theorem. We take the auxiliary probability $P(A \cdot B' \mid B)$ and decompose it by Theorem II in the two possible multiplicative fashions:

$$P(A \cdot B' \mid B) = P(A \mid B) \, P(B' \mid B \cdot A)$$
$$= P(B' \mid B) \, P(A \mid B \cdot B').$$

The two products on the righthand side are automatically equal. Provided the probability $P(B' \mid B)$ is not zero, we may solve for the probability $P(A \mid B \cdot B')$ and thereby arrive at Bayes' theorem.

The usefulness of Bayes' theorem depends on one's ability to determine the probabilities appearing in the square brackets, for these probabilities provide the proportionality factor between the probability of A on the original hypothesis B and the probability on the augmented hypothesis $B \cdot B'$.

An illustration—especially a possibly controversial one—would not be out of order. The presence of water vapor in the Martian atmosphere was established in 1969. Did the addition of this information to scientific knowledge about Mars alter the probability of the inference that there is life on Mars, given the then-existing knowledge? To judge from newspaper accounts, one would say yes, there was an increase in the probability. If a sceptic demurs, one can offer him an analysis of the matter with Bayes' theorem, as follows.

Let us first make some correspondences between statements in our illustration and the symbols appearing in Bayes' theorem.

A: "There is life on Mars."

B: A set of statements summarising scientific knowledge prior to the establishment of statement B'.

B': "There is water vapor in the Martian atmosphere."

The immediate question is whether the ratio of probabilities in square brackets in Bayes' theorem differs from unity. Let us look first at the denominator. Given only B, statement B' was not a certain inference; hence we have the strict inequality

$$P(B' | B) < 1.$$

But given $B \cdot A$, statement B' is a certain inference, for life (as we know it) requires water; so we have

$$P(B' | B \cdot A) = 1.$$

We conclude that the ratio is larger than one, and hence that the probability of there being life on Mars did indeed increase on the addition of the water-vapor information.

The illustration may appear to be nothing more than a vindication of common sense, but note well that we have used our theory of probability to analyze a scientific question that could not even be posed if one were to take a strict frequency interpretation of probability.

2.7. AN EXAMPLE FOR FUTURE REFERENCE

By now we have amassed considerable symbolic machinery for expressing and manipulating relations among probabilities. As an exercise, we apply it to a situation that we will meet somewhat less abstractly in the next chapter. The results we derive here will be needed when we tackle the second major problem of probability theory, that of assigning initial probabilities.

Suppose that from a hypothesis B we draw three inferences: A_1, A_2, and A_3. We specify that, as a set, they are, given the hypothesis, both exhaustive and mutually exclusive. With individual probabilities $P(A_j | B)$ somehow assigned, consistent with the normalization condition, we might sketch the abstract situation as in figure 2.7.1. Next to the abstract situation, we have a possible concrete case.

If we should have to assign probabilities to such a set of inferences, we might not want to assign them directly, but rather might prefer to split up the problem. We could group the inferences, for instance, as the disjunction $(A_1 \vee A_2)$ and as A_3, and make assigning probabilities to these inferences the first step in the solution of the problem. For this option there is a corresponding diagram, as in figure 2.7.2. If the rules for manipulating probabilities are complete and consistent, we ought to be able to express all the probabilities appearing in this diagram in terms of those appearing in the preceding one, and vice versa. This is indeed possible.

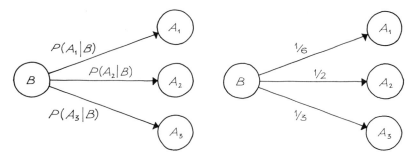

FIGURE 2.7.1
The arrow from hypothesis to inference is labeled with the corresponding probability.

To compute the probability $P(A_1 \vee A_2 \mid B)$, we use the theorem in equation (2.4.1) and the specified mutually-exclusive property of A_1 and A_2, given B:

$$P(A_1 \vee A_2 \mid B) = P(A_1 \mid B) + P(A_2 \mid B) - \text{(zero)}. \qquad (2.7.1)$$

Finding the probabilities on the arrows emanating from $(A_1 \vee A_2)$ is hardly more difficult. Let us evaluate the probability $P(A_1 \mid (A_1 \vee A_2) \cdot B)$ first; the other will follow by an identical line of reasoning.

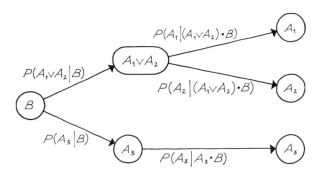

FIGURE 2.7.2
This abstract diagram is logically equivalent to the abstract diagram in figure 2.7.1. Each arrow is labeled with the probability that the inference is correct, given that *all* the preceding statements along the sequence of arrows are true.

We may usefully look on $P(A_1 \mid (A_1 \vee A_2) \cdot B)$ as being the probability of A_1 when an initial hypothesis B is augmented by the addition of the statement $(A_1 \vee A_2)$. Taking this view, we may apply Bayes' theorem by first making correspondences with the symbols in that theorem—A_1 corresponds to A, $(A_1 \vee A_2)$ corresponds to B'—and then writing out the ensuing relation as

$$P(A_1 | B \cdot (A_1 \vee A_2)) = P(A_1 | B) \times \left[\frac{P(A_1 \vee A_2 | B \cdot A_1)}{P(A_1 \vee A_2 | B)} \right].$$

The probability in the numerator of the square brackets is unity, for the hypothesis $B \cdot A_1$ ensures that the statement A_1 is correct and thus that at least one of the inferences A_1, A_2, is correct. So we quickly arrive at the intermediate result

$$P(A_1 | B \cdot (A_1 \vee A_2)) = \frac{P(A_1 | B)}{P(A_1 \vee A_2 | B)} \geqslant P(A_1 | B). \qquad (2.7.2)$$

The inequality, arising because the numerical value of the probability $P(A_1 \vee A_2 | B)$ cannot be greater than one, is an incidental benefit. Knowledge that the statement $(A_1 \vee A_2)$ is true, as well as B, can only increase (or leave unaltered) the probability that the inference A_1 is correct. This conclusion is quite generally true.

Since we already have an expression for the probability in the denominator, this part of the calculation is completed. With the aid of equation (2.7.1) we may write explicitly

$$P(A_1 | (A_1 \vee A_2) \cdot B) = \frac{P(A_1 | B)}{P(A_1 | B) + P(A_2 | B)}. \qquad (2.7.3)$$

For the second arrow from $(A_1 \vee A_2)$, that going to A_2, we need merely replace A_1 in the above equation everywhere by A_2 and vice versa.

There remains only the arrow from A_3 to A_3. Since the statement A_3 is certain to be true on the hypothesis $A_3 \cdot B$, the probability $P(A_3 | A_3 \cdot B)$ is simply unity. This completes the reexpression, and we can use our abstract results to fill in the diagram for the concrete case, as displayed in figure 2.7.3.

Our results also provide the means for reversing the reexpression of diagrams, that is for expressing the probabilities appearing in the first abstract diagram in terms of those in the second.

2.8. THE EXPECTATION VALUE ESTIMATE

A theory of probability should provide means for estimating a numerical quantity when we are not sure of the correct value. Let us approach this question by means of two suggestive cases. For the first case, we return to a problem we considered in Chapter 1, the problem of estimating the pressure

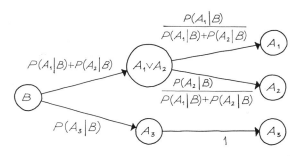

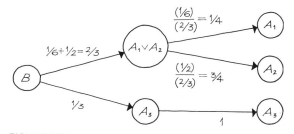

FIGURE 2.7.3
The reexpression of the probabilities appearing in
figure 2.7.2 in terms of those appearing in
figure 2.7.1, for both the abstract and the
concrete case.

produced by helium confined to a cubical box. We established the result that,
for the state ψ_j, the pressure p_j is given by $\frac{2}{3}E_j/V$. Since we had excluded
states whose energy fell outside the limits $\tilde{E} \pm 0.01\tilde{E}$ and associated equal
weight with those whose energy was within these limits, we quite reasonably
estimated the pressure as

$$p_{\text{est}} = \frac{2}{3}\frac{\tilde{E}}{V} \pm 0.01\left(\frac{2}{3}\frac{\tilde{E}}{V}\right).$$

It is this last step that we now want to examine more carefully. Let us see how
it might appear in symbolic form.

In setting up the statistical problem, we had the information originally
given us and in addition supposed both that a quantum-mechanical treatment
was appropriate and that there was only a single particle in the box. This is
then the hypothesis, which we call statement B. Then we considered a set of
inferences. Each inference was, effectively, a statement of the form, "Our

physical system is appropriately described by the wave function ψ_j." Let us denote such an inference by the symbol A_j. The corresponding probability may be written as $P(A_j | B)$. Though without explicitly saying so, we took the inferences to be mutually exclusive and exhaustive on the hypothesis. Together these properties imply that the probabilities should satisfy the normalization condition.

We can now put our verbal statements about the probabilities into mathematical form. If the number of energy eigenstates with energy within the range $\tilde{E} \pm 0.01\tilde{E}$ is designated by n', then we may express the probabilities as

$$P(A_j | B) = \begin{cases} 0 & \text{if } E_j \text{ is outside the energy limits,} \\ 1/n' & \text{if } E_j \text{ is within the energy limits.} \end{cases}$$

This numerical form weights equally the states within the energy limits, excludes the others, and satisfies the normalization condition.

An estimate for the pressure may now be written in the form

$$p_{\text{est}} = \sum_{\substack{\text{states within} \\ \text{the energy limits}}} \frac{2}{3}\frac{E_j}{V} P(A_j | B) = \sum_{j=1}^{\infty} \frac{2}{3}\frac{E_j}{V} P(A_j | B).$$

In the first stage, the summation goes over n' probabilities, each with a value $1/n'$. This will produce a result of $\frac{2}{3}\tilde{E}/V$, or very nearly so. A shift away from exactly that result might occur if there are more states in the higher half of the energy range,

$$\tilde{E} < E_j \leqslant \tilde{E} + 0.01\tilde{E},$$

than in the lower half; if so, we would—dare one say "intuitively"?—want the estimate of pressure to be a bit larger than $\frac{2}{3}\tilde{E}/V$. This is a refinement with which we did not concern ourselves earlier. It emerges rather naturally in this attempt to put our previous reasoning into symbolic form. In the second stage, the summation has been extended over all energy eigenstates. Those states which should not contribute to the pressure estimate actually do not, for the probability $P(A_j | B)$ is zero for them.

We have here a specific example of reasoning from a set of probabilities to the value we estimate for a function, here the function $\frac{2}{3}E_j/V$. The procedure amounts to summing over all the possible values of the function, each value being weighted with the probability that, in the situation under consideration, it is the correct value of the function.

Before we try to generalize the situation, let us look at the notion in another context. Suppose someone is handed a box of wooden matches and asked to estimate the number inside. He is allowed to shake the box and does so. From the rattle he can judge that the number of matches is two or more, but not greater than six, say. Let us suppose he is of a statistical bent and assigns a probability to each inference A_n that the box contains precisely n matches. These probabilities we write as $P(A_n | R)$, with R standing for the hypothesis that he has rattled the box of matches.

The inferences are mutually exclusive on the given hypothesis, for the box cannot contain both two and five matches, say. If the inferences run from $n = 0$ to $n = \infty$ over all integers, they are exhaustive, for the box does contain some definite number of matches that must lie in this range. The distribution of assigned probabilities might look like that in figure 2.8.1. That distribution is properly normalized:

$$\sum_{n=0}^{\infty} P(A_n | R) = 0 + 0 + 0.1 + 0.25 + 0.35 + 0.2 + 0.1 + 0 + \ldots$$
$$= 1.$$

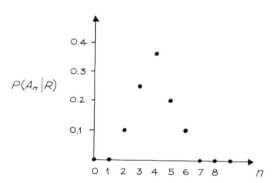

FIGURE 2.8.1

The distribution of probabilities for the box of wooden matches.

On the basis of this set of probabilities, the inference A_4, that the box has precisely four matches, would be the best single bet. The inferences A_3 and A_5 run a close second; so one should not be surprised if the box turned out to have three or five matches.

Let us see what emerges if we estimate the number of matches by first weighting each possible n with its probability and then summing. (This procedure may be reminiscent of the procedure of taking an "average.")

$$\begin{pmatrix} \text{An estimate of} \\ \text{the number of} \\ \text{matches} \end{pmatrix} = \sum_{n=0}^{\infty} nP(A_n \mid R)$$

$$= 0(0) + 1(0) + 2(0.1) + 3(0.25) + \ldots$$

$$= 3.95.$$

This does not lead to an integral estimate for n, but it does strongly suggest that $n = 4$ is the best integral guess.

Indeed, imagine that we had not seen the distribution of probabilities from which this estimate of 3.95 was calculated. Suppose we knew only the value of the estimate and the procedure by which someone had calculated it. If we were then required to make a single guess at the number of matches, we would surely pick the number 4. Whether this choice would be a reliable guess would depend on the distribution of probabilities. If we knew in addition that there is a peak around the estimated value of 3.95 (or 4), as here, then we would also have considerable confidence in a guess based on the mathematical estimate.

The two examples indicate the *usefulness* of an estimate based on weighting the possible numerical values of a function—pressure p_j or number of matches n—with their respective probabilities, and then summing. If we accept the line of reasoning leading to this kind of estimate as being in accord with common sense, we can generalize the situation and define precisely a procedure for estimating the value of a function when we can assess only probabilities that the function takes on particular values.

Let us write the given function as $f(x)$ and suppose that, on the hypothesis B, the variable x can have the discrete values $x_1, x_2, \ldots, x_n$. The number n may be finite or infinite. (We will frame only the definition for the discrete case now. The extension to continuous values of x is not overly difficult, but is unnecessary at this point.) To designate the inference that the variable x has the value x_j, we use the symbol X_j, and we specify that the n inferences in the set are, on the hypothesis B, exhaustive and mutually exclusive. To each inference X_j corresponds a probability $P(X_j \mid B)$. How the probabilities are assigned is of no concern here, though they must satisfy the normalization condition. Following the line of reasoning used in the two examples, we define an estimate—written $\langle f(x) \rangle$—of the value of the function $f(x)$ by the relation

$$\langle f(x) \rangle \equiv \sum_{j=1}^{n} f(x_j)P(X_j \mid B). \qquad (2.8.1)$$

The virtue of this definition is not abstract, but lies in its usefulness and in its close connection with common sense. The estimate $\langle f(x) \rangle$ is commonly called the *expectation value* of $f(x)$. It is important to remember that $\langle f(x) \rangle$ depends on *both* the function $f(x)$ *and* the probabilities $P(X_j | B)$.

2.9. ESTIMATES OF DEVIATIONS

It is worthwhile to return to the estimation of the number of matches and to ask how far the actual number in the box is *likely* to deviate from the single guess $n = 4$ or the expectation value $\langle n \rangle = 3.95$. With the distribution of probabilities in front of us, a reasonable estimate would be plus or minus one unit (one match). Two units are possible, but rather improbable.

There will be times when the distribution of probabilities will not be given quite as explicitly and simply as in this match case, but in which we will nonetheless be able to compute expectation values with relatively little effort. We should try to devise procedures for estimating deviations based solely on the calculation of expectation values.

Relative to the expectation value $\langle n \rangle$, an individual deviation in n may be written as $(n - \langle n \rangle)$. The deviations may be both positive and negative. For this reason, the expectation value of simply $(n - \langle n \rangle)$ is not a useful measure of the deviations to be anticipated:

$$\langle (n - \langle n \rangle) \rangle = \sum_{n=0}^{\infty} (n - \langle n \rangle) P(A_n | R)$$

$$= \sum_{n=0}^{\infty} n P(A_n | R) - \langle n \rangle \sum_{n=0}^{\infty} P(A_n | R)$$

$$= \langle n \rangle - \langle n \rangle \times 1 = 0.$$

We must first do something to $(n - \langle n \rangle)$ to ensure that negative deviations make a positive contribution to the estimate. This can be done by working with the absolute values of the deviations. For the given probabilities, this leads to

$$\langle |n - \langle n \rangle| \rangle = \sum_{n=0}^{\infty} |n - \langle n \rangle| \, P(A_n | R) = 0.865$$

as an estimate of the deviation from $\langle n \rangle$ to be anticipated. It conforms closely to our estimate by eye.

Often the use of an absolute value is awkward, but another, usually more convenient, measure is available. One can ensure that negative deviations contribute positively by first squaring the deviations and then computing an expectation value; afterwards the square root of the entire expression is to be taken. This means that we first compute

$$\langle (n - \langle n \rangle)^2 \rangle,$$

which may be simplified by expanding the square:

$$
\begin{aligned}
\langle (n - \langle n \rangle)^2 \rangle &= \langle n^2 - 2n\langle n \rangle + \langle n \rangle^2 \rangle \\
&= \langle n^2 \rangle - 2\langle n \rangle \langle n \rangle + \langle n \rangle^2 \\
&= \langle n^2 \rangle - \langle n \rangle^2.
\end{aligned}
$$

It is the reduction to calculation of merely $\langle n^2 \rangle$ and $\langle n \rangle$ that makes this a convenient procedure. A little arithmetic leads to

$$\langle (n - \langle n \rangle)^2 \rangle = 16.85 - (3.95)^2 = 1.2475.$$

When we take the square root, the result is

$$\langle (n - \langle n \rangle)^2 \rangle^{1/2} = 1.116$$

as an estimate of the anticipated deviation. This is sometimes called the *root mean square* estimate of the deviation. In general it gives a somewhat larger estimate than the absolute-value method, but both agree quite well with our estimate by eye from the distribution of probabilities itself. Because of the *convenience* of the root mean square method, we will generally use it from now on.

An expectation value estimate should not be accepted uncritically, for if the probability distribution is broad, the expectation value, though still giving the best single estimate, is not likely to provide a particularly good estimate. To test for the confidence one should have, one can compute anticipated deviations (by any one of several procedures). If the estimate of the anticipated deviations is relatively small, good; if not, then the expectation value estimate should not be ascribed great reliability.

None of the estimates of anticipated deviations can ensure, by their relative smallness, that a large deviation is impossible. A measure like the root mean square deviation, which does sample the entire distribution of probabilities but produces only a single number, cannot exclude the possibility of a large deviation with a very small probability of occurrence. To test for this, one needs to go back to the probabilities themselves, though one can get a clue from deviation estimates such as $\langle (n - \langle n \rangle)^4 \rangle^{1/4}$, which emphasizes more strongly the large deviations. It is usually true, however, that a root mean

square estimate of the anticipated deviations, plus some insight into the problem, suffice for a good assessment of the confidence to be placed in an expectation value estimate.

2.10. PROBABILITIES AND FREQUENCIES

We have adopted a meaning for the term "probability" that is totally independent of any notion of "frequency of occurrence," either imagined or
experimental. Rather, we have spoken about the probability of an inference,
given a hypothesis, and by that we have meant the degree of rational belief
in the correctness of the inference, given the hypothesis. Starting with this
understanding of the meaning of the term "probability," and leaning on
Cox's work for the two crucial theorems, we have developed an extensive
theory of probability. In this section we will discuss "frequencies of occurrence" in the terms of that theory.

There is no dispute with the claim that "frequencies of occurrence" have
a place in a theory of probability. If we imagine tossing a coin N times, we
have every right to be interested in the ratio of the number of heads to the
total number of tosses. Similarly, if we imagine examining individual molecules
in a gas, we could well be interested in the ratio of the number of molecules with a speed between a fixed pair of limits to the total number of
molecules we examine.

These two examples typify situations in which observations or experiments
are repeated many times, and the occurrence or nonoccurrence of some
specified property is established for each of those instances. Any one of a
variety of numerical values may emerge as the actual ratio of the number of
occurrences to the total number of observations. For example, if a conventional coin is flipped N times, the ratio of the number of times a head appears
to the total number of tosses may be any one of the numbers

$$\frac{0}{N}, \frac{1}{N}, \frac{2}{N}, \ldots, \quad \text{or} \quad \frac{N}{N}.$$

We begin the analysis by specifying a hypothesis and a set of inferences.
Suppose we are concerned with the occurrence of some property α. It might
be the property that a head appears in the flipping of a coin or that a molecular
speed lies between two fixed limits. Being concerned with ratios in N observations or experiments, we might frame a hypothesis, denoted by H'_N, containing
the following statement: "A total of N separate observations concerning

the property α are made." The statement would mean, for instance, that the coin is flipped N times, and the occurrence or nonoccurrence of a head is noted each time.

Associated with the N observations, we have a set of N inferences, each of the form, " In the jth observation, the property α is observed to occur." Such an inference we write as A_j, with the index j having the range $1 \leqslant j \leqslant N$. In the coin example, the inference A_2 might be made more specific as the statement, " In the second flip of the coin a head is observed to appear."

The preceding puts us in a position to consider N distinct probabilities, each of the form $P(A_j | H'_N)$. Immediately some questions arise. Is the numerical value of the probability to be the same for all observations, all values of the index j? For example, does the relation

$$P(A_2 | H'_N) = P(A_1 | H'_N)$$

hold? Is the probability of the occurrence of property α in the jth observation independent of whether or not the property has occurred in the preceding observations? For example, does the relation

$$P(A_2 | H'_N \cdot A_1) = P(A_2 | H'_N)$$

hold?

For the same coin flipped in the same way each time, the probability of the occurrence of a head in any given flip would be taken to be the same as the probability in any other flip. That probability would also be taken to be independent of whether a head had appeared in the preceding toss or had not. One would regard the N flips of the coin as N totally independent flips. Such decisions would not necessarily be made with, say, tests for the effectiveness of a new drug. Suppose twenty people with colds are to be given the new medication. One might initially assess the probability of a one-day cure to be the same for all twenty patients. But if one were then to observe that the first nineteen were indeed cured in a single day, quite reasonably one would increase the probability of a cure for the twentieth person over the initial probability. The twentieth cure would be an almost certain inference, given nineteen preceding cures. In a formal sense, we would have the relation

$$P(A_{20} | H'_{20} \cdot A_1 \cdot \ldots \cdot A_{19}) > P(A_{20} | H'_{20})$$

for the cold cures, with A_j being the inference that the jth person is cured in a single day.

Let us be modest in the scope of the situation we propose to analyze and specifically exclude cases like the cold-cure tests. We codify the restrictions by introducing a new hypothesis, denoted simply by H_N, which contains the following statements: "A total of N separate observations concerning the property α are made. The probability of the occurrence of property α in any one observation is the same as the probability in any other of the N observations. The probability of the occurrence of property α in any one observation is independent of whether property α occurs in any or all of the preceding observations."*

With the more restrictive hypothesis H_N, we have arranged matters so that the probability $P(A_j | H_N)$ has the same numerical value for all j. For the sake of brevity, let us write this property as

$$P(A_j | H_N) = p \quad \text{for all } j. \tag{2.1.1}$$

This defines the number p, which is the probability that property α occurs in any single one of the N observations.

For convenience later, let us introduce a second concise designation:

$$P(\sim A_j | H_N) = 1 - P(A_j | H_N) = 1 - p \equiv q. \tag{2.10.2}$$

Thus we define q as the probability that property α does not occur in any specific single observation.

The second of the two statements by which we have restricted the original hypothesis implies that relations such as

$$P(A_j | H_N \cdot A_i) = P(A_j | H_N), \quad \text{for } i < j, \tag{2.10.3a}$$

$$P(A_j | H_N \cdot \sim A_i) = P(A_j | H_N), \quad \text{for } i < j, \tag{2.10.3b}$$

now hold. These relations say that the probability that property α occurs in the jth observation is independent of whether or not the property has been observed in the ith observation, provided $i < j$. Indeed, any consistent conjunction of the statements

$$A_1, \ldots, A_{j-1}, \sim A_1, \ldots, \sim A_{j-1},$$

may be added to H_N as hypothesis without altering the probability that property α occurs in the jth observation. This property of the probabilities will play an essential role in the succeeding derivations.

By now we have adequately established the background, and we may begin to frame and to answer questions about the entire set of N observations.

---

* One may question whether H_N is strictly a "statement" and hence is technically admissible as a hypothesis. Nonetheless, the use of H_N provides a concise means of describing the situation that we will analyze.

First a still modest question, but one whose answer has more general relevance: what is the probability that in the first n observations property α *does* occur and that in the remaining $N - n$ observations property α *does not*?

The inference in this situation has the formal expression

$$A_1 \cdot \ldots \cdot A_n \cdot \sim A_{n+1} \cdot \ldots \cdot \sim A_N,$$

the conjunction of n affirmations of the occurrence of property α and $N - n$ denials, in the indicated observational order. Hence the formal expression for the corresponding probability is

$$P(A_1 \cdot \ldots \cdot A_n \cdot \sim A_{n+1} \cdot \ldots \cdot \sim A_N \,|\, H_N),$$

but we must surely be able to simplify this. Starting with the insertion of parentheses and the multiplicative decomposition permitted by Theorem II, we may reduce the probability in the following fashion:

$$P(A_1 \cdot \ldots \cdot A_n \cdot \sim A_{n+1} \cdot \ldots \cdot \sim A_N \,|\, H_N)$$
$$= P((A_1 \cdot \ldots \cdot \sim A_{N-1}) \cdot \sim A_N \,|\, H_N)$$
$$= P(A_1 \cdot \ldots \cdot \sim A_{N-1} \,|\, H_N) \, P(\sim A_N \,|\, H_N \cdot (A_1 \cdot \ldots \cdot \sim A_{N-1}))$$
$$= P(A_1 \cdot \ldots \cdot \sim A_{N-1} \,|\, H_N) \, P(\sim A_N \,|\, H_N)$$
$$= P(A_1 \cdot \ldots \cdot A_n \cdot \sim A_{n+1} \cdot \ldots \cdot \sim A_{N-1} \,|\, H_N) \, q.$$

The crucial step (from the third to the fourth line) makes use of the property that, by the hypothesis H_N, the probability that inference $\sim A_N$ is correct is independent of whether or not the preceding inferences are correct; that is,

$$P(\sim A_N \,|\, H_N \cdot (A_1 \cdot \ldots \cdot A_n \cdot \sim A_{n+1} \cdot \ldots \cdot \sim A_{N-1})) = P(\sim A_N \,|\, H_N),$$

an extension of the relations given explicitly by equations (2.10.3a, b). The steps have achieved a simplification by extracting the last inference, $\sim A_N$, from the original inference and by replacing it with $P(\sim A_N \,|\, H_N)$ in a multiplicative fashion.

The steps may be repeated. With each repetition one more inference is removed from the original inference. It is replaced in a multiplicative fashion by either p or q, by p if the inference removed was of the form A_j or by q if it was of the form $\sim A_j$. The end result of $N - 1$ such operations is a complete reduction to the concise form

$$P(A_1 \cdot \ldots \cdot A_n \cdot \sim A_{n+1} \cdot \ldots \cdot \sim A_N \,|\, H_N) = p^n q^{N-n}. \qquad (2.10.4)$$

Now suppose we construct an over-all inference in the following, more general way. For each observation j, we specify either A_j or $\sim A_j$, that is, an occurrence or a nonoccurrence of property α. The only restriction is that we specify a total of n inferences of the form A_j and a total of $N - n$ inferences of the form $\sim A_j$. The over-all inference is to be formed by their conjunction. The inference with which we have just worked is one example, but now the order in the specified sequence might be different, for example,

$$\sim A_1 \cdot A_2 \cdot \ldots \cdot A_n \cdot A_{n+1} \cdot \sim A_{n+2} \cdot \ldots \cdot \sim A_N.$$

The latter differs in that A_1 has been changed to $\sim A_1$, and $\sim A_{n+1}$ to A_{n+1}, preserving the number of affirmations and denials at n and $N - n$, respectively.

In more colloquial terms, we are considering an over-all inference that specifies in *some single ordered way* n occurrences of property α and $N - n$ nonoccurrences. To the question, what is the probability of such an over-all inference, we have the answer: the probability is given by the right hand side of equation (2.10.4). The reduction procedure leading to equation (2.10.4) did not depend upon the order in which the n affirmations and $N - n$ denials appeared in the set of N observations. Therefore all sequences specifying, in some definite order, n occurrences and $N - n$ nonoccurrences have the same probability.

This is the first of the significant mathematical results of this section. For lack of an adequate notation, we must write it in the following cumbersome yet explicit fashion:

$$P\left(\begin{matrix} \text{a } specific \text{ sequence} \\ \text{of } n \text{ occurrences and} \\ N - n \text{ nonoccurrences} \end{matrix} \;\middle|\; H_N \right) = p^n q^{N-n}. \tag{2.10.5}$$

When we ask merely about the ratio of the number of occurrences to the total number of observations, we do not care about the precise order in which the occurrences and nonoccurrences are present. Rather, we are concerned with the probability of the following inference: "The property α is observed to occur in n observations out of the total number N of observations." Let us designate this inference by B_n. There are usually many ways, many distinct observational sequences, in which the inference B_n may be a correct inference.

We begin the task of calculating the probability $P(B_n | H_N)$ by relating the inference B_n to the specific sequences of n occurrences and $N - n$ non-occurrences. To avoid a proliferation of symbols we write the relation as

$$B_n = \left(\begin{matrix} \text{the disjunction of the over-all inferences} \\ \text{for all specific sequences of } n \\ \text{occurrences in } N \text{ observations} \end{matrix}\right). \tag{2.10.6}$$

As an example, we take the tractable case of $N = 2$ and $n = 1$. Then for specific sequences of $n = 1$ occurrences we would have $(A_1 \cdot \sim A_2)$ and $(\sim A_1 \cdot A_2)$. The expression for B_1 would be

$$B_1 = (A_1 \cdot \sim A_2) \vee (\sim A_1 \cdot A_2).$$

This says, effectively, that one occurrence of property α out of two observations may happen in either of two ways: property α occurs in the first observation but not the second, or property α occurs in the second observation but not the first.

The over-all inferences for the different specific sequences of n occurrences in N observations are mutually exclusive on the hypothesis H_N. For example, the inferences $(A_1 \cdot \sim A_2)$ and $(\sim A_1 \cdot A_2)$ are mutually exclusive. So we may use a line of deduction already employed in Section 2.5 to reduce $P(B_n \mid H_N)$ in the following way:

$$P(B_n \mid H_N) = P \left(\begin{array}{l} \text{the disjunction of the over-all} \\ \text{inferences for all specific} \\ \text{sequences of } n \text{ occurrences} \\ \text{in } N \text{ observations} \end{array} \middle| H_N \right)$$

$$= \sum_{\substack{\text{sum over all specific sequences of} \\ n \text{ occurrences in } N \text{ observations}}} P \left(\begin{array}{l} \text{the over-all inference for} \\ \text{the specific sequence of } n \\ \text{occurrences in } N \text{ observations} \end{array} \middle| H_N \right)$$

$$= \left(\begin{array}{l} \text{the number of different} \\ \text{specific sequences of } n \\ \text{occurrences in } N \\ \text{observations} \end{array} \right) p^n q^{N-n}. \tag{2.10.7}$$

The step to the last line is permitted by equation (2.10.5). With this intermediate result, we have reduced the determination of the probability $P(B_n \mid H_N)$ to a counting problem: we must count the number of different specific sequences of n occurrences in N observations.

In a later chapter we will again meet this counting problem, and so the labor devoted to the solution here will be labor doubly well spent. Imagine the results of N observations displayed along a line, with an upward arrow ↑ denoting occurrence of property α in a specific observation and a downward arrow ↓ denoting nonoccurrence. The sole way in which N observations can yield zero occurrences and N nonoccurrences is presented in the first sketch with all arrows down:

This situation becomes the first entry in table 2.10.1.

TABLE 2.10.1

Tabular entries in the solution of the counting problem: for N observations concerning property α, how many different observational sequences are there which give n occurrences?

Number of occurrences	Number of different observational sequences
0	1
1	N
2	$N(N-1)/2!$
3	$N(N-1)(N-2)/!$
$\vdots$	$\vdots$
n	$N!/n!(N-n)!$
$\vdots$	$\vdots$
N	1

To display a situation with one occurrence and $N-1$ nonoccurrences we may imagine one arrow flipped up:

$$\uparrow\downarrow\downarrow\downarrow\downarrow\downarrow\downarrow\ldots.$$

It need not be the first observation that yields an occurrence. We may flip up any one of the N downward arrows. So there are N different sequences of one upward arrow and $N-1$ downward arrows.

Next we consider the situation with two occurrences and $N-2$ nonoccurrences. Starting from the display with all arrows downward, we must flip up two arrows. For the first of the two we may pick any one of the N downward arrows. Having chosen the first, we may pick the second from among the $N-1$ remaining downward arrows. This gives us $N(N-1)$ ordered pairs of choices. Not all of these, however, lead to different sequences of two occurrences and $N-2$ nonoccurrences. Flipping up first the arrow for observation 2, say, and then that for observation 5, say, leads to the same sequence of observational results as first flipping up the arrow for observation 5 and then that for observation 2, as indicated in the sketch:

first   second
$\downarrow\uparrow\downarrow\downarrow\uparrow\downarrow\downarrow\ldots$ ⎫ different ordered pairs
$\downarrow\uparrow\downarrow\downarrow\uparrow\downarrow\downarrow\ldots$ ⎬ of choices but the same
second   first ⎭ observational sequence.

To get the correct number of distinct observational sequences with two occurrences and $N-2$ nonoccurrences we must divide $N(N-1)$ by 2.

Now we imagine three arrows flipped up for three occurrences and $N - 3$ nonoccurrences. The reasoning is analogous. For the first of the three arrows, we may choose any one of the N downward arrows. Having chosen the first, we may pick the second from among the remaining $N - 1$ downward arrows. And having chosen the first and second, we may pick the third from among the remaining $N - 2$ downward arrows. This gives $N(N - 1)(N - 2)$ ordered triplets of choices. Again, these do not all lead to different observational sequences. The ordered triplet of choices, arrow 2, then arrow 5, and finally arrow 7, leads to the same observational sequence as do the ordered triplets

$$(5, 2, 7), (5, 7, 2), (7, 5, 2), (7, 2, 5) \text{ and } (2, 7, 5).$$

The ordered triplets differ in the permutation of the three numbers among themselves: $3! = 1 \cdot 2 \cdot 3 = 6$. To get the correct counting of different observational sequences with three occurrences, we must divide the number of ordered triplets of choices, $N(N - 1)(N - 2)$, by the permutations among three chosen arrows, $3!$.

Rather than plod onward with four arrows upward, and then five, six, ..., we may use the preceding analysis to generate the general result for n arrows upward. The number of ordered choices of n flippings in which we can get n arrows up is

$$N(N - 1)(N - 2) \cdots (N - [n - 1]).$$

Not all of these lead to different observational sequences of n arrows up. We must divide by the number of permutations among a chosen set of n arrows. Thus the number of different observational sequences with n occurrences and $N - n$ nonoccurrences is

$$\frac{N(N - 1)(N - 2) \cdots (N - [n - 1])}{n!}.$$

The expression can be tidied up by noting that

$$N(N - 1)(N - 2) \cdots (N - [n - 1]) = \frac{N!}{(N - n)!}.$$

With the aid of this more convenient form, we may write

$$\begin{pmatrix} \text{the number of different} \\ \text{specific sequences of} \\ n \text{ occurrences in} \\ N \text{ observations} \end{pmatrix} = \frac{N!}{n!(N - n)!}. \qquad (2.10.8)$$

When 0! appears, as for $n = N$, we agree to call it unity: $0! = 1$. This convention is necessary if we are to have a single expression valid for all n.

Having solved the counting problem, we may return to equation (2.10.7) and insert the solution. The probability of n occurrences and $N - n$ non-occurrences of property α, *regardless* of the sequence in which the occurrences happen in the observations, is then

$$P(B_n \mid H_N) = \frac{N!}{n!(N - n)!}\, p^n q^{N-n}. \qquad (2.10.9)$$

This result is central to the discussion that follows.

In figure 2.10.1 two examples are given of the probability $P(B_n \mid H_N)$ as a

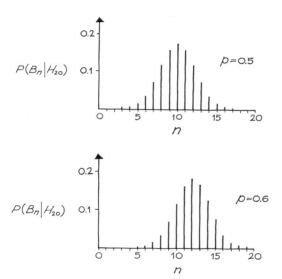

FIGURE 2.10.1
Two examples of the probabilities $P(B_n|H_N)$ as a function of n, the number of occurrences.

function of n. In the upper diagram the probability of the occurrence of property α at a single observation is taken to be $p = 0.5$. This implies the equality $q = p$ for the upper diagram; so the numerical value of the factor $p^n q^{N-n}$ in $P(B_n \mid H_N)$ does not change with n. The shape of the distribution is determined solely by the combination of factorials. This diagram would apply in a discussion of the occurrence of heads in twenty flips of a conventional coin. In the lower diagram, with $p = 0.6$, the distribution is somewhat asymmetric, which is the more typical case.

The inferences B_n, forming a set of inferences with n ranging from zero to N, are exhaustive and mutually exclusive on the hypothesis H_N. Thus the sum of the probabilities should satisfy the normalization condition. To confirm this, we must perform the following summation:

$$\sum_{n=0}^{N} P(B_n \mid H_N) = \sum_{n=0}^{N} \frac{N!}{n!(N-n)!} p^n q^{N-n}. \qquad (2.10.10)$$

The trick for evaluating this summation—and for working with $P(B_n \mid H_N)$ in general—consists of recognizing the close connection with the binomial theorem. For any two numbers, p and q, and a positive integer N, the binomial theorem gives an expansion of $(p+q)^N$ as

$$(p+q)^N = q^N + Npq^{N-1} + \frac{N(N-1)}{2!} p^2 q^{N-2} + \cdots + p^N$$

$$= \sum_{n=0}^{N} \frac{N!}{n!(N-n)!} p^n q^{N-n}. \qquad (2.10.11)$$

Remarkably, each probability $P(B_n \mid H_N)$ corresponds uniquely to a term in the expansion of $(p+q)^N$ by the binomial theorem. For this reason the probability distribution $P(B_n \mid H_N)$ is often called the *binomial probability distribution*.

Now that we have established a connection with the binomial theorem, we may perform the summation in equation (2.10.10) as follows:

$$\sum_{n=0}^{N} P(B_n \mid H_N) = (p+q)^N = 1.$$

The last step is a consequence of equations (2.10.1, 2). The probabilities as derived are indeed properly normalized.

We have now a set of probabilities with which to discuss frequencies in considerable detail. The special cases of $p = 1$ and $p = 0$ are genuinely trivial. Indeed, for them one need not have bothered with probability theory, for one can strictly deduce the corresponding frequencies, 1 and 0. The truly germane case is $p \neq 0, 1$, on which we focus our attention.

When the probability of occurrence at a single observation is neither 0 nor 1, each probability $P(B_n \mid H_N)$ has a nonvanishing value. Consequently, for each of the possible ratios of the number of occurrences to the total number of observations,

$$\frac{0}{N}, \frac{1}{N}, \frac{2}{N}, \ldots, \frac{n}{N}, \ldots, \frac{N}{N}, \qquad (2.10.12)$$

there is a finite probability that the ratio will actually be found if N observations are made. The theory does not tell us what the actual ratio will be.

Rather, it provides us, for each possible ratio, with the probability that that ratio will emerge.

The connection between the probability p and the frequencies of occurrence will become more evident if we calculate the expectation value estimate of the frequency. We proceed from the definition of expectation value given in equation (2.8.1), and write

$$\left\langle \frac{n}{N} \right\rangle = \sum_{n=0}^{N} \left(\frac{n}{N} \right) P(B_n \mid H_N)$$

$$= \sum_{n=0}^{N} \left(\frac{n}{N} \right) \frac{N!}{n!(N-n)!} \, p^n q^{N-n}. \qquad (2.10.13)$$

The indicated summation is similar to that which we performed with the aid of the binomial theorem. Only the presence of an extra factor of the summation variable n makes the summation more difficult. But again there is a trick. For the sake of performing the summation itself, we may consider p and q as fully independent variables, that is, as not constrained by the relation $q = 1 - p$. Then we note that the extra factor of n may be taken together with the factor p^n and expressed as a partial derivative:

$$np^n = p \frac{\partial}{\partial p} (p^n).$$

And finally the summation may be done explicitly as follows:

$$\sum_{n=0}^{N} \left(\frac{n}{N} \right) \frac{N!}{n!(N-n)!} \, p^n q^{N-n}$$

$$= \sum_{n=0}^{N} \frac{1}{N} \frac{N!}{n!(N-n)!} \left[p \frac{\partial}{\partial p} (p^n) \right] q^{N-n}$$

$$= \frac{1}{N} p \frac{\partial}{\partial p} \left[\sum_{n=0}^{N} \frac{N!}{n!(N-n)!} \, p^n q^{N-n} \right]$$

$$= \frac{1}{N} p \frac{\partial}{\partial p} [(p+q)^N]$$

$$= p(p+q)^{N-1}.$$

The temporary independence of q and p is first used in the step in which the partial differentiation with respect to p is pulled out in front of the summation. It is used again in performing the differentiation leading to the last expression. Having accomplished the summation for the general case with arbitrary values for p and q, we may now insert the particular relation, namely $q = 1 - p$. The factor $(p+q)^{N-1}$ becomes simply unity. Upon

157584

inserting the result into equation (2.10.13), we arrive at

$$\left\langle \frac{n}{N} \right\rangle = \sum_{n=0}^{N} \left(\frac{n}{N} \right) P(B_n \,|\, H_N) = p. \qquad (2.10.14)$$

The probability p gives directly the expectation value estimate of the frequency. We must bear in mind that, in an actual set of N observations, any one of the many possible frequencies of occurrence may be found. The full range, from $0/N$ to N/N, was given in expression (2.10.12). For each such frequency, each such ratio of the form n/N, we have a definite, non-zero probability, namely $P(B_n \,|\, H_N)$, dependent on n and N as well as on p. But when we calculate the expectation value of the frequency, we do find a particularly simple result: the expectation value estimate of the frequency of occurrence is given by the probability that property α is observed to occur in a single observation.

To extend our understanding of the connection between probabilities and frequencies, let us examine the matter of deviations. Suppose we were to make a set of N observations concerning the property α. Some definite ratio of occurrences would emerge. How close to the expectation value of the ratio is the observed ratio likely to be? And how should the extent of the deviation we anticipate depend on N, the total number of observations in the set?

To some extent these questions can be answered by looking at graphs, of which figure 2.10.2 presents an example, but an analytic expression will

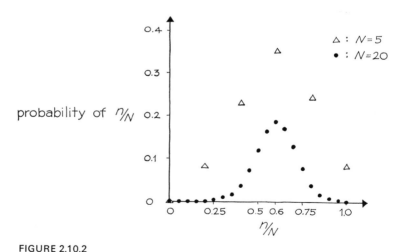

FIGURE 2.10.2

For p fixed at $p = 0.6$, this diagram presents the probabilities of the ratios n/N for two values of the number N of observations. The peak in the distribution remains fixed at a value equal to p. It is the width of the distribution of probabilities that changes, decreasing significantly with increasing N.

provide more detailed information. Let us calculate the root mean square estimate of the deviation

$$\Delta\left(\frac{n}{N}\right) \equiv \left\langle\left(\frac{n}{N} - \left\langle\frac{n}{N}\right\rangle\right)^2\right\rangle^{1/2}$$

$$= \left[\left\langle\left(\frac{n}{N}\right)^2\right\rangle - \left(\left\langle\frac{n}{N}\right\rangle\right)^2\right]^{1/2}.$$

Since the value of $\langle n/N \rangle$ is already known, we need determine only the expectation value of the square of the ratios. The formal expression for the latter is

$$\left\langle\left(\frac{n}{N}\right)^2\right\rangle = \sum_{n=0}^{N} \left(\frac{n}{N}\right)^2 \frac{N!}{n!(N-n)!} p^n q^{N-n}.$$

The tricks used before to accomplish the summation will work here. First we isolate $n^2 p^n$ and rewrite that factor as

$$n^2 p^n = n\left(p \frac{\partial}{\partial p}\right)p^n = \left(p \frac{\partial}{\partial p}\right)\left[\left(p \frac{\partial}{\partial p}\right)p^n\right]$$

$$\equiv \left(p \frac{\partial}{\partial p}\right)^2 p^n.$$

The expression in the second line may be considered merely a concise abbreviation for the sequence of differentiations and multiplications indicated in the first line. Now, regarding p and q as fully independent variables, we may perform the summation as follows:

$$\sum_{n=0}^{N} \left(\frac{n}{N}\right)^2 \frac{N!}{n!(N-n)!} p^n q^{N-n}$$

$$= \sum_{n=0}^{N} \frac{1}{N^2} \frac{N!}{n!(N-n)!} \left(p \frac{\partial}{\partial p}\right)^2 p^n q^{N-n}$$

$$= \frac{1}{N^2} \left(p \frac{\partial}{\partial p}\right)^2 [(p+q)^N]$$

$$= \frac{1}{N^2} p[N(p+q)^{N-1} + pN(N-1)(p+q)^{N-2}].$$

At this point, having finished with the partial derivative manipulations, we may once again set q equal to $1-p$. Upon tidying up, we have the result:

$$\left\langle\left(\frac{n}{N}\right)^2\right\rangle = p^2 + \frac{1}{N} p(1-p).$$

For the root mean square estimate of the deviation, we find

$$\Delta\left(\frac{n}{N}\right) = \frac{\sqrt{p(1-p)}}{N^{1/2}}. \tag{2.10.15}$$

We must remember that frequencies of occurrence are ratios whose value must lie between 0 and 1, end points included. So for any given N and some fixed p, we may plot the probability of each possible ratio along an abscissa with a range of 0 to 1. The graph will show a peak at (or near) the value p. For such a graph, the import of equation (2.10.15) is that the spread in the points will decrease as $1/N^{1/2}$ when N is increased. Such a decrease is qualitatively evident in figure 2.10.2.

The result for $\Delta(n/N)$ may be expressed in terms of what we should expect to observe in a set of N observations. We have no guarantee that the observed ratio of occurrences will be equal, or nearly equal, to $\langle n/N \rangle$, itself equal to p. Nonetheless, if N is very large, then it is highly improbable that the observed ratio will depart significantly from $\langle n/N \rangle$.

There are cases in which the analysis of the preceding paragraph may be used in the opposite direction, that is, to assess the probability p that property α will occur in a single observation from an observed frequency of occurrence in a large (though always finite) number of observations. One must be cautious, for there are possible pitfalls. I paused after writing the last sentence and flipped a penny twenty times. The observed ratio of occurrences of heads (8) to the total number of observations (20) was 0.4. I might assess the probability of the occurrence of a head at a single toss as $p = 0.4$. If I knew nothing else, in particular, knew nothing about symmetry for the two sides of the coin, this would be a reasonable assessment. Because of a coin's symmetry, however, I feel that p should be closer to 0.5. So I flipped the coin a further twenty times, finding six heads; taking the forty tosses altogether, I might assess p to be 0.35. The coin is not necessarily "biased"; a third run of twenty tosses yielded precisely ten heads. Rather, we are dealing with relatively small N, for which relatively large fluctuations in observed ratios are quite probable. Generally one needs large N for a reliable assessment of p from an observed frequency, and even then one has *no guarantee* that going to still larger N will not alter the assessment significantly. There is, moreover, a big difference between assessing p from an observed frequency only and assessing from that information plus other relevant knowledge, such as a coin's high degree of symmetry. All this is, however, in the nature of a (rather loose) aside. In statistical mechanics one typically does not have information about observed frequencies of occurrence; so this means of assessing or assigning probabilities is largely irrelevant.

2.11. A REDUCTION PROCEDURE

As an introduction to the technical problem to be posed and solved in this section, let us consider an election for public office. Suppose we are ardent supporters of a candidate; let us denote by W the inference, "Our candidate wins the election." The statement that there is an election, the results of public opinion polls, and the like, form a hypothesis E. Let us suppose that the political analysts provide a set of probabilities $P(W \cdot V_j | E)$. The symbol V_j denotes the inference, "Our candidate gets j votes in the election." This is, admittedly, a trifle artificial, but let us suppose that the political analysts do provide such a set of probabilities, with j ranging from zero to infinity. Our real interest lies with the probability $P(W|E)$, the probability that our candidate wins, regardless of how many votes he gets. Question: can we somehow extract $P(W|E)$ from the given set of probabilities?

It would seem that the answer to the question should be affirmative, and indeed it is. First we make the existence of some connection evident by applying Theorem II:

$$P(W \cdot V_j | E) = P(W|E)\ P(V_j | E \cdot W).$$

Now we note that the inferences V_j form a set of inferences that are exhaustive and mutually exclusive on the hypothesis $E \cdot W$. In the election our candidate will necessarily receive a number of votes between zero and infinity (exhaustive property) but only one such number of votes (mutually exclusive property). Hence if we sum the above equation over all j, the righthand side will be $P(W|E)$ times a sum whose value is unity. So we may indeed extract the desired probability:

$$P(W|E) = \sum_{j=0}^{\infty} P(W \cdot V_j | E).$$

The essential element in the reduction procedure is the summation over a set of inferences that are exhaustive and mutually exclusive on the basis of a particular hypothesis. We made use of that property of the inferences V_j with respect to the hypothesis $(E \cdot W)$ which appears after Theorem II has been applied. At the cost of considerably more effort, one may show that the same result emerges provided merely that the exhaustive and mutually exclusive property holds with respect to the initial hypothesis (here E) as "particular hypothesis."

We may summarize the reduction procedure as follows: whenever we have a set of probabilities of the form $P(A' \cdot A_j | B)$, where j takes on a range of values, we may extract the probability $P(A' \mid B)$ as

$$P(A' \mid B) = \sum_j P(A' \cdot A_j \mid B)$$

provided that the inferences A_j form a set of inferences exhaustive and mutually exclusive on either the hypothesis B or the hypothesis $B \cdot A'$ or both. This procedure will play an essential role in Chapter 4 and again in Chapter 6.

REFERENCES for Chapter 2

Our development of probability theory is based on Richard T. Cox's little book, *The Algebra of Probable Inference* (Baltimore: The John Hopkins Press, 1961). Most of the material in the first and third sections of the book appeared much earlier in a paper by Cox in the *American Journal of Physics*, **14**, 1 (January-February 1946). The derivation of the two theorems from the two axioms is carefully presented. One can, moreover, see clearly the stages of which conventions about the scale for measuring probabilities are needed and adopted. Some results that we derived in this chapter, in particular, in Section 2.10, and some that we will derive in Chapter 3, are developed by Cox in greater generality. The slim volume contains a wealth of insight into probability theory, all presented in a modest style. Professor Cox's book was reviewed by E. T. Jaynes, *American Journal of Physics*, **31**, 66 (1963). The review is well worth reading for its clear statement of Cox's achievement in establishing consistency and uniqueness for a theory of probability, when probability theory is broadly viewed as a branch of logic.

Sir Harold Jeffreys, in his *Scientific Inference* (Cambridge, Eng.: Cambridge University Press, 2nd ed., 1957), is primarily concerned with the problem of inference and induction, but the second chapter is devoted to a brief (and necessary) development of his probability theory; an extensive treatment is to be found in his *Theory of Probability* (Oxford: Oxford University Press, 3d ed., 1967). There is much here of general interest to the natural scientist; particularly germane is the chapter entitled "Statistical Mechanics and Quantum Theory."

John Maynard Keynes, much better known as an economist of international stature, and less well-known as a very successful manager of finances for King's College of the University of Cambridge, was a probability theorist of note. His *Treatise on Probability* (London: Macmillan, 1921, republished in paperback in 1962 by Harper and Row) is a landmark in the history of the subject. Part I, on "Fundamental Ideas," provides a masterful introduction to the subtleties of what probability is to mean and the theory to encompass.

Pierre Simon, Marquis de Laplace, wrote his *Philosophical Essay on Probabilities* (New York: Dover, reprinted in 1951), as a popular exposition of his theory of probability, basing it in part on lectures given in 1795. His intent was to avoid mathematics where possible and yet to present the principles and general results. Close to two centuries later, the essay remains fascinating and intellectually fruitful reading. In it one finds Laplace's famous remark: "Probability theory is only common sense reduced to calculation."

Erwin Schrödinger presented his views about the foundations of probability theory in two articles, *Proc. Royal Irish Acad.*, **51**, section A (1947), 51 and 141.

PROBLEMS

2.1. On merely the mild assumption that the hypothesis B is sufficiently relevant to the inference A that a meaningful probability $P(A \mid B)$ exists, prove the following two results and suggest a verbal interpretation of each:

$$P(A \cdot \sim A \mid B) = 0, \qquad P(A \vee \sim A \mid B) = 1.$$

2.2. On the minus sign in the theorem of Section 2.4. Take the inferences D_1 and D_2 and the hypothesis R of Section 2.3 and calculate the probability $P(D_1 \vee D_2 \mid R)$ in two ways: first by reasoning in a colloquial manner, after noting that the case D_1 indicates the case D_2, and then by using the theorem of Section 2.4. Do the two results agree? Is the minus sign vindicated?

2.3. More on Bayes' theorem. In discussing Bayes' theorem, we explicitly excluded the possibility $P(B' \mid B) = 0$. Suppose that $P(B' \mid B)$ *is* zero. What do we do then? (Hint: ask yourself whether $B' \cdot B$ is then acceptable as a hypothesis.)

2.4. Suppose the probability that heads appears when a coin is tossed once is $p = 0.5$, and suppose the coin is tossed 20 times. What is the probability that heads appears during the first ten tosses and tails during the remainder? What is the probability that heads appears 10 times anywhere among the 20 tosses? Why are the two probabilities numerically different?

2.5. Suppose two standard dice are rolled a single time. Assume that the dice are tossed independently and that the probability of the appearance of any particular face on a die is the same as that of any other face on the die. Call the preceding the hypothesis H. The *sum* of the numbers that appear necessarily falls between 2 and 12, endpoints included. Let S_n represent the inference, "The sum of the numbers appearing is n."

a. Calculate the probabilities $P(S_n \mid H)$ and display them in a graph.

b. Compute the expectation value of the sum of the numbers appearing.

c. Estimate the deviation to be anticipated from the expectation value estimate, both by eye and by some reasonable analytical method, and justify the method you use.

2.6. An object that must land with one of two sides down has been tossed 100 times. One side has a single dot painted on it; the other, two dots. The object is not necessarily symmetric. Denote the inference, "The side with one dot will land down on the next toss," by A_1, and analogously for two dots.

a. On the hypothesis H that the first paragraph of the problem is true and that the object is to be tossed once again, are the inferences mutually exclusive? Exhaustive? What value must the sum of the probabilities, $P(A_1 | H)$ and $P(A_2 | H)$, have?

b. Would you be willing to assign definite numerical values to the two probabilities above? If so, what values? Justify the answers you give.

c. Now we add to the hypothesis the information that the average number of the dots landing down in the 100 throws is a specific number C. (The number C does lie between 1 and 2, but it is not necessarily 1.5.) Call the augmented hypothesis H^*. Answer the questions raised in parts a and b for the case of the augmented hypothesis.

(Note: *some* of the questions asked in this problem do not have unique answers; that is, there is scope for disagreement among "reasonable" answers.)

2.7. Random walk in one dimension. This problem examines the proverbial drunkard and how far he is likely to stagger from the lamppost. Suppose the drunkard is constrained (by fences, perhaps) to stagger along the side walk—so that the problem is one-dimensional—and that he takes steps of equal length l (somewhat unlikely). We specify that the drunkard is as likely to step to the right as to the left on the next step, regardless of what his previous steps were. He starts from the lamppost at $x = 0$ and takes N steps. All this is given as hypothesis. Much of the mathematical analysis required for the following questions has already been performed in Section 2.10. You need only extract it.

a. Support the claim that the probability p of a step to the right (increasing x) is $p = 1/2$.

b. What is the probability that, after N steps, the drunkard is a distance $x = Nl$ from the lampost? $x = -Nl$?

c. What is the expectation value $\langle x \rangle$ of this distance from the lamppost? Why is the value not unreasonable?

d. The expectation value of $|x|$ would not be zero, but is rather difficult to calculate. To get an estimate of how far the drunkard is likely to be distant, in the absolute sense, calculate the root mean square estimate, $\langle x^2 \rangle^{1/2}$. How does this vary with N?

e. There is another route to the $N^{1/2}$ dependence. The drunkard's distance after N steps is some sequence of steps to the right and left: $x = S_1 + S_2 + S_3 + \ldots$, where S_n denotes the nth step, and $S_n = +l$ or $-l$ with equal probability. Square the expansion for x and argue that, since the first step is as likely to be $-l$ as $+l$, and likewise for the others, the expectation value of the cross terms will be zero. Then sum the expectation values, take a square root, and arrive at $N^{1/2}l$ as an estimate of his (absolute) distance.

3

THE ASSIGNMENT OF PROBABILITIES

We must now choose the criteria that will guide us in assigning probabilities in statistical mechanics, in specifying the degree of rational belief in an inference, given a relevant hypothesis. This is a crucial step, for once the criteria have been adopted, the framework for a theory of statistical mechanics will have been constructed. But the assignment of probabilities is a notoriously difficult problem.

We can readily appreciate that the extent of the difficulty depends strongly on the context, on the nature of the hypothesis and inference. When the hypothesis is as simple as "A standard die is tossed once," the probability of the inference, "An even number will appear," can easily be agreed on, even though the process of arriving at the assignment may be difficult to articulate and to justify. The inference, "It will rain tomorrow," is equally simple, but for the meteorologist the hypothesis is enormously more complex.

The latter is generally a combination of both measurement and past experience. It is much easier to arrive at the probability of $\frac{1}{2}$ for the die case than to reason to a probability of 3 out of 10, say, for the probability of rain tomorrow.

The hypotheses and inferences that we will meet in statistical mechanics will not be as simple as those for rolling dice, nor as complex as those for predicting the weather. Some of the information in the hypothesis will remain in qualitative form, such as an assertion that quantum mechanics is applicable, but other, initially quantitative parts we will be able to cast into mathematical restrictions on an acceptable assignment of probabilities. Since the restrictions will fail to determine uniquely the probabilities of the inferences under consideration, the introduction of a new principle will be necessary, a principle that will enable us to select a single assignment of probabilities out of the many assignments that are compatible with the hypothesis and the restrictions it implies. What is needed is some quantitative criterion for the " best " choice of probability assignment from the vast array of options open to us.

We will find that criterion for " best " choice, that guiding principle, in a notion central to the discipline of information theory. For this reason the first part of the chapter will be devoted to introducing and discussing the notion of the "amount of missing information" that is associated with a specified assignment of probabilities. At first the notion will be quite fuzzy, but as we work on it, the notion will become more sharply defined. After that, we will turn to the central task of the chapter and set up a scheme, based on two criteria, for assigning probabilities.

3.1. SOME NECESSARY BACKGROUND

In developing statistical mechanics, we will consider a set of inferences, $A_1, A_2, \ldots, A_n$, and some hypothesis B, which will consist of the information given about the physical situation and perhaps the statement that a quantum-mechanical treatment is appropriate. At times the number of inferences in the set will be finite, at others infinite, but since the inferences will always, on the given hypothesis, be exhaustive and mutually exclusive, we may, from the outset, limit ourselves to the problem of assigning probabilities to inferences that meet these conditions.

There are some immediate limitations on acceptable values for the numbers representing the probabilities. As a consequence of our conventions on the numerical scale, the value of a probability $P(A_j \mid B)$ must lie between 0 and 1, endpoints included:

$$0 \leqslant P(A_j \mid B) \leqslant 1. \tag{3.1.1}$$

A further restriction is imposed by the property that the inferences are exhaustive and mutually exclusive on the given hypothesis. This property implies—as the derivation leading to equation (2.5.4) showed—that the probabilities must satisfy the relation

$$\sum_{j=1}^{n} P(A_j \mid B) = 1. \tag{3.1.2}$$

This condition, which we have called the normalization condition, is a restriction on the probabilities taken as a set.

The two restrictions, equations (3.1.1) and 3.1.2), represent all that one can say in complete generality about the numerical values of the probabilities for a set of inferences that are exhaustive and mutually exclusive on the given hypothesis. A name for the result of assigning numerical values that satisfy the two restrictions will be useful. Let us call such an assignment a *probability distribution*. In some way probabilities summing to unity have been distributed over the set of inferences.

Brevity is a virtue. To simplify the writing of equations involving probabilities, let us adopt a shortened notation and write merely P_j for $P(A_j \mid B)$, that is, adopt the abbreviation

$$P_j \equiv P(A_j \mid B). \tag{3.1.3}$$

Whenever we write merely P_j, we must remember that somewhere in the background is a particular hypothesis B and an explicit inference A_j. For the probabilities of concern in the remainder of the book, the hypothesis and inferences will be such that the inferences are exhaustive and mutually exclusive on the given hypothesis. Provided we bear these points in mind, we may safely use the P_j notation.

In the new notation, the two restrictions on the probabilities become

$$0 \leqslant P_j \leqslant 1,$$

$$\sum_{j=1}^{n} P_j = 1.$$

This suffices to establish the conceptual and notational background for the chapter.

3.2. THE NOTION OF A MEASURE OF MISSING INFORMATION

To introduce the notion of the expectation value of a function, we analyzed the process of estimating the number of matches in a wooden matchbox. The same situation can provide an introduction to the topic of this section and the following four. We are presented with a definite set of probabilities for the inferences that the matchbox has a specific (integral) number of matches. The probability distribution is that of figure 2.8.1. At this stage we do not know for certain how many matches there actually are in the box. The case of four matches happens to be favored over all others, but it has a probability of only 0.35, far from that of certainty.

If we were told more about the matchbox, we could alter the probability distribution, perhaps increasing the probability that there are four matches, perhaps not. That would depend on what we were told. If we were given sufficient information, we could determine precisely the number of matches in the box. In a formal sense we would reduce the probability distribution to the ultimate case in which some single inference has unit probability, and all others have zero. That this is possible in principle (as well as in practice) follows from the property that the inferences in the set are exhaustive and mutually exclusive on the hypothesis. At least one of the inferences is correct, but not more than one of them may be correct. In the beginning, of course, we do not know which inference is correct; hence the various degrees of rational belief, each less than that of certainty, in the correctness of the inferences.

The relevant point is this: *given sufficient information*, we can reduce an initially specified probability distribution for a set of inferences (exhaustive and mutually exclusive on some initial hypothesis) to the ultimate case wherein some single inference is certain to be correct and the others certain to be incorrect.

This provokes some questions. How much is " sufficient information "? In particular, if we are provided with only the initial probability distribution, can we assess quantitatively the amount of " missing information," the addition of which would enable us to reduce the probability distribution to the ultimate case?

At first reading it is far from obvious that the second question—on which we focus attention—could ever be answered in the affirmative. One immediately wonders whether a qualitative notion like " information " or, better, " missing information " can be made quantitative. But let us assume for the moment that a notion of " missing information " can, in at least some limited sense, be made into a quantitative concept, and let us see where this assumption leads. Ultimately we will have to look back and see whether we have achieved anything reasonable and useful.

The amount of missing information of a given probability distribution is a notion that arose out of work in communication theory. That field can provide a suggestive example before we confront the task of making the notion mathematically precise.

In common English prose, the various letters of the alphabet do not occur with the same probability. The letter e predominates, followed by t, o, a, n, i, r, s, ... in order of descending probability. Certain combinations frequently occur, such as ee (in "seem," "proceed") and th (in "thought," "throw"). At least in words the letter q is invariably followed by the letter u.

Suppose a friend thinks of a word from the English language and asks us to name the fourth letter in it. Twenty-six inferences for the actual fourth letter deserve consideration: a, b, c, ..., z. On the basis of the known frequency of occurrence of the twenty-six letters in English prose, we can make a reasonable probability assignment for each of these exhaustive and mutually exclusive inferences. The probabilities are presented in table 3.2.1. Now we pose the question: how much more information do we need in order to determine the letter?

TABLE 3.2.1

The probabilities for the letter-guessing example. The probabilities are based on the known frequencies of occurrence of the individual letters in common English prose.

Letter	Probability	Letter	Probability	Letter	Probability
a	0.078	j	0.001	s	0.065
b	0.013	k	0.004	t	0.090
c	0.029	l	0.036	u	0.028
d	0.044	m	0.026	v	0.010
e	0.131	n	0.074	w	0.015
f	0.028	o	0.082	x	0.003
g	0.014	p	0.022	y	0.015
h	0.059	q	0.001	z	0.001
i	0.069	r	0.068		

Since we do not yet have a quantitative measure of missing information, we cannot give a definite answer to the question. We can, however, play with the question and see whether any reasonable notions emerge.

The probability distribution has a ragged profile, with strong preferences for e, t, o, a, n. This would seem to be an asset in our attempt to determine the correct letter. If we were to proceed by asking our friend, "Is it —?", we would be likely to start with the letter with the highest probability. Because of the steepness of the probability distribution when it is arranged in descending order, it is unlikely that we would have to ask more than eight or ten times. We are likely to feel that the amount of information we need in order to establish certainty about the letter in this situation is less than would be needed if the probabilities were all equal. We would probably have to ask, "Is it —?", more often in the latter case. In a still qualitative sense, the amount of additional information that we need in order to establish certainty for some one inference does seem to depend on the profile of the probability distribution from which we start.

It is worth noting that, even with such a steeply sloped distribution, a letter with rather low probability (such as the v in "university," with a probability 0.010) might be the correct inference. Any measure of missing information ought to depend on the probability of each inference in the set.

We might be clever and ask about a group of letters all at once, for instance, "Is it any one of the five letters with the highest probability?" If yes, we might ask about those five letters one by one. The letter might, however, be among the bottom 21 inferences. In whichever way the question about the top five group is answered, some information has been provided. More information is needed—and could be provided by answers to more questions— in order to proceed from the correct group to the correct letter. In the very beginning, however, we do not know whether the top group or the bottom is the correct group. Any consistent estimate of the total information we need— based on only the given initial probability distribution—ought to encompass both possibilities.

This is about as far as we need go here. "Amount of missing information" is still a fuzzy concept, but we can try, with qualitative reasoning as a guide, to give it a meaning both precise and reasonable.

3.3. CONDITIONS ON A QUANTITATIVE MEASURE OF MISSING INFORMATION

Let us state the problem as precisely as we can. We consider a set of n probabilities, P_1, P_2, ..., P_n, for inferences that are exhaustive and mutually exclusive on the given hypothesis. We seek a function of these probabilities that will provide a quantitative measure of the missing information associated with the probability distribution. Let us designate the function by

$$MI(P_1, P_2, \ldots, P_n).$$

The function is to provide a numerical measure of the amount of additional information that is needed to reduce the given probability distribution to an ultimate distribution, in which some single inference has unit probability and the others have zero probability. Since this measure is to characterize the given probability distribution, it should depend only on the specified probabilities and *not* in any way on which inference the addition of information might indicate as the correct one. This point cannot be overemphasized. After all, in the beginning we do not know which inference is the correct one (though one but only one actually is correct). For any chosen inference in the set, we know only the probability that it is correct. It is from this initial situation, of partial knowledge and partial ignorance, that we are trying to assess the amount of additional information needed to determine the correct inference.

As the first step in establishing a measure of missing information, we will specify a number of conditions any such function must satisfy if it is to be acceptable. Here we can give quantitative content to any qualitative ideas we may have about what would be a reasonable measure of missing information. In the second step we will look for a mathematical function that satisfies the conditions. If we have chosen the conditions well, there will be one and only one mathematical function that satisfies them. We will have a solution, and it will be unique.

Since we will arrive at the measure selected by Claude E. Shannon, and by his line of reasoning, it might be more honest to say that the steps outlined above characterize Shannon's process of selection as presented in *The Mathematical Theory of Communication*. In either view, the central point is that qualitative reasoning is used to give precise quantitative meaning to a notion that, initially, is only qualitatively defined.

We proceed to set up three conditions that the function $MI(P_1, P_2, \ldots, P_n)$ must satisfy.

If we change the probabilities slightly and continuously (while maintaining the normalization), we certainly want the numerical value of the function MI to change continuously. The probability distribution is changing continuously and so should the amount of information needed to establish certainty for some one inference. This is surely a modest and reasonable condition. We frame it as the first condition.

CONDITION I:
$MI(P_1, P_2, \ldots, P_n)$ shall be a continuous function of the probabilities $P_1, P_2, \ldots, P_n$.

Now we consider how we would want a measure of missing information to respond to an increment in the number of inferences. If the inferences are specified to be equally probable and we increase the number of inferences, we are likely to feel that our uncertainty about the correct inference increases. With more possibilities we need more information to establish certainty, say, by ruling out possibilities one by one. If we take the case of $n = 2$ being incremented by one to $n = 3$, we can represent the situation as in figure 3.3.1. Then a minimal requirement on MI can be put into mathematical form as the second condition.

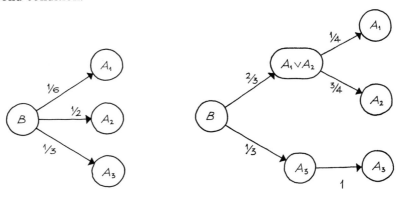

FIGURE 3.3.1
It seems reasonable to demand the strict inequality.

CONDITION II:

If all the probabilities are equal, that is, $P_j = 1/n$ for all j, then $MI(1/n, 1/n, \ldots, 1/n)$ shall be a monotonic increasing function of the positive integer n.

The third condition is a consistency requirement. It is considerably less obvious than the two preceding conditions, though we have had a hint of it in the letter-guessing example. We will first work with a simple case, and then generalize for the final mathematical form of the condition.

Let us imagine assessing the amount of missing information of a probability distribution by supplying additional information, the latter resulting ultimately in certainty for some single inference. We want the amount of missing information thus assessed to be independent of the steps by which we choose to supply information. This is the consistency requirement. Also, the amount of missing information should *not* depend on which inference the addition of information indicates as the correct one. This we must demand if the measure of missing information is to characterize the initial probability distribution.

The simple case is ready-made for us: we take over the exercise in manipulating probabilities of Section 2.7. The two diagrams with the concrete example are reproduced in figure 3.3.2. The inferences A_1, A_2, and A_3 are specified to be exhaustive and mutually exclusive on the hypothesis B. The corresponding probabilities are taken in this example to be

$$P_1 = P(A_1 \mid B) = \tfrac{1}{6},$$
$$P_2 = P(A_2 \mid B) = \tfrac{1}{2},$$
$$P_3 = P(A_3 \mid B) = \tfrac{1}{3}.$$

The amount of missing information of this probability distribution is expressed formally by $MI(\tfrac{1}{6}, \tfrac{1}{2}, \tfrac{1}{3})$. This represents the amount of information we would have to supply if we wanted to determine directly the correct A_j.

We might, however, want to proceed in a two-step fashion. We might group the inferences as $(A_1 \lor A_2)$ and as A_3, and imagine first supplying the information needed to find out which is the correct group. The two groups form a set of inferences exhaustive and mutually exclusive on the hypothesis B, and from figure 3.3.2 we may read off the respective probabilities. For the determination of the correct group we would have to supply an amount of information given by $MI\left(\tfrac{2}{3}, \tfrac{1}{3}\right)$.

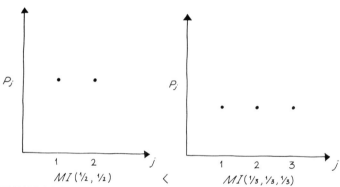

FIGURE 3.3.2
Diagrams to illustrate the effect of the grouping of inferences.

But this is only the first stage in the imagined supplying of information. We have yet to determine the correct inference among A_1, A_2, and A_3. If $(A_1 \lor A_2)$ should happen to be the correct group, the formal expression for the information still needed would be $MI(\tfrac{1}{4}, \tfrac{3}{4})$. But the other group, here A_3, might be the correct one. Then the additional information would be $MI(1)$.

Since we do not know initially which group will be the correct group, we cannot assert that $MI(\tfrac{1}{4}, \tfrac{3}{4})$, say, rather than $MI(1)$, will be the amount of information needed to proceed from the group stage to the final inference stage. All we really know is the probability that the first group will be the correct group, and correspondingly for the second. So it seems reasonable that we should consider the information needed in general in order to proceed from the groups to a final inference to be given by a sum over the two groups, each contribution weighted with the probability of the group's occurrence. This means an amount

$$\tfrac{2}{3} MI(\tfrac{1}{4}, \tfrac{3}{4}) + \tfrac{1}{3} MI(1).$$

In effect, this is the expectation value estimate of the amount of information we have to supply in the second step. With merely the initial probability distribution and the rules for manipulating probabilities at our disposal, we really cannot say anything more definite. This uncertainty about the information needed in the second step is inherent in the initial probability distribution. If we are to make a statement, based solely on the initial probability distribution, about the total information needed, we must consider both $MI(\tfrac{1}{4}, \tfrac{3}{4})$ and $MI(1)$. Weighting them in a sum with their respective probabilities is about the most natural way of handling this situation.

The total missing information, as evaluated in the two-step process, may then be taken as the sum of the information needed to determine a group plus the probability-weighted sum of the information needed to determine the correct inference, given the group. The consistency requirement is the demand that the amount of information expressed in this way shall agree numerically with the amount when imagined to be supplied directly. This gives rise, in our example, to the equation

$$MI(\tfrac{1}{6}, \tfrac{1}{2}, \tfrac{1}{3}) = MI(\tfrac{2}{3}, \tfrac{1}{3}) + \tfrac{2}{3} MI(\tfrac{1}{4}, \tfrac{3}{4}) + \tfrac{1}{3} MI(1).$$

The generalization is not difficult. We consider n inferences $A_1, A_2, \ldots, A_n$, each with its initial probability $P_1, P_2, \ldots, P_n$, and arrange them into n' groups. The first k of the inferences, say, are grouped together as a single inference $(A_1 \vee A_2 \ldots \vee A_k)$ with probability P_1' equal to $(P_1 + P_2 + \ldots + P_k)$. This expression for the probability is valid because the inferences forming the group are mutually exclusive, and is merely an extension of equation (2.7.1.) Then the inferences $k + 1$ through $k + m$ are lumped into a second inference $(A_{k+1} \vee \ldots \vee A_{k+m})$ with probability P_2' equal to $(P_{k+1} + \ldots + P_{k+m})$. We continue the process until we have placed each of the n inferences into some one of the n' groups and have arrived at a set of primed probabilities for the groups: $P_1', P_2', \ldots, P_{n'}'$. The procedure is displayed in figure 3.3.3. To determine the correct group we would have to supply an amount of information given by $MI(P_1', P_2', \ldots, P_{n'}')$.

Now we analyze the information needed to go from the intermediate groupings to the final inferences. The probability that A_1 is the correct inference, given both the original hypothesis and the statement that group 1 is the

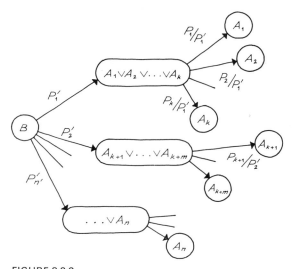

FIGURE 3.3.3
The n inferences $A_1, A_2, \ldots, A_n$ have been collected
into n' groups. Incompleted arrows indicate
"more of the same."

correct group, is P_1/P_1'. This expression for the probability as a ratio follows, by extension, from equations (2.7.2, 3). Under these assumptions, we have, for the k inferences in group 1, the probabilities $P_1/P_1', \ldots, P_k/P_1'$, respectively. If group 1 happened to be the correct group, we would have to supply an amount of information given by $MI(P_1/P_1', \ldots, P_k/P_1')$ in the process of determining which A_j, for $1 \leqslant j \leqslant k$, was the correct inference. An amount expressed analogously would be required if group 2 were the correct group, and so on for all n' groups. Since we do not know initially which group will be the correct one, we should consider the information needed to proceed from the group stage to the final inference stage to be given by a sum over all n' groups, each weighted with the probability of its occurrence. This means an amount

$$ P_1' \, MI(P_1/P_1', \ldots, P_k/P_1') + \ldots + P_{n'}' \, MI(\ldots, P_n/P_{n'}') $$

for the second step.

In the two-step procedure the total missing information is given by the sum of the information needed to determine the correct group plus the probability-weighted sum of the information needed to determine the correct final inference, given the group. The consistency requirement is formulated by demanding that the total missing information evaluated in this two-step fashion shall agree numerically with the total missing information if supplied directly. We state this in mathematical terms as the third condition.

CONDITION III:
The function *MI* shall, for all possible groupings, satisfy the following consistency relation:

$$MI(P_1, P_2, \ldots, P_n) = MI(P'_1, P'_2, \ldots, P'_{n'})$$
$$+ P'_1 \, MI(P_1/P'_1, \ldots, P_k/P'_1) + \cdots$$
$$+ P'_{n'} \, MI(\ldots, P_n/P'_{n'})$$

where $P'_1 = (P_1 + P_2 + \ldots + P_k)$, and so on, there being n' such groups.

The three conditions, and especially the last, are abstract, but in an essential sense this is unavoidable. It is a consequence of the approach we *must* take, for we do not yet have a concrete function *MI* whose properties we could describe. Rather, we have to set up conditions that an acceptable function *MI* must satisfy, and this is necessarily an abstract procedure. Remarkably, we have actually imposed conditions sufficiently restrictive so that the function *MI* follows unambiguously by deductive reasoning. We proceed now to the derivation of the mathematical form of *MI*.

3.4. THE DERIVATION

First we note that the imposition of Condition I permits us to work in the derivation exclusively with probabilities that, numerically, are rational numbers, the ratios of two integers. (Though not obvious now, this will be helpful.) The extension of the function *MI* to probabilities that are irrational numbers follows by continuity. So we need consider only a set of n probabilities in which an individual probability P_j is expressible as

$$P_j = \frac{m_j}{\sum_{l=1}^{n} m_l}, \tag{3.4.1}$$

where m_j is a positive integer or zero. The summation in the denominator of the ratio ensures that the normalization condition is fulfilled. At this point the problem is that of determining the form of $MI(P_1, P_2, \ldots, P_n)$ when the probabilities are restricted to being rational numbers.

In the next move we use Condition III to reduce the problem to that of determining MI for the special case of equally probable inferences. Purely for the sake of the derivation, we imagine a number Σm_l of equally probable, exhaustive, and mutually exclusive inferences, and we regard the probabilities in equation (3.4.1) as a result of grouping in the assessment of the missing information of that imagined probability distribution. If we abbreviate the value of the summation by N,

$$N \equiv \sum_{l=1}^{n} m_l, \tag{3.4.2}$$

each of the N equally probable inferences will be assigned a probability of $1/N$. The diagram presented in figure 3.4.1 will help here.

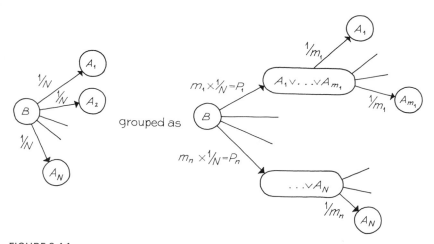

FIGURE 3.4.1
As a tactical move, we imagine a set of N equally probable, exhaustive, and mutually exclusive inferences and regard the probabilities given in equation (3.4.1) as resulting from a grouping of the N inferences. Equally probable inferences are collected in groups of $m_1, m_2, \ldots, m_n$ members. As previously, incompleted arrows indicate "more of the same."

For the imagined N equally probable inferences, Condition III declares that we may represent the amount of missing information by two formally different but numerically equal expressions:

$$MI(1/N, \ldots, 1/N) = MI(P_1, \ldots, P_n) +$$
$$+ P_1\, MI(1/m_1, \ldots, 1/m_1) + \ldots$$
$$+ P_n\, MI(1/m_n, \ldots, 1/m_n). \tag{3.4.3}$$

The expression on the left directly indicates the amount of missing information of the probability distribution for the imagined N equally probable inferences. The dots help to represent N entries of value $1/N$ each in MI.

The righthand side indicates the same amount of missing information, but in an imagined two-step process. From the N equally probable inferences, a group is formed by the disjunction of m_1 inferences. The probability of the group is $m_1 \times (1/N)$, which is precisely P_1 as given in equation (3.4.1). Then another group is formed by disjunction of m_2 inferences; this group has probability P_2. The procedure is continued until each of the N inferences is encompassed by some one of the n groups. The amount of information needed to determine the correct group is $MI(P_1, \ldots, P_n)$. This expression is, of course, the quantity in which we are interested. The rest is just part of the derivation. Then comes the step from the group stage to the final inference stage. The probability of any one of the equally probable inferences in the m_1 group, given that one of the inferences in the group is correct, is $1/m_1$. If the m_1 group is the correct group, the amount of additional information needed would be $MI(1/m_1, \ldots, 1/m_1)$, with m_1 entries in the expression. Since we do not know whether the m_1 group is the correct group or not, we must weight its contribution with the probability of the m_1 group, that is, with the probability P_1. The other groups are treated analogously.

In this analysis the MI function, evaluated for equally probable inferences, appears many times. We can give the equation presented above a neater appearance if we introduce an abbreviation. If we have n equally probable inferences, that is, the probability is $1/n$ for each inference, we define the abbreviation $F(n)$ by

$$F(n) \equiv MI(1/n, \ldots, 1/n). \tag{3.4.4}$$

There are n entires of value $1/n$ each in the function MI.

With the aid of the abbreviation, equation (3.4.3) becomes

$$F(N) = MI(P_1, \ldots, P_n) + \sum_{j=1}^{n} P_j F(m_j). \tag{3.4.5}$$

From the appearance of this equation we can judge that to determine MI in the general case, we need merely determine the form of the function $F(n)$, perhaps by means of a special case. This we proceed to do.

Since the preceding equation must hold for all rational probabilities as expressed in equation (3.4.1), it must hold if we choose all the m_j to be equal to some given integer, the integer m, say. Then matters degenerate to a very special case:·

$$N = \sum_{l=1}^{n} m_l = nm$$

and

$$P_j = \frac{m}{N} = \frac{m}{nm} = \frac{1}{n}.$$

The expression $MI(P_1, \ldots, P_n)$ becomes $F(n)$. The summation reduces to merely $F(m)$, for, when $m_j = m$ for all j, we have

$$\sum_{j=1}^{n} P_j F(m_j) = \sum_{j=1}^{n} P_j F(m) = F(m) \sum_{j=1}^{n} P_j = F(m).$$

Thus in this special case, equation (3.4.5) becomes

$$F(nm) = F(n) + F(m). \tag{3.4.6}$$

We have reduced our original problem to the solution of this relatively simple equation. An immediate solution is given by the natural logarithmic function:

$$F(n) = K \ln(n), \tag{3.4.7}$$

with K a positive constant. Equation (3.4.6) does not restrict the constant K, but Condition II requires that the constant be positive. This follows because $F(n)$ is identical to $MI(1/n, \ldots, 1/n)$ and the latter is, by Condition II, required to be a monotonic *increasing* function of the positive integer n. Hence the constant K must be chosen positive.

What specific numerical value is chosen for K is relatively unimportant, for the value merely sets the scale for the measure of missing information. But since one is free to choose the value of K, one can use the natural logarithm (base $e \simeq 2.7$) without restricting the generality of the solution, for logarithms to different bases differ only by constant factors. (Since this conclusion may not be immediately clear, problem 3.1 offers an assist in seeing it.) We may defer the choice of a specific positive value for K until later.

By the investigation of a special case, we have discovered the mathematical form of the function $F(n)$. If we return now to equation (3.4.5), we can solve for the general function MI:

$$MI(P_1, \ldots, P_n) = F(N) - \sum_{j=1}^{n} P_j F(m_j)$$

$$= -\sum_{j=1}^{n} P_j[F(m_j) - F(N)]$$

$$= -\sum_{j=1}^{n} P_j[K \ln m_j - K \ln N]$$

$$= -K \sum_{j=1}^{n} P_j \ln\left(\frac{m_j}{N}\right) = -K \sum_{j=1}^{n} P_j \ln P_j.$$

For the last step we need only remember that m_j/N is merely an abbreviated version of the righthand side of equation (3.4.1), the equation giving P_j.

There have been some tactical moves of a distracting nature, but we have solved the problem posed in the first paragraph of this section. The amount of missing information of a probability distribution with individual probabilities $P_j \equiv P(A_j | B)$, provided the inferences are exhaustive and mutually exclusive on the hypothesis, is expressed by

$$MI(P_1, \ldots, P_n) = -K \sum_{j=1}^{n} P_j \ln P_j. \qquad (3.4.8)$$

Arriving at this solution has not been easy. Yet one must ask, are there any other solutions? The answer is no. The three seemingly broad conditions we imposed are actually so stringent that the problem of finding a function MI which satisfies them has only a single solution (except for the choice of the positive constant K). For a proof one need, really, only use Condition II more extensively. Shannon's proof of uniqueness is presented in Appendix 2 of *The Mathmatical Theory of Communication*.

3.5. SOME PROPERTIES OF THE FUNCTION *MI*

Through qualitative reasoning we have arrived at a quantitative measure of the missing information of a given probability distribution. Since the actual function MI is, at first sight, rather strange, it would be well to examine it more closely in some simple cases. In this manner we can discover some of its properties—beyond those imposed by the three conditions—and see whether they correspond to what we might reasonably expect of such a quantitative measure.

The simplest case in which we would ever employ probabilistic reasoning is a situation with only two inferences, A_1 and A_2, from some hypothesis. Then we have two probabilities, P_1 and P_2, and the amount of missing information is

$$MI(P_1, P_2) = -K\left[P_1 \ln P_1 + P_2 \ln P_2\right]. \qquad (3.5.1)$$

We may ask, how does MI vary as we vary P_1 and P_2? Actually, we cannot vary the two probabilities independently, for they are constrained by the normalization condition, $P_1 + P_2 = 1$. We use this to eliminate P_2 in MI:

$$MI = -K\left[P_1 \ln P_1 + (1 - P_1) \ln (1 - P_1)\right].$$

The righthand side we can plot as a function of P_1, with P_1 running from zero to one, as has been done in figure 3.5.1. The expression is never negative— certainly a desirable property for a measure of missing information—and reaches a maximum when $P_1 = P_2$. That the equal probability case should correspond to the maximum is something we might have expected, for we can never be in a worse situation than that.

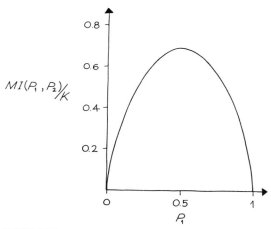

FIGURE 3.5.1

The behavior of $MI(P_1, P_2)$ as a function of P_1.

For $P_1 = 1$, the measure implies zero missing information. We would justifiably be disturbed if this were not so, for $P_1 = 1$ means that we are certain that A_1 is the correct inference, and we need no further information to establish certainty. The case $P_1 = 0$ is analogous, for then $P_2 = 1$; and we are again certain of the inference. This is auspicious.

Before going on we should clarify a point that the construction of the graph presented. When $P_1 = 0$, the direct contribution of P_1 to $MI(P_1, P_2)/K$ is $-0 \ln 0$. Since $\ln 0$ is $-\infty$, we are faced with the indeterminate form $0 \times \infty$. How should some meaning be assigned to this? The answer lies in Condition I, which asserts that MI shall be a continuous function of the probabilities. So we should examine the limit as P_1 goes to zero.

Let us consider more generally the function $-x \ln x$ for x in the range $0 \leqslant x \leqslant 1$. This is the universal form of the individual terms in MI/K. To establish the limit as x goes to zero, we turn the indeterminate form $0 \times \infty$ into the form ∞/∞ and apply L'Hôpital's rule:

$$\lim_{x \to 0}(-x \ln x) = \lim_{x \to 0} \frac{(-\ln x)}{(1/x)} = \lim_{x \to 0} \frac{(-1/x)}{(-1/x^2)} = \lim_{x \to 0} x = 0. \qquad (3.5.2)$$

The zero limit justifies the assignment of zero made in the graph. In the interval $0 \leqslant x \leqslant 1$, the function $-x \ln x$ has the behavior indicated in figure 3.5.2. The function is never negative and vanishes at both endpoints.

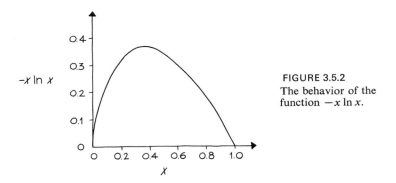

FIGURE 3.5.2
The behavior of the function $-x \ln x$.

A consequence of the properties of $-x \ln x$ is that no term in the function MI is ever negative. Any inference with a probability not equal to zero or one makes a positive contribution and leads to a need for additional information. If an inference is impossible on the hypothesis ($P_j = 0$), then it contributes nothing to MI and rightly so, for we need no further information to exclude it. If an inference is certain on the hypothesis ($P_j = 1$), then it makes a zero contribution to MI. But in this case the normalization condition requires all other probabilities to be zero, and hence they, too, make no contribution. The value of MI is zero, and agreeably so, for we need no further information to determine the correct inference. Zero is the minimum numerical value of MI and occurs only when we are certain which inference is correct.

The investigation so far leads naturally to the question of what, in the general case, the maximum value of MI is. We rather expect that maximum to occur when all inferences are equally probable, but let us see. We ask for the maximum value of

$$MI = -K \sum_{j=1}^{n} P_j \ln P_j$$

subject to the normalization condition

$$\sum_{j=1}^{n} P_j = 1.$$

This is the kind of mathematical problem—the determination of an extremum subject to one or more conditions—for which the Lagrange multiplier technique was devised. We will find that technique quite useful in a crucial derivation in Chapter 4. In order to gain familiarity with the technique, let us begin using it here, where we have a good idea of what the result will be. Appendix B offers a quick geometric introduction to Lagrange multipliers, an introduction that provides an insight into why they work.

Let us follow the pattern set up in Appendix B and begin by writing a function Λ defined as

$$\Lambda = \Lambda(P_1, P_2, \ldots, P_n, \lambda)$$

$$\equiv -K \sum_{j=1}^{n} P_j \ln P_j + \lambda \left[1 - \sum_{j=1}^{n} P_j \right],$$

with λ being a Lagrange multiplier. Now we demand, as a condition for an extremum, that the partial derivatives with respect to each P_j, and with respect to λ, be zero:

$$\frac{\partial \Lambda}{\partial P_j} = -K(\ln P_j + 1) - \lambda = 0 \qquad \text{for each } j,$$

$$\frac{\partial \Lambda}{\partial \lambda} = 1 - \sum_{j=1}^{n} P_j = 0.$$

On solving for P_j, we find

$$P_j = \exp\left(-\frac{\lambda}{K} + 1 \right).$$

Since the righthand side is independent of j, all n probabilities must have the same value. Now we impose the last equation, which is just the normalization condition:

$$1 = \sum_{j=1}^{n} P_j = n \times \exp\left(-\frac{\lambda}{K} + 1 \right).$$

This implies that the exponential is simply $1/n$. (The value of λ itself is thus determined but is of no interest to us.) Hence *MI* has an extremum when $P_j = 1/n$ for all j, and we can certainly recognize this as a maximum. When we insert the probabilities into *MI*, we arrive at

$$(MI)_{\max} = +K \ln n. \tag{3.5.3}$$

For a fixed number n of inferences, the amount of missing information of the probability distribution reaches its maximum value when the inferences are equally probable. In a few steps we can demonstrate a related property: if we pick any pair of probabilities and make those two probabilities more nearly equal, holding the other probabilities fixed, then the new value of MI is greater than the old value.

For the sake of definiteness, let us take P_1 and P_2 as the two probabilities in the pair. Then we must hold the probabilities $P_3, \ldots, P_n$ fixed. That, together with the normalization condition, restricts the way in which we may make P_1 and P_2 more nearly equal. Their sum must remain constant:

$$\sum_{j=1}^{n} P_j = 1 \quad \text{implies} \quad P_1 + P_2 = 1 - \sum_{j=3}^{n} P_j \equiv C.$$

With the constant C as defined, we may write P_2 in terms of P_1 as

$$P_2 = C - P_1.$$

Then, with all probabilities fixed except those of the pair P_1, P_2, we have

$$MI = -K\left[P_1 \ln P_1 + (C - P_1)\ln(C - P_1) + \sum_{j=3}^{n} P_j \ln P_j \right].$$

The question is this: if we vary P_1 in such a way as to make P_1 and P_2 more nearly equal, does MI indeed increase?

To answer the question, we take the derivative of MI with respect to P_1, holding $P_3, \ldots, P_n$ (and hence C) fixed:

$$\frac{dMI}{dP_1} = -K\left[\ln P_1 + 1 - \ln(C - P_1) - 1 \right] = -K \ln\left(\frac{P_1}{C - P_1}\right).$$

Having calculated the derivative, we may replace $(C - P_1)$ by P_2 and arrive at the intermediate result:

$$\frac{dMI}{dP_1} = -K \ln\left(\frac{P_1}{P_2}\right).$$

If we let ΔP_1 represent a small change in P_1, we have for the corresponding change in MI the expression

$$\Delta MI = \frac{dMI}{dP_1} \Delta P_1 = \left[-K \ln\left(\frac{P_1}{P_2}\right) \right] \Delta P_1.$$

And now we consider the two cases of interest:

1. $P_1 < P_2$ initially. One must take ΔP_1 positive in this case in order to make the two probabilities more nearly equal. With $P_1 < P_2$, the argument of

the logarithm is less than one and so the logarithm itself is negative. Thus the expression as a whole is positive.

2. $P_1 > P_2$ initially. Now one must take ΔP_1 negative in order to make the two probabilities more nearly equal. With $P_1 > P_2$ the logarithm is positive. The leading negative sign compensates for the negative value of ΔP_1. So once again the expression as a whole is positive.

Only for the sake of definiteness did we single out the pair consisting of P_1 and P_2. We have a general proof of the specified property: if we pick any pair of probabilities and make those two probabilities more nearly equal, holding the other probabilities fixed, the new value of MI is greater than the old value. As the probabilities become more nearly equal, indicating less preference among the inferences, the amount of information needed to establish the correct inference increases. Had we wished to, we could have used this property to prove that MI reaches its maximum value when all inferences are equally probable.

3.6. RECAPITULATION AND REMARKS

Let us summarize the context and salient results of the past few sections.

The context of the entire discussion may be presented in a few sentences. We consider a set of probabilities P_1, P_2, $\ldots$, P_n for a set of inferences A_1, $A_2, \ldots, A_n$. The inferences are specified to be, on the given hypothesis, exhaustive and mutually exclusive. Though necessarily some single inference is correct and the others incorrect, we do not know which inference is correct. For each inference we know only the probability that it is the correct inference. The probability distribution expresses the state of knowledge concerning the correct inference.

In this context we look for a function of the probabilities—written as $MI(P_1, P_2, \ldots, P_n)$—which will be a quantitative measure of the additional information needed to reduce the given probability distribution to an ultimate probability distribution, in which some single inference has unit probability and the others have zero probability. The function MI is to measure the "missing information" of the given probability distribution, in the sense that it is to express quantitatively the amount of information that must be supplied before we will know the correct inference. The assessment of the amount of missing information is to be based solely on the initially given probability distribution, not in any way on which inference may emerge as the correct inference.

This outlines the problem we set. Following Shannon, we imposed three conditions on the function *MI*. These determine the function *MI* uniquely (modulo the value of the positive scale factor K) as

$$MI(P_1, P_2, \ldots, P_n) = -K \sum_{j=1}^{n} P_j \ln P_j, \qquad K > 0.$$

The conditions and the derivation provided us with a function whose properties we could investigate. Of particular significance are the bounds on the value of *MI* for a fixed number n of inferences:

$$0 \leqslant MI(P_1, P_2, \ldots, P_n) \leqslant K \ln n.$$

The measure of missing information is zero if and only if the probability distribution is an ultimate one, in which the correct inference is known with certainty. The measure reaches its maximum value if and only if the inferences are equally probable. For a situation between these extremes, a shift in the values of a pair of probabilities toward equality for the two probabilities leads to a larger value of *MI*.

This is satisfying, for these are properties which we may reasonably expect of a measure of missing information. We may conclude that we have established a function, dependent only on the initially given probability distribution, which does in a reasonable fashion measure the amount of "missing information," the amount of additional information needed to determine the correct inference.

Before we go on, we should consider an alternative set of words to describe what *MI* measures. In one view, a probability distribution expresses the (usually limited) extent to which one knows which inference is correct. In another view, a probability distribution indicates the extent to which one is uncertain about the correct inference. For instance, the more nearly equal the probabilities are, the more uncertain one is of the correct inference.

In a reasonable sense, the more one needs to be told about a situation in order to determine the correct inference, the more uncertain one was in the beginning. The notions of "missing information" and of "uncertainty" are two ways of describing a state of ignorance about the correct inference. Once one has established the function *MI* as a measure of missing information, one may usefully look on it as expressing also one's "uncertainty" about the correct inference when one knows only the probability distribution. This is not a view of the function *MI* that we will need, but it is well worth noting and will occasionally be mentioned.

Finally, the purely quantitative aspect of *MI* warrants emphasis. The function *MI* measures the *amount* of missing information. It says nothing about the *kind* of information that is missing. Whether the additional information would come from polling voters or from examining molecules in a gas is irrelevant to *MI*. We have here a thoroughly quantitative notion, but a fruitful one, as the next section will show.

3.7. CRITERIA FOR THE ASSIGNMENT OF PROBABILITIES

At last we turn to the primary objective of this chapter: the establishment of criteria for the assignment of probabilities. We should bear in mind that here, as previously in the chapter, we are restricting consideration to probabilities for a set of inferences that are exhaustive and mutually exclusive on the given hypothesis. The numerical value of each probability must lie between zero and one, endpoints included; the sum of the probabilities in the set must be unity. A set of numerical assignments conforming to these restrictions we agreed to call a probability distribution. Now we must seek criteria that will enable us, when given a hypothesis and a set of inferences, to decide on the most appropriate, the " best," probability distribution.

The first criterion is rather obvious.

CRITERION I:
The assigned probabilities should reflect and reproduce the information that we actually have (or explicitly assume).

This criterion is implemented by transcribing our information, given in the hypothesis or implied by it, into mathematical form as restrictions on an acceptable probability distribution. In any given case, the mathematical form will depend on the nature of the information. Some examples will clarify the practical meaning of this criterion.

Pictures are always helpful. We could use a single helium atom in a box, but even that is unnecessarily complex. The existence of several quantum-mechanical states with the same energy makes the pictorial representation unnecessarily cumbersome. So let us take a particle constrained to one-dimensional motion. For this case the energy levels and quantum-mechanical states are in one-to-one correspondence. Let us arrange the states in order of increasing energy and label each with an index j, as in ψ_j, with the index having the range $1 \leqslant j \leqslant \infty$. The energy of state ψ_j is $E_j = (j)^2 E_1$. The proportionality factor E_1 happens to be $\pi^2 \hbar^2 / 2mL^2$ for a " one-dimensional box" of length L.

Suppose our hypothesis contains the information that the actual energy E of the system is known to have an upper bound of $16 E_1$, that is, $E \leqslant 16 E_1$. If our inferences are of the form, "The state ψ_j provides the appropriate description of the physical system" (A_j), then we set $P_j = 0$ for $j > 4$. For this stage in the assigning of probabilities, we have the picture in figure 3.7.1. For $j \leqslant 4$ we cannot yet make any definite statements about the probabilities (except that they must be nonnegative and must sum to one). Hence we have the erratic distribution of points and the question marks.

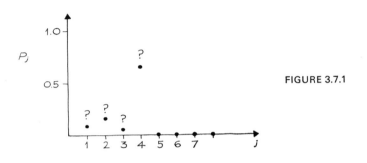

FIGURE 3.7.1

The hypothesis may be augmented to give a different situation. If the information that energy eigenstates 1 through 3 are equally likely were included in the hypothesis, as well as the upper bound on the energy, we would set

$$P_1 = P_2 = P_3 \quad \text{and} \quad P_j = 0 \text{ for } j > 4.$$

This information would not determine the probability distribution completely. The normalization condition can, however, be used to some advantage at this point. It implies a connection between P_4 and the first three probabilities:

$$P_1 = P_2 = P_3 = \tfrac{1}{3}(1 - P_4).$$

The probability distribution depends on the undetermined value of P_4. For the picture we have that in figure 3.7.2, where the first three points are at the same height, but question marks are still needed to indicate an incomplete assignment. Since the value of P_4 may be anywhere between zero and one, there is a continuous infinity of acceptable probability distributions.

Again we can revise the hypothesis. Suppose now that the expectation value of the energy is specified, as $5 E_1$, as well as the energy upper bound:

$$\langle E \rangle = 5 E_1 \quad \text{and} \quad E \leqslant 16 E_1.$$

Knowledge that the first three states are equally probable is *not* included in this new hypothesis.

The information given is distinctly helpful in determining a probability distribution, but it does fall far short of determining a unique probability assignment. The upper bound again implies $P_j = 0$ for $j > 4$. The specification of the energy expectation value can then be presented, as a restriction on the choice of probabilities, in a concise mathematical form:*

$$\sum_{j=1}^{\infty} E_j P_j = \sum_{j=1}^{4} (j)^2 E_1 P_j = 5E_1.$$

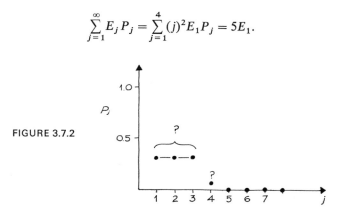

FIGURE 3.7.2

When we include the normalization condition, we find that we have only two conditions on the four probabilities that are not necessarily zero. Figure 3.7.3. indicates six of the infinitely many probability distributions compatible with the information given. In the fifth of these assignments, distribution (e), the points have been required to lie along a straight line. If one wants to ascribe a non-zero probability to each of the four inferences, it is not always possible to have them lie along a straight line. Here a little calculation shows that one can achieve both aims only if $\langle E \rangle$ lies within certain limits:

$$\tfrac{20}{6} E_1 < \langle E \rangle < \tfrac{70}{6} E_1.$$

The point of this is merely that to assign the probabilities along a straight line is sometimes incompatible with other, quite reasonable desires.

---

* Logically, one cannot *know* an expectation value before the probabilities have, at least in principle, been determined. Nonetheless, one may have information in the hypothesis that enables one to *force* an expectation value to assume a specific numerical value. Chapter 4 will present a situation with cogent justification for such action. The example here should be understood in the "force" sense.

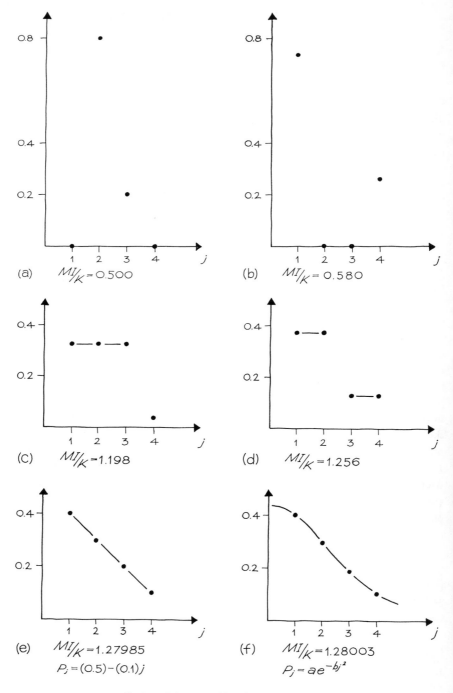

FIGURE 3.7.3 A display of six, out of infinitely many, probability distributions compatible with the given information.

Just as the varieties of information are legion, so are the mathematical forms that we use to incorporate information into the probability assignment. In general, we use what information we have in order to impose mathematical restrictions on the initially undetermined probability distribution. The more information we have, the more we can fence in the probabilities. But usually we do not have enough information for the corresponding restrictions to determine the probabilities uniquely. In the typical situation, several different probability distributions are compatible with the information we have. To determine a *unique* set of probabilities, we must introduce a second criterion. The step is unavoidable.

Suppose we have imposed on the probabilities the mathematical restrictions that our information implies. In other words, we have applied Criterion I. Still we are confronted with a host of possible assignments, all compatible with what we know. We may call those distributions the *compatible probability distributions*. From the vast array of compatible probability distributions, we must choose one. How? Let us consider doing this by a process of elimination, employing qualitative reasoning in search of a general principle.

Some of the compatible probability distributions will exhibit sharp peaks in regions where others show no such spectacular behavior. The former assert high probability for the corresponding inferences, but our information does not really warrant such a strong preference. If the information did, all compatible distributions would show such peaks. The graphs in figure 3.7.3 illustrate such distributions. Qualitatively, it would seem reasonable to prefer those compatible distributions without the jagged profiles. They do not imply more than we can justify.

Some compatible assignments may give zero probability to certain inferences, whereas others do not. Since our information does not require the zeros, why treat those inferences so shabbily? Share the probability. Qualitatively, it would seem reasonable to prefer those compatible distributions that do not neglect an inference by assigning to it zero probability.

The gist of this brief sortie into choice by elimination is that compatible probability distributions with *unrequired* peaks or zeros express, we suspect, more than our knowledge warrants. They make unjustifiable assertions. A compatible distribution without such unwarranted preference or neglect would seem a *better* choice, in what is truly a matter of free choice for us.

We may correlate these remarks and feelings with the amount of missing information exhibited by various compatible probability distributions. For a specific example, we can refer to figure 3.7.3 and the six (out of infinitely many) compatible distributions it displays.

Distribution (a) exhibits an unrequired peak and two unrequired zeros. The value of the missing information of this probability distribution is the smallest of those of the six distributions displayed. The next distribution, (b), shows an unjustifiable preference for the two end inferences, with complete neglect of the central two. Its value of MI is only a shade larger than that of distribution (a).

The next two distributions we are likely to regard as more reasonable. At the very least, neither distribution totally neglects an inference. Nonetheless, it would be difficult to justify the assigning of equal probabilities to a triplet, as in (c), or to two pairs, as in (d). Let us note that the value of MI has increased substantially over the value for the least-favored distribution, (a).

In distribution (e) the probabilities have been compelled to fall along a straight line. No inference is neglected, and no pairs or triplets are given preferential treatment. One is likely to feel that this is the most reasonable of the distributions examined thus far. Also, it has the greatest value for MI of those distributions. We should remember, though, that a straight-line assignment is not always possible if one wishes to avoid neglecting any inference that need not be neglected. The last distribution, (f), represents a probability assignment in the form of a (decaying) exponential dependence on the energy of the state appearing in the inference. Here the exponential dependence is hardly distinguishable from the straight-line distribution, but this distribution does have a slightly larger value of MI, indeed, the largest of those for the six compatible distributions displayed on the page.

What we see here is a correlation between the qualitative reasonableness of a compatible distribution and the quantitative value of the missing information of the distribution. The suggestion is near at hand that the larger the value of MI of a compatible distribution, the "better" that distribution is. The natural extrapolation would be: the "best" of the compatible probability distributions is the one with the largest value of MI.

The same conclusion may be reached by a more formal line of reasoning. We are concerned with making a choice from the set of compatible probability distributions. Each of these distributions has built into it all the information provided in the hypothesis or implied by it. Yet we find that some compatible distributions show a smaller amount of missing information than others.

Let us imagine singling out for special consideration two of the compatible distributions, that distribution with the largest value of missing information and then some other. The second distribution is closer to being an ultimate probability distribution, closer to indicating some single inference as the correct one, than is the compatible distribution with the largest value of MI. But where is the information that could justify this *specific* closer approach to the selection of some single inference? Such information does not exist in the data we possess. If we wish to avoid introducing a bias, if we wish to avoid unwarranted preference for some inferences and unwarranted neglect of others, we must choose the compatible probability distribution with the largest value of the missing information.

Taking the foregoing to be reasonable, we may state the second and final criterion.

CRITERION II:

From among all the probability distributions that satisfy Criterion I, that is, from among the compatible probability distributions, we choose, as best, the one that exhibits the largest amount of missing information.

With the statement of this final criterion, we have accomplished the objective of the chapter. But before going on to application in statistical mechanics, we should briefly look at the strengths and weaknesses of the two criteria.

There is likely to be little dispute over the reasonableness of the first criterion. One must somehow make use, in assigning probabilities to inferences, of the information provided by the hypothesis. If one looks at probability theory as a theory of rational guessing, then of course one wants to use whatever information one has in making one's guesses. That in a mathematical theory one would use that information to place restrictions on otherwise undetermined numbers (the P_j's) is quite natural. In this way one specifies the set of compatible probability distributions.

In the typical case, the set of compatible distributions contains many, even infinitely many, distributions. To select some one compatible distribution as, in some sense, the best distribution, the unbiased distribution, one needs to introduce a further guiding principle. Here one really is free to adopt any criterion that one feels can be justified. There is no uniquely determined choice. For our choice of principle—Criterion II—one cannot unequivocally state that it is the correct choice or the optimum choice. Though it might seem strange at first sight, it has many of the hallmarks of reasonableness, and in a situation where one must make a choice, it is at the very least a good choice. Although one cannot assert that it is the optimum choice, this is no real weakness, for there are no objective means by which to determine the optimum guiding principle.

Yet there are some potential weaknesses. Primarily, one can question the generality of the scheme. Can all the relevant information in a hypothesis, or implied by it, be transcribed into mathematical restrictions on an acceptable probability distribution? If not, it may be difficult (or impossible) to define the set of compatible probability distributions. And after one has established the set of compatible distributions, can one be sure that Criterion II will yield a unique answer? There might be several distributions with a common "largest" value of the missing information.

These two questions cannot be answered with an unqualified "yes." The complete generality of the scheme is far from established. But one can say that the scheme is comprehensive enough to enable one to frame a successful theory of statistical mechanics, and that is our primary concern.

REFERENCES for Chapter 3

Leon Brillouin explores a vast array of very relevant topics in his *Science and Information Theory* (New York: Academic Press, 1962). Included are aspects of information theory, statistical mechanics, thermodynamics, and quantum theory, often with an emphasis on principle and philosophic content. The reader may, however, experience some difficulty in sorting out the pieces and arranging them into a coherent picture in his mind.

Shannon's development of information theory appeared in two issues of the *Bell System Technical Journal* (July and October 1948). The material is most accessible in Claude E. Shannon and Warren Weaver, *The Mathematical Theory of Communication* (Urbana: University of Illinois Press, 1949). Of the structure of information theory presented in Shannon's part of this volume, only a small part is germane to our needs in establishing criteria for the assignment of probabilities, primarily, the material in Shannon's Chapter 1, Section 6, and in his Appendix 2. That the three conditions lead to a *unique* expression for *MI* is proved in that appendix.

PROBLEMS

3.1. The choice of logarithmic base. In Section 3.4 the claim was made that using the natural logarithms in the solution for the unknown function $F(n)$ entailed no loss of generality. Yet the use of logarithms to the base 10, yielding a solution $F(n) = K' \log_{10} n$, with K' any positive constant, appears to give a different solution. Show that this is merely a matter of appearances. (Hint: As the major step, show that the choice $K' = K \ln 10$ yields precisely the expression adopted in the text.)

3.2. Solution for $F(n)$. For the solution to the (functional) equation of Section 3.4, we took $F(n) = K \ln n$. A positive integer may be decomposed in one and only one way as a product of prime numbers. Hence a truly different solution for $F(n)$ is this: the number of factors needed for writing n as a product of primes. Convince yourself that this provides a solution to the equation, and then consider, on the basis of Condition II, whether it is a solution *acceptable* for *MI*.

A proof that the logarithmic solution for $F(n)$ is the only solution acceptable under Condition II is given by Amnon Katz, pp. 14–17 of his book (cited in the references for Chapter 4). That proof suffices to establish the uniqueness of the final solution for the function *MI*.

3.3. Use a Lagrange multiplier to find the extreme values of the function $f(x, y) = xy$ when x and y are constrained to satisfy the equation $x^2 + y^2 = 1$. Compare with the results that you get by using the constraint equation to eliminate y in the product xy in terms of x. (Incidentally, note that the Lagrange multiplier turns out in this case to equal the extreme values of the function. Lagrange multipliers do, typically, have some meaningful interpretation.)

3.4. Associated with each of the following two sets of probabilities is a certain amount of missing information: (1) three inferences with probabilities $1/3, 1/3, 1/3$; and (2) four inferences with probabilities $3/4, 1/8, 1/8, 0$. Which set has the larger value of *MI*?

THE STATISTICAL
DESCRIPTION OF EQUILIBRIUM

The preceding chapters have provided the concepts in probability theory needed for a systematic development of statistical mechanics, the discipline concerned with estimating the properties of a physical system, given only limited information about the actual state of affairs. Now we use those concepts to analyze a situation of paramount importance: a physical system in equilibrium at a well-defined temperature. Indeed, the bulk of this book will be devoted to particular instances of such systems, which are not only legion but also the subject of intensive research. A gas provides the most familiar example. When the intermolecular forces may be neglected, the relation among pressure, temperature, and volume is relatively simple, but what about the situation at high density, where one must certainly take intermolecular forces into account? Great advances have been made in this field, but the extreme case—condensation into a liquid at a sharply defined temperature—remains in a less than fully satisfactory theoretical state. Paramagnetic and ferromagnetic substances provide other instances, as does liquid helium, with its peculiar properties at low temperature.

First we must specify rather precisely what we mean by equilibrium and by temperature. Fortunately, one is likely to feel reasonably familiar with both;

so we can build on that background. Then we will decide on the set of inferences to be used in discussing a system known to be in equilibrium. Finally we will establish the " best " probability distribution for the inferences in the general situation of equilibrium at a well-defined temperature. The last will require two preliminary calculations of modest complexity. The end result, however, will be comfortingly simple.

Before launching into the chapter, we should agree on the use of a pair of adjectives: microscopic and macroscopic. Frequently we will need to characterize a physical system, a volume of space, or a physical property as being either microscopic or macroscopic. Let us agree that microscopic will denote " very small ": containing a few atoms only, on the order of atomic spatial dimensions, or referring to individual particles, as the case may be. Macroscopic will, in contrast, mean " rather large ": containing many, many particles, of spatial extent visible to the naked eye, or referring to all of the many particles at once, again as the case may be.

4.1. THE MEANING OF EQUILIBRIUM

Most of us have, at one time or another, watched cream being poured into a cup of black coffee. At first the black coffee is laced with streamers of cream, but in a short time—even if not stirred—the coffee takes on a uniform " coffee brown " appearance. If the person dallies over his coffee, the once hot coffee becomes " cold "; it takes on the temperature of the surrounding room.

These are two familiar instances of an approach to equilibrium. In the first instance it is " equilibrium " within the cup: the rather uniform distribution of cream in the coffee. In the second it is " equilibrium " with the surroundings: the common temperature of coffee and room. The system settles down to a condition in which, to our human senses at least, there is no further change. It is likely that these are changes on the microscopic level, but on the macroscopic level there is no discernible change with time.

It is this kind of ultimate condition that we will mean when we apply the term " equilibrium ". A precise, all-encompassing definition is difficult to frame, but the general meaning of " equilibrium " is that a system has settled down to the point where its *macroscopic* properties are unchanging with time. In practice there is seldom any doubt whether a system has reached equilibrium or not. And there is always a test of sorts: do the macroscopic properties, given further " aging " of the system, change discernibly?

The time required for the attainment of equilibrium varies immensely. For coffee stirred with a few motions of the spoon, effective uniformity of color is achieved within seconds. The motion of coffee within the cup may take a minute or so to die out. To achieve a common temperature of coffee and surroundings may take fifteen minutes or more. Some chemically reactive mixtures, if not catalyzed, will effectively take an infinite time to reach chemical equilibrium. A mixture of hydrogen and oxygen is a favorite example. At room temperature the two gases coexist without appreciable production of H_2O. The introduction of a spark radically alters the approach to equilibrium. For a system of nuclear spins, as in a crystal of lithium flouride, the spins may come into equilibrium among themselves in a time of order 10^{-5} seconds, but equilibrium between the spins and the crystal lattice (the vibrational motion of the nuclei and bound electrons) may take as long as five minutes.

A theory of the approach to equilibrium is extraordinarily difficult. For some special systems a quite reasonable description does exist, but the theories of the general case leave something to be desired in the way of rigor or transparency or both. As far as possible we will sidestep this thorny problem, and deal with systems that have attained equilibrium. That physical systems, if allowed to settle down, do somehow reach equilibrium is an observational fact.

We should concede that a condition of equilibrium does not preclude the existence of fluctuations, even in macroscopic properties. If we think of a gas as a host of molecules in chaotic motion, we are led to expect variations in the force that the molecules, by impact, exert on the walls of the container. The pressure may show some temporal variation. The attribute "equilibrium" for the gas implies that the fluctuating pressure centers around some typical pressure and that the latter does not change with time.

A remark: the "equilibrium" about which we are talking is sometimes called *thermal equilibrium*, in distinction to the equilibrium of ordinary mechanics. The latter means no change on both the macroscopic and the microscopic level, that is, no change whatsoever.

Knowledge that a physical system has reached equilibrium carries a profound implication for the statistical description of the system: our estimates (of the properties and behavior of the system) should be independent of time. For example, the expectation value estimate of a pressure or total magnetic moment should be constant in time. To be sure, fluctuations in the actual system may be inevitable. Yet, with only the severely restricted information that we typically have, we cannot predict the actual temporal sequence of fluctuations. That limitation, together with the property that equilibrium specifically excludes long-term changes with time, implies that we should demand that our estimates be independent of time.

4.2. AN OPERATIONAL DEFINITION OF TEMPERATURE

"Temperature is hotness measured on some definite scale."* This is not a bad start. In a room we have a feeling of being hot or cold and an idea that the temperature is high or low. A look at the thermometer generally confirms this. Indeed, the purpose of a thermometer is to provide a quantitative measure of "hotness." There is, however, a vast array of possible thermometers: mercury-filled, alcohol-filled, bimetallic, carbon resistor, and so on. Being precise about temperature requires the choice of a definite substance or apparatus that responds in a reproducible way to what we qualitatively think of as "changes in temperature." It requires also the choice of a definite scale.

The first step in our approach to a precise operational definition of temperature is a matter of historical review. The work of Boyle (in the mid-seventeenth century) and of Charles and Gay-Lussac (around 1800) gave us the "perfect-gas law." All gases, provided they are sufficiently dilute, exhibit the same, simple relationship between observed pressure, volume, and some suitably defined temperature. The modern form of the relation, with T designating temperature and N the number of molecules, is the proportionality

$$p_{obs} V \propto NT. \tag{4.2.1}$$

A subscript "obs" has carefully been appended to the general symbol p for pressure to denote the experimentally observed pressure. There is, after all, at least a logical distinction between an observed pressure and a pressure calculated as an expectation value from some probability distribution. The latter we will write as $\langle p \rangle$.

The condition that the gas be "sufficiently dilute" can, in practice, be achieved to a high degree of approximation. Ordinarily we can take the condition to mean that the typical intermolecular separation must be large relative to the molecular size. For air at sea level, the ratio is about 10 to 1, and already at this dilution the deviations from the "perfect" gas behavior are small.

---

* For this simple but perceptive comment, I am indebted to Eric Rogers and his *Physics for the Inquiring Mind* (Princeton, N.J.: Princeton University Press, 1960). His Chapter 27 has something to say to both the uninitiated and the sophisticated.

One might be suspicious about the reservation that the perfect gas law holds for some "suitably defined" temperature. In practice the relation holds quite well if one uses for the temperature T the reading on an ordinary mercury Centigrade thermometer plus 273.15. On such a thermometer the two fixed points of the scale are specified by the assignments:

0 degrees Centigrade for the "temperature" of a mixture of melting ice and water exposed to air at atmospheric pressure; and

100 degrees Centigrade for the "temperature" of steam in equilibrium with liquid water at atmospheric pressure.

Then the region of glass tube between the two fixed points is marked off into 100 divisions of equal length. These divisions define the intermediate temperatures. When the mercury expands to scratch number 21 (typical of a room), the "temperature" is 21 degrees Centigrade. In the simplicity of this one tends to miss the essence of the procedure: certain operational steps have been chosen—freely—in order to provide a quantitative measure of "hotness." In any event, something like the above definition of temperature is surely what was used by those who discovered the perfect gas law.

Galileo is credited with the invention of the thermometer. Around 1595 he developed a simple device, essentially an air-filled bulb having a long tube whose free end was held under water. The device indicated temperature changes through the expansion or contraction of the trapped air and the consequent changes in the height of the water in the tube. Though affected by barometric pressure, Galileo's thermometer anticipated by centuries the definition shortly to be adopted by us. To Fahrenheit goes the credit for the first reliable mercury and alcohol thermometers, the era being the early 1700's. The early choices of fixed points for a temperature scale were bizarre but not unreasonable. "Cellar temperature," as found, for instance, 84 feet deep in the cellar of the Paris observatory, was used as late as the mid-eighteenth century. Newton took melting snow and the human body. In 1742 Celsius proposed a scale with zero at the boiling point of water and 100 at melting ice. Seven years later, this scale was turned "proper end up" by a colleague, Strömer, to give the common Centigrade scale. Thus mercury Centigrade thermometers were certainly available to Charles and Gay-Lussac. A fascinating historical account is given by Henry Carrington Bolton in this *Evolution of the Thermometer, 1592–1743* (Easton, Pa.: Chemical Publishing, 1900).

The major virtue, for us, of the perfect-gas law is that the law may be used to construct an operational definition of temperature. This means using the combination $p_{obs}V/N$ to define the temperature T. The reasonableness of this procedure has two bases: (1) the procedure would give a definite meaning to temperature in good accord with our qualitative notions; and (2) the temperature thus defined would be independent of the type of gas used. The truth of the second assertion follows from the experimental evidence on which the empirical gas-law relationship is based.

If one uses a "gas thermometer" to define temperature, the defining relation becomes

$$p_{obs} V \equiv kNT, \tag{4.2.2}$$

where k is a proportionality constant that must be specified to complete the definition. The numerical value of k is a matter of convention, as are its units, for the unit of temperature remains to be specified.

The definition is completed by assigning a unit of measurement to temperature and by specifying that the "temperature" of a certain substance under certain conditions, as measured by a gas thermometer, shall have a certain numerical value in the chosen unit. The unit is the *kelvin*, the modern form of the old "degree Kelvin," and is abbreviated as simply K. The designation honors the British physicist Lord Kelvin, one of the most brilliant exponents of thermodynamics. The kelvin is not expressible in terms of mass, length, and time. It is an entirely independent unit, like, in some systems of units, the Coulomb unit of charge.

With the unit of temperature T established by convention, the units of the proportionality constant k are fixed. Dimensional balance in equation (4.2.2) implies that k has the same units as does energy divided by temperature. In the cgs system, which we shall use in this book, the units become ergs per kelvin.

The "certain substance under certain conditions" is water when the solid, liquid, and gas phases are in equilibrium with one another. The advantage of this choice is that there is only one value of pressure and one of temperature —as determined independently, say, with a mercury thermometer!—for which this three-way equilibrium exists. The specified condition—the *triple point* of water—provides an extremely reproducible fiducial point for a temperature scale. By international convention, the temperature of water at its triple point is taken to be 273.16 kelvin *by definition*. (As measured on a mercury Centigrade scale, the "temperature" would be about 0.01 degrees Centigrade.)

These conventions complete the definition of temperature as defined and measured—they are really synonymous here—by a gas thermometer. The numerical value of the constant k is now a matter of experimental determination. In principle it is determined by measuring the values entering the combination $p_{obs}V/N$ for a gas thermometer in contact with water at its triple point and then reading off k as that number divided by 273.16. In practice the value of k is found by indirect methods; the numerical result is

$$k = 1.381 \times 10^{-16} \text{ ergs per kelvin.}$$

In honor of Ludwig Boltzmann, who, with Maxwell and Clausius, pioneered in the development of statistical mechanics in the latter half of the nineteenth century, the constant k is called *Boltzmann's constant*. As introduced here, Boltzmann's constant is just a proportionality constant, but it gains in significance when we view it as a link between macroscopic quantities (here p_{obs}, V, and T) and microscopic ones (here the number N of molecules).

The diagram in figure 4.2.1 sketches the essentials of a constant-volume gas thermometer. The volume of the gas is kept constant, despite temperature changes, by raising or lowering the entire tube on the right until the mercury in the lefthand tube matches a fixed point. The effect on the gas of "temperature changes" is then solely a matter of gas pressure changes. These register as changes in the height of the mercury column in the righthand tube. In effect, the gas pressure is measured by balancing it against the pressure exerted by the mercury height difference on the right.

In principle the thermometer is calibrated by putting the gas bulb in contact with water at its triple point and thus determining the pressure, $(p_{obs})_{273.16}$, associated with a temperature of 273.16 K for the V and N of the gas. In effect, one has determined the constant V/kN for the thermometer. Temperature is then measured by establishing a common temperature for the gas in the bulb and the object whose temperature is to be determined—by putting

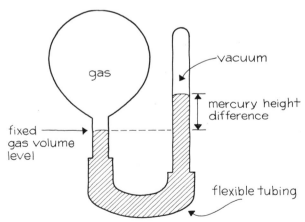

FIGURE 4.2.1
A schematic view of the essentials of a constant-volume gas thermometer.

them in contact and waiting for them to attain mutual equilibrium—and then finding the new pressure. The gas thermometer definition of temperature in equation (4.2.2) yields the temperature of the object as

$$T = p_{\text{obs}} \frac{V}{kN} = p_{\text{obs}} \times \frac{273.16}{(p_{\text{obs}})_{273.16}}.$$

The operational definition of temperature provided by a gas thermometer has some obvious limitations. For example, if the "temperature" becomes too low, the gas condenses into a liquid or, at the very least, exerts so small a pressure that the measurements are difficult. Some extension of the definition beyond the range where the use of a gas thermometer is possible or convenient is necessary.* There are many ways of doing this, ways that give good agreement when the domains of applicability overlap. We need not go into this here, for later there will emerge from statistical mechanics a natural extension of the definition of temperature to all domains in which the concept of temperature appears reasonable. Indeed, we will need to return to the subject of temperature in Chapter 7 and again in Chapter 11. One should look on the gas-thermometer definition as merely an operational definition of temperature, valid as far as it goes, incomplete, yet a good means for getting us started in statistical mechanics.

4.3. A LINK BETWEEN TEMPERATURE AND ENERGY

Later in this chapter we will want to make knowledge of the temperature of a system a vital factor in the statistical description of the system. Because statistical mechanics is a microscopic, mechanical theory, whereas temperature is a different kind of concept, an intrinsically macroscopic concept that is bound (at least so far) to an operational definition, the incorporation of information about temperature into a probability distribution will require some association of temperature with the mechanical notion of energy. In establishing a link, we can take full advantage of our explicit definition of temperature with the gas thermometer.

So let us look at the "inside" of a gas thermometer. We are likely to feel that in a hot gas the molecules move more rapidly than in a cool gas. In our

---

* In fairness to the gas thermometer, it must be said that temperatures as high as the melting point of metallic palladium, $1,822(\pm 2)$ K, have been measured with a nitrogen constant-volume thermometer. Helium gas has been used at least as low as 10 K.

gas-thermometer scheme, an increase in temperature at fixed volume is manifested by an increase in the observed pressure. With pressure being due to molecular impacts on the wall, the pressure increase can arise only from more collisions per second or harder ones or (actually) both. This increased activity entails higher speed, and hence more energy, for the molecules, and thus implies some link between temperature and energy. We can establish a precise connection by relating both temperature and energy in an explicit manner to the pressure.

By virtue of our operational definition, the numerical value of the temperature of the gas is given by

$$T \equiv \frac{p_{obs} V}{Nk},$$

and for convenience later, we rearrange this expression as

$$p_{obs} = \frac{NkT}{V}. \tag{4.3.1}$$

Now we deal with the energy-pressure relation, which arose already in Chapter 1, and we take over one result from that early work. In the quantum-mechanical treatment of a single helium atom in a cubical box, we arrived at an expression for the pressure, $(p_j)_{\text{one atom}}$, exerted on the box walls when the single atom was properly described by a state $(\psi_j)_{\text{one atom}}$ with energy $(E_j)_{\text{one atom}}$:

$$(p_j)_{\text{one atom}} = \frac{2}{3} \frac{(E_j)_{\text{one atom}}}{V}. \tag{4.3.2}$$

(The subscripts have been added in order to reserve unadorned p_j and E_j for the corresponding quantities when we deal with the entire system of N molecules in a realistic gas thermometer.) The analysis leading to this result concerned work and energy; in an expansion of the container, the pressure does some work, and hence the energy of the gas must decrease. Dimensional arguments alone could lead to equation (4.3.2), except for the factor of 2/3, which requires some details about the dependence of the energy eigenvalues on the volume.

In a realistic gas thermometer, there is certainly more than "a single helium atom." So we need to ask whether this relation between pressure and energy will apply when a host of molecules is present. The reply is affirmative, provided the gas is dilute enough so that we may neglect the mutual potential energies of the molecules. Then the relation between an energy eigenvalue for the entire gas of N molecules and the volume will be as before:

$$E_j \propto \frac{1}{V^{2/3}}.$$

After all, when we neglect the mutual interactions, the energy of the entire gas is strictly the sum of the kinetic energies of the individual molecules. (To avoid complications with intramolecular energy, we take the "molecules" to be monatomic.) Since each of the individual energies has a $V^{-2/3}$ dependence on the volume, so must their sum, the total energy. Thus the work-energy analysis of Section 1.2 may be applied to yield the relation

$$p_j = -\frac{\partial E_j}{\partial V} = +\frac{2}{3}\frac{E_j}{V}$$

for the pressure p_j when the entire gas is appropriately described by an energy eigenstate ψ_j with energy E_j.

The direct proportionality of p_j to E_j is the major asset of this connection. It ensures that expectation values of pressure and total energy will necessarily be proportional:

$$\langle p \rangle = \frac{2}{3}\frac{\langle E \rangle}{V}. \tag{4.3.3}$$

The values in any given case will depend on the probability distribution used to calculate the expectation values, but we are assured of the indicated proportionality. It is an eminently reasonable one, for we anticipate that an increase in $\langle E \rangle$ should entail an increase in $\langle p \rangle$.

With these preliminary results established, we may frame a temperature-energy relation for a dilute-gas thermometer. When we know the pressure or temperature of the gas thermometer, we do not know enough to determine a unique quantum-mechanical state for the gas. Rather, we must deal with a set of states, and must describe the gas in a statistical fashion with a probability distribution. On general grounds we may place a restriction on that probability distribution. We may require that the expectation value estimate $\langle p \rangle$ of the pressure be equal to the observed pressure:

$$\langle p \rangle = p_{\text{obs}}. \tag{4.3.4}$$

This is surely a reasonable requirement. It will gain further acceptability when we find that the statistical theory predicts that discernible deviations in pressure from the expectation value $\langle p \rangle$ are highly improbable when the

number of gas particles is large, as in a realistic gas thermometer. The requirement leads to a self-consistent theory.

The imposition of equation (4.3.4) permits us to deduce a connection
between the statistically calculated energy and the temperature. Upon
substituting for $\langle p \rangle$ from equation (4.3.3) and for p_{obs} from equation (4.3.1),
we find

$$\frac{2}{3} \frac{\langle E \rangle}{V} = \frac{NkT}{V}$$

or

$$\langle E \rangle = \tfrac{3}{2} NkT. \tag{4.3.5}$$

This is the link—perhaps an obvious one—between temperature and energy
for the dilute-gas thermometer. The link emerges as a statistical connection.
For us the import of equation (4.3.5) is this: knowledge of the temperature
of the dilute-gas thermometer enables us to impose (on the probability
distribution describing the gas) the requirement that an expectation value
calculation of the total energy yield $\tfrac{3}{2} NkT$.

At this point, where we see temperature as being *statistically* related to a
mechanical concept like energy, we should note again that temperature is
intrinsically a *macroscopic* notion. A molecule by itself may have a well-
defined position or velocity or energy but, by itself, it can have no well-
defined temperature. Only when the molecule is part of a macroscopic system
(at a well-defined temperature) may we associate a temperature with the
molecule, and even then we are indulging in a loose usage. It is the macroscopic system that has the measurable temperature.

4.4. STATES IN THE STATISTICAL DESCRIPTION OF EQUILIBRIUM

This is the last of the background sections. In it we will establish the set of
inferences to be used in the statistical description of a physical system known
to be in equilibrium. The succeeding sections will be addressed to establishing
the probabilities to be assigned those inferences, given one or another specific
hypothesis.

The problem before us can be stated in considerable generality. We seek a
set of inferences to be used in estimating the properties of some physical
system that is in equilibrium. The system might be our ubiquitous box of gas,
though perhaps no longer dilute. It might be a paramagnetic sample in a

magnetic field. The possibilities are legion, but there is always some single system of interest to us.

It goes almost without saying that we know the environment in which the system is located. This may mean the volume and shape of a container, the direction and magnitude of an externally applied magnetic field, and so on. Since these are macroscopic quantities, we may regard them as having definite values; quantum-mechanical uncertainties are (presumably) unimportant. We assume that the number of particles in the system and the behavior of the interparticle forces are specified. Even when the interparticle forces are not precisely known, one generally assumes a plausible form for them, perhaps with some adjustable parameters. Then one compares the statistical estimates with experiment and adjusts the parameters for best fit. So an assumption that the interparticle forces are at least specified, if not literally known, is acceptable.

The approach to the set of inferences for this kind of situation will be a quantum-mechanical one. This approach will enable us to derive later the classical expressions as the limit of the more generally applicable quantum-mechanical ones. Since we will certainly not possess in the hypothesis enough information to specify a unique wave function for describing the system, we must consider many possible quantum-mechanical states. For each of these states, we must assign a probability to the inference that it appropriately describes the actual system. But the first question to be answered is, which states should we use in the inferences?

There are several reasons for deciding to use states that are energy eigenstates of the entire system. One reason is simplicity. We have seen that, for the gas thermometer, there is a direct connection between temperature and the expectation value of the energy. As part of the general development, we will need to analyze the gas thermometer and derive a probability distribution for describing it. The use of energy eigenstates will make the restriction on the probability distribution (arising from the temperature-energy relation) easy to handle.

There is, however, a far more basic reason for using energy eigenstates: they guarantee time-independent quantum-mechanical estimates of physical quantities. Such time-independence is a property we want when describing a system known to be in equilibrium. To see the mathematical origins of this desirable property, we must look more closely at a system's energy eigenstates.

Let us designate the energy operator for the entire system by $\mathcal{H}$. Then an energy eigenstate ψ_j is determined by the equation

$$\mathcal{H}\psi_j = E_j\psi_j, \tag{4.4.1}$$

where E_j is an energy eigenvalue of the entire system. The shape and volume of the container or the value of the external magnetic field are needed for a specification of the energy operator and explicitly appear in it. We saw this previously for the single helium atom in a box, where the size of the cubical box determined the spatial dependence of the potential energy. In general, some macroscopic environmental parameters will appear in the energy operator. By convention these parameters, which are of a "mechanical" nature, are called *external parameters*. They appear again in the energy eigenvalues. After all, it is to be expected that the environment will partially determine the energy values of the system. For example the presence or absence of a magnetic field must have some effect on the possible energy values. While we are on the subject of dependence, let us note that the energy operator $\mathcal{H}$ is not dependent on time. This follows from the constancy of the environment in which the system is located.

Several distinct quantum-mechanical states may have the same energy; that is, the spatial properties of the wave functions may be different but the total energy the same. If the system is macroscopic, such a situation is rather unlikely, at least when all interactions are included. As before, the index j on ψ_j and E_j designates a specific quantum-mechanical state, not merely a specific energy level.

Two factors are crucial to the time-dependence (or, rather, time-independence) of quantum-mechanical estimates from energy eigenstates: (1) the evolution in time of the energy eigenstates themselves; and (2) the manner in which estimates are formed from wave functions.

The evolution is determined by the time-dependent Schrödinger equation. For a general wave function $\psi(t)$, that equation reads

$$i\hbar \frac{\partial \psi(t)}{\partial t} = \mathcal{H}\psi(t).$$

If we write an energy eigenstate at time t_0 as $\psi_j(t_0)$, the evolution in time leads to a state $\psi_j(t)$, given by

$$\psi_j(t) = e^{-i(t-t_0)E_j/\hbar}\psi_j(t_0). \tag{4.4.2}$$

This expression both satisfies the Schrödinger equation and reduces at time $t = t_0$ to the proper initial wave function. The significant aspect of this result is that all the time-dependence is situated in phase factor containing only constants and the time.

Now we examine the manner in which estimates are formed from wave functions. Quantum mechanics instructs us to compute the probabilities for the positions of particles by forming the square of the absolute value of the wave function. In such an operation the time-dependence of a wave function like that in equation (4.4.2) disappears entirely. Time-independent position probabilities result—precisely what we want for a description of equilibrium.

Quite generally, estimates in quantum mechanics depend on the product of a wave function with the complex conjugate wave function. Hence the time-dependence of an energy eigenstate, which arises only in a phase factor, will not lead to time-dependent results.

The upshot of the analysis is that the use of energy eigenstates does guarantee time-independence for estimates of physical quantities, a characteristic we demand in describing a system known to be in equilibrium.

An aside: let us note that there is a vast difference between (1) no time variation of predictions and (2) the prediction of no time variations. Suppose we knew that the electron in an isolated hydrogen atom were in the ground state, the lowest energy eigenstate. Estimates or predictions of the electron's position would be based on the absolute square of the ground-state wave function. Since that square would be independent of time, there would be no time variation in the predictions of position. But certainly we would not predict no time variation in the electron's position. The electron would be moving; the non-zero kinetic energy in the ground state would provide ample basis for that modest conclusion. The relevance of all this is the following: although we demand no time variation in our estimates, we do not thereby deny the existence of fluctuations in time. *When* in time the fluctuations will occur is certainly beyond prediction—the data are inadequate. We must do the best we can on the basis of what we do know.

So we focus our attention on the set of energy eigenstates of the system, each ψ_j being determined by the eigenvalue equation displayed in equation (4.4.1). There will be many such states. Let us specify that the wave functions we use are mutually orthogonal:

$$\int \psi_j^* \psi_k \, d(\text{variables}) = 0 \quad \text{if} \quad k \neq j.$$

Here ψ_j^* is the complex conjugate of the wave function ψ_j. The single integral sign symbolizes integration over all variables. That mutual orthogonality may be demanded without loss of generality is a basic quantum-mechanical result. (If the states have different energies, the orthogonality follows automatically. Those states with equal energies can be combined in such a way that the resulting wave functions are orthogonal.)

For inferences about the system, let us take statements of the form, "The energy eigenstate ψ_j provides the appropriate description of the physical system." There will be as many inferences in this set as there are distinct energy eigenstates of the system. Let us denote the total number by n; then the index j will run from $j = 1$ to $j = n$. Whether the number n is finite or infinite is unimportant for now. (If n is infinite, we do assume that the energy eigenstates are merely denumerably infinite, as is typical for a spatially confined system.) On the hypothesis that the system is in equilibrium, we may draw two general conclusions about the set of inferences.

The inferences in the set are mutually exclusive. The mutual orthogonality of the wave functions ensures this, for that implies that if the system is in state ψ_j, there is zero probability of its being in state ψ_k when k is different from j.

For our statistical purposes we may take the set of inferences to be exhaustive. This conclusion is more difficult to justify, though the following may be said in support of it. The system of concern to us is known to be in equilibrium, to show no systematic temporal variation in its properties. Only energy eigenstates have the property of guaranteeing time-independent predictions and estimates of the properties of a system. So we may take the view that some energy eigenstate is the appropriate state (given merely the limited information we possess). Since the inferences cover, one by one, all the energy eigenstates, we may take the set of inferences to be exhaustive.

Let us tie together the sailent points of this section. We are concerned with describing in quantum-mechanical terms some specific physical system known to be in equilibrium, and with estimating its properties. We take as generally known the environment (the external parameters), the number of particles, and the interparticle forces. Later, information about energy or temperature will be added to the general knowledge. Even with the additional information, the information available is inadequate to enable one to specify a single quantum-mechanical state as providing *the* appropriate description of the system. We need to consider many states and to introduce probabilities. Only the energy eigenstates have the property of guaranteeing time-independent estimates of physical properties. So we must seek to describe the system with energy eigenstates. We take the inferences to be of the form, "The energy eigenstate ψ_j provides the appropriate description of the physical system." Given the general hypothesis about equilibrium and the properties of the energy eigenstates, we may take the inferences in the set to be mutually exclusive and exhaustive.

4.5. THE PROBABILITY DISTRIBUTION WHEN $\langle E \rangle$ IS SPECIFIED

The preceding section has set up the general framework for the statistical description of a system known to be in equilibrium. All that remains is to establish the probabilities to be assigned to the inferences. The probabilities depend on the full hypothesis, for that determines the set of compatible probability distributions.

Since the task of establishing the probability distribution for a system whose temperature is known is difficult, because, as already mentioned, temperature is not a "mechanical" parameter, in this section we will perform a preliminary calculation. We will assume that we know enough about the actual physical system that we may reasonably force an expectation value estimate of the total energy to yield a specific known value. This kind of calculation is directly applicable to the gas thermometer, as we noted in Section 4.3, and that essential application will be made in the section following this one.

To each of the inferences we must now assign a number P_j, the probability that, on the basis of our information, it is the correct inference. Thus far there is only one restriction on the set of probabilities. Because the inferences are exhaustive and mutually exclusive, the probabilities, however assigned in detail, must satisfy the normalization condition:

$$\sum_{j=1}^{n} P_j = 1. \qquad (4.5.1)$$

To restrict further the probabilities, we invoke the assumption that we may force an expectation value calculation of the total energy to yield a specific known value. Calling that known value the constant $\tilde{E}$, we demand that the probabilities satisfy the relation

$$\sum_{j=1}^{n} E_j P_j = \tilde{E}. \qquad (4.5.2)$$

This exhausts our fund of information and thus our source of restrictions on the probabilities. To select the "best" probability distribution out of the infinitely many compatible distributions, we must turn to Criterion II of Section 3.7. To avoid the introduction of bias we must choose, from among the compatible distributions, the one that exhibits the largest missing information. In mathematical terms, this means we seek the probability distribution that maximizes

$$-K \sum_{j=1}^{n} P_j \ln P_j$$

subject to the two restrictions, equations (4.5.1) and (4.5.2). The Lagrange multiplier technique (discussed in Appendix B) provides the most convenient method for solving this mathematical problem; we will use it.

Let us introduce two Lagrange multipliers, λ_0 and λ_1, and write

$$\Lambda = \Lambda(P_1, P_2, \ldots, P_n, \lambda_0, \lambda_1)$$
$$\equiv -K \sum_j P_j \ln P_j + \lambda_0 \left[1 - \sum_j P_j \right] + \lambda_1 \left[\tilde{E} - \sum_j E_j P_j \right].$$

We look for an extremum by requiring that the partial derivatives of Λ with respect to all the probabilities and to both λ_0 and λ_1 be zero:

$$\frac{\partial \Lambda}{\partial P_j} = -K(\ln P_j + 1) - \lambda_0 - \lambda_1 E_j = 0,$$

$$\frac{\partial \Lambda}{\partial \lambda_0} = 1 - \sum_j P_j = 0,$$

$$\frac{\partial \Lambda}{\partial \lambda_1} = \tilde{E} - \sum_j E_j P_j = 0.$$

On solving the first of these equations for the probability P_j, we find

$$P_j = e^{-(1 + \lambda_0/K)} e^{-(\lambda_1/K)E_j}.$$

Since the factor λ_1/K multiplying the energy is usually denoted by β, let us follow convention and reexpress the result as

$$P_j = e^{-(1 + \lambda_0/K)} e^{-\beta E_j}.$$

The second equation is just the normalization condition:

$$\sum_{j=1}^{n} e^{-(1 + \lambda_0/K)} e^{-\beta E_j} = 1.$$

This permits us to determine the constant exponential factor in terms of a sum of $\exp(-\beta E_j)$ over all j. A tidy result begins to emerge:

$$P_j = \frac{e^{-\beta E_j}}{\left(\sum_{i=1}^{n} e^{-\beta E_i} \right)}. \tag{4.5.3}$$

If we arrange the states ψ_j in order of increasing energy, the probability distribution exhibits a very smooth profile. It neglects no states, save those with infinite energy (if such there be), to which it assigns vanishingly small probability, for β must then be positive to ensure convergence of the sum.

Though an exponential dependence is perhaps unexpected, the probability distribution does conform to qualitative notions of a distribution that avoids unwarranted assertions.*

There remains the third equation, which now takes the form

$$\frac{\sum_{j=1}^{n} E_j e^{-\beta E_j}}{\left(\sum_{i=1}^{n} e^{-\beta E_i}\right)} = \tilde{E}. \tag{4.5.4}$$

This equation serves to determine the parameter β in terms of the known constant $\tilde{E}$. Provided the summations can be done (or adequately approximated) and the ensuing equation solved for β in terms of $\tilde{E}$ and the energy eigenvalues, one has a complete determination of the desired probability distribution.

Incidentally, one might have thought that the probability distribution P_j would exhibit a peak somewhere near the specified energy $\tilde{E}$, rather than be a monotonic function of E_j. One must be careful to distinguish a probability distribution for *states* of the system from a probability distribution for various values of the *energy* of the system. As we will discuss in more detail later, one can get the latter probability distribution from the former by grouping together states having comparable energy. The number of states with energy in a small range of given size generally grows rapidly with the energy at which that range is centered. Whereas the probability for an individual state decreases with energy, the number of states to be considered increases with energy. The result of the competition is a sharp peak near $\tilde{E}$ in the probability distribution for various values of the energy, quite in accord with common sense. This discussion is a bit premature but may allay a natural worry.

4.6. FURTHER ANALYSIS OF THE GAS THERMOMETER

The results of the preceding section are directly applicable to a statistical description of the gas in a gas thermometer, for which the connection between the "known constant $\tilde{E}$" and the temperature T is established through their

---

* We have not checked in detail that the *extremum* found by the Lagrange multiplier technique is actually a *maximum*. The foregoing paragraph makes it plausible that we do have a maximum, and further support is provided by one of the problems. A proof that the solution derived here does maximize the missing information subject to the restrictions is given by E. T. Jaynes in his Brandeis lectures, cited in the references for this chapter. We have the one and only solution to the constrained maximization problem.

mutual relationship to the expectation value of the energy:

$$\langle E \rangle = \tilde{E} \quad \text{and} \quad \langle E \rangle = \tfrac{3}{2}NkT$$

imply

$$\tilde{E} = \tfrac{3}{2}NkT. \tag{4.6.1}$$

So the parameter β in the probability distribution for the states of the gas is determined by equation (4.5.4) with the appropriate value of $\tilde{E}$ inserted:

$$\frac{\sum\limits_{j=1}^{n} E_j e^{-\beta E_j}}{\left(\sum\limits_{i=1}^{n} e^{-\beta E_i}\right)} = \tfrac{3}{2}NkT. \tag{4.6.2}$$

Inspection of the equation is enough to tell us that β for the gas thermometer is a function of the temperature. If we change the temperature T on the right-hand side (by imagining a higher temperature for the gas, say), the value of the lefthand side must change accordingly. The only quantity on the left that may change is β, for the eigenvalues are independent of temperature. So β must be a function of the temperature.

To determine the manner in which β depends on temperature, we must evaluate the summations appearing in equation (4.6.2) and then compare them with the righthand side. The energy eigenvalues are those for the entire system of N particles, and the sum goes over all energy eigenstates. If we were to try to evaluate the summations for the number of atoms typical of a gas thermometer, we would soon be lost in the algebra. Even for the case of $N = 1$ the summations are not trivial. With the aim of reducing the calculations to manageable proportions, let us deal with $N = 1$. That the ensuing result is valid also for $N \gg 1$ can be made plausible by noting that details about the number of atoms are contained in the probabilities through the energy eigenvalues E_j and the extent of the summation. One may have β independent of N without violating the clear requirement that the probability distribution must in some way depend on the number of atoms. Later, in Section 9.6, we will confirm the validity of the connection between β and T made here.

So we take the simplest case, $N = 1$, a single atom in a cubical box, and look for the relation between β and T. For the energy eigenstates and corresponding energies, we have the results from Section 1.2. An energy eigenvalue E_j is given by

$$E_j = \frac{\pi^2 \hbar^2}{2m V^{2/3}} (n_x^2 + n_y^2 + n_z^2) \equiv \gamma(n_x^2 + n_y^2 + n_z^2).$$

The abbreviation γ has been introduced merely to shorten subsequent expressions. To each j corresponds a particular set of the positive integers n_x, n_y, n_z. A summation over all j is here a summation over n_x, n_y, n_z as these range over all positive integers, one to infinity. This means, incidentally, that "n" in the j summation is actually infinity.

Now we may go back to equation (4.6.2) and write down the formidable summations relating β and T:

$$\frac{\sum\limits_{n_x=1}^{\infty} \sum\limits_{n_y=1}^{\infty} \sum\limits_{n_z=1}^{\infty} \gamma(n_x^2 + n_y^2 + n_z^2)e^{-\beta\gamma(n_x^2+n_y^2+n_z^2)}}{\sum\limits_{n'_x=1}^{\infty} \sum\limits_{n'_y=1}^{\infty} \sum\limits_{n'_z=1}^{\infty} e^{-\beta\gamma(n'_x{}^2+n'_y{}^2+n'_z{}^2)}} = \tfrac{3}{2}kT.$$

The complicated expression may be greatly simplified. First we use the symmetry in n_x, n_y, n_z in the numerator. The final contribution of each of three terms *outside* the exponential will be identical. Provided we multiply by 3, we may drop n_y^2 and n_z^2 where they appear *outside* the exponential, leaving just n_x^2:

$$\frac{3 \sum\limits_{n_x} \sum\limits_{n_y} \sum\limits_{n_z} \gamma n_x^2 e^{-\beta\gamma(n_x^2+n_y^2+n_z^2)}}{\sum\limits_{n'_x} \sum\limits_{n'_y} \sum\limits_{n'_z} e^{-\beta\gamma(n'_x{}^2+n'_y{}^2+n'_z{}^2)}} = \tfrac{3}{2}kT.$$

Once we have done this, we may usefully factor the exponentials and cancel, top and bottom, the summations in n_y and n_z. In more detail for the numerator:

$$3 \sum\limits_{n_x} \sum\limits_{n_y} \sum\limits_{n_z} (\gamma n_x^2 e^{-\beta\gamma n_x^2}) \times (e^{-\beta\gamma n_y^2}) \times (e^{-\beta\gamma n_z^2})$$

$$= 3\left(\sum\limits_{n_x} \gamma n_x^2 e^{-\beta\gamma n_x^2}\right) \times \left(\sum\limits_{n_y} e^{-\beta\gamma n_y^2}\right) \times \left(\sum\limits_{n_z} e^{-\beta\gamma n_z^2}\right).$$

The step follows by imagining that the n_z summation is done first with n_x, n_y, held fixed, then the n_y summation is done with n_x held fixed, and finally the full summation is completed with the n_x summation. An analogous factorization holds for the denominator.

After the cancellation of the n_y, n_z summations, a somewhat more tractable result emerges:

$$\frac{3 \sum\limits_{n_x} \gamma n_x^2 e^{-\beta\gamma n_x^2}}{\sum\limits_{n'_x} e^{-\beta\gamma n'_x{}^2}} = \tfrac{3}{2}kT. \qquad (4.6.3)$$

Even these two summations cannot be evaluated in terms of simple functions. If the volume is large, the value of γ will be small, and so the value of the exponential will change relatively little during each increment in n_x (at least in the summation region where the exponential still has a significant magnitude). Such will be the case for a realistic gas thermometer. So we may satisfactorily approximate the summation in the denominator by an integral:

$$\sum_{n_x=1}^{\infty} e^{-\beta \gamma n_x^2} \simeq \int_{\eta=0}^{\eta=\infty} e^{-\beta \gamma \eta^2} \, d\eta. \qquad (4.6.4)$$

The approximation procedure is illustrated in figure 4.6.1, which shows also

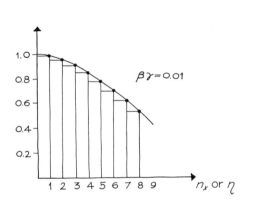

FIGURE 4.6.1
The approximation of the summation by an integration. The points indicate the values of the summand in equation (4.6.4), and the smooth curve gives the integrand in that equation. To compare summation and integration, let us regard summation as multiplication of each value of the summand by one and then a summation of the terms, so that the numerical value of the sum is represented by the area of the bars in the diagram. The area under the curve closely approximates that area already for $\beta\gamma = 0.01$. If the volume (present in γ) is large, then $\beta\gamma$ is very small, and the error introduced by integration is small and hence negligible.

why the lower limit of the approximating integral is taken as $\eta = 0$. (The symbol η is merely a dummy variable of integration and has no special significance.)

The extra factor of n_x^2 in the numerator presents no real problem, for we may write that summation as the partial derivative with respect to β of a summation like that in the denominator,

$$3 \sum_{n_x} \gamma n_x^2 e^{-\beta \gamma n_x^2} = 3 \sum_{n_x} -\frac{\partial}{\partial \beta} e^{-\beta \gamma n_x^2}$$

$$= -3 \frac{\partial}{\partial \beta} \left(\sum_{n_x} e^{-\beta \gamma n_x^2} \right),$$

and then use the integral approximation for the sum in parentheses.

Equation (4.6.3) now takes the form

$$\frac{-3\frac{\partial}{\partial\beta}\left(\int_0^{\infty} e^{-\beta\gamma\eta^2}\, d\eta\right)}{\int_0^{\infty} e^{-\beta\gamma\eta^2}\, d\eta} = \tfrac{3}{2}kT. \tag{4.6.5}$$

With the aid of the table of integrals in Appendix A, we find

$$\frac{-3\frac{\partial}{\partial\beta}\left(\tfrac{1}{2}\sqrt{\pi}(\beta\gamma)^{-1/2}\right)}{\tfrac{1}{2}\sqrt{\pi}(\beta\gamma)^{-1/2}} = +\tfrac{3}{2}\beta^{-1} = \tfrac{3}{2}kT$$

and thus

$$\beta = \frac{1}{kT}. \tag{4.6.6}$$

For several reasons, this is a satisfying result. It is simple, and it is the kind of result one might expect on the basis of merely a dimensional argument. The quantity β appears in an exponent, $\exp(-\beta E_j)$, and an exponent must be dimensionless; therefore β must have the dimensions of $(\text{ergs})^{-1}$. Since we know that β is a function of the temperature and since temperature carries the kelvin unit, there must be some dimensional factor that, in effect, converts from the temperature unit to energy units. Boltzmann's constant k, with dimensions of ergs per kelvin, is a likely candidate. This would lead to

$$\beta = \left(\begin{matrix}\text{dimensionless}\\ \text{factor}\end{matrix}\right)\frac{1}{kT}$$

on dimensional grounds alone.

In addition, the form $\beta = 1/kT$ for the temperature dependence of β gives a satisfactory temperature dependence to the entire probability distribution. We expect that as the temperature is increased, the probability of the higher energy states will increase, at least relative to the lower energy states. For the ratio of the probabilities of states j and i, we have

$$\frac{P_j}{P_i} = \frac{e^{-\beta E_j}}{e^{-\beta E_i}} = e^{-\beta(E_j - E_i)} = e^{-(E_j - E_i)/kT}.$$

If $E_j - E_i > 0$, then an increase in temperature (and hence a reduction of β) leads to an increase in the ratio, and hence to the anticipated result.

4.7. THE PROBABILITY DISTRIBUTION WHEN T IS KNOWN

The objective of this final section is to establish the probability distribution for a general system in equilibrium when the system's temperature is known. We will find that relatively little mathematics is needed. The major element is a carefully structured argument based on results and criteria that we have already obtained.

Temperature is not a "mechanical" parameter and therefore fits only awkwardly into a theory of statistical *mechanics*. Nonetheless, since the temperature is frequently one of the few things known about a physical system, we must find a way to work from a knowledge of temperature to a probability distribution. Let us first look closely at the condition "... when the system's temperature is known." A temperature can be "known" only if it has been measured. The implication is that we should pay attention to the measurement process itself.

Temperature measurement, in the defining scheme we have adopted, involves certain operations with a calibrated gas thermometer. Specifically, the system of interest is placed in contact with the gas thermometer, and after a wait until a mutual equilibrium has been attained, an observation of the pressure of the gas forming the thermometer enables us to specify the temperature of the system. We now "know" the temperature of the system of interest. If the thermometer and the system of interest are subsequently separated, we still say that we "know" the temperature of the system. The justification for our saying this is the belief that if the two were again put in contact and equilibrium attained, the observed gas pressure, and hence the temperature, would be the same as before. (There is a natural restriction here: one must not tamper with the system of interest. If one were to heat it with a Bunsen burner, one would expect a change in temperature.)

From the brief description we can extract the two crucial aspects of our "knowing" the temperature of a system: (1) there is (or was) a mutual equilibrium between the system of interest and the thermometer; and (2) observations made on the thermometer lead to the numerical determination of the temperature of the system of interest. A knowledge of the temperature provides surprisingly little information, and what information there is, is only macroscopic, though it does reflect the microscopic state of affairs.

In order to determine a probability distribution for the system, we must make use of the "mutual equilibrium" property. Let us look first for a statistical description of the system of interest and the thermometer, in contact, considered as a single, joint system. Once we have a probability distribution for the joint system, we can readily derive from that a probability distribution for the system of interest.

The joint system is in equilibrium, so we should describe it with energy eigenstates of the joint system. Thus the first task is that of constructing (at least mentally) those states. The contact between the two components—the system of interest and the gas thermometer—means that there is a mutual interaction. Indeed, the existence of such an interaction is necessary in principle if the two components are to reach a *mutual* equilibrium. Nevertheless, in practice the interaction is typically small. So to a very good approximation we may specify an energy eigenstate of the joint system by specifying an energy eigenstate ψ_j of the system of interest and an energy eigenstate $\psi'_{j'}$ of the gas thermometer. The energy of the state of the joint system, thus specified, would be $E_j + E'_{j'}$. The smallness of the "contact" interaction provides the justification for leaving out the very small energy associated with the interaction itself. We should note also that one seldom knows much about the interaction between thermometer and system of interest, beyond the fact that the interaction is present and is small. Even if we wished to treat the interaction in detail, we would be prevented from doing so.

An inference concerning the joint system would be of the form, "The joint system is appropriately described by the joint energy eigenstate formed from the state ψ_j of the system of interest and the state $\psi'_{j'}$ of the gas thermometer." Let us use the symbol $P_{jj'}$ to denote the probability, given the hypothesis in the background, that this inference is correct. The hypothesis contains information about the constituents of the system of interest, their mutual interactions, the environment, similar data for the gas thermometer, and the results of the temperature measurement. The range of the index j will be $1 \leqslant j \leqslant n$, where n is the total number of energy eigenstates of the system of interest. A corresponding relation, $1 \leqslant j' \leqslant n'$, will hold for the index j', where n' is the number of energy eigenstates of the thermometer. As before with only a single system, we may consider the inferences in the set to be mutually exclusive and exhaustive. Thus the probabilities must satisfy the normalization condition:

$$\sum_j \sum_{j'} P_{jj'} = 1.$$

Before going on, let us note that the characteristic inference here may be regarded as the conjunction of two other, yet similar, inferences. The inference of the preceding paragraph about the joint system is equivalent to the conjunction, "The system of interest is appropriately described by the state ψ_j" *and* "The thermometer is appropriately described by the state $\psi_{j'}'$." Since the existence of equilibrium ensures that both the inferences about the system of interest and those about the thermometer form sets in which the inferences are mutually exclusive and exhaustive, the reduction procedure of Section 2.11 may be applied. Once we have the probability distribution for the joint system, we may derive the distribution for the system of interest alone by summing over the states of the thermometer, that is, by summing over all j'. So a determination of the distribution for the joint system will amply satisfy our needs.

Now we turn to the heart of the derivation: the physical and statistical arguments that will determine the "best" probability distribution for the joint system. Let us note first that the probability distribution for the thermometer alone should remain unchanged if the joint system is physically separated into its two components. After all, the thermometer is in equilibrium and will remain so; also, there is no significant change in the information we possess. We already know what the probability distribution for the gas thermometer itself should be under such circumstances. That we derived in Sections 4.5 and 4.6 as

$$\begin{pmatrix} \text{the probability that the} \\ \text{thermometer is appropriately} \\ \text{described by its energy eigen-} \\ \text{state } \psi_{j'}' \text{ with energy } E_{j'}' \end{pmatrix} = \frac{e^{-\beta E'_{j'}}}{\left(\sum_{i'} e^{-\beta E'_{i'}}\right)}, \qquad (4.7.1)$$

with the parameter β found to be $1/kT$. (Remember that in the present section the variables referring to the thermometer carry primes.) So we are in a position to place a significant constraint on the probability distribution for the joint system. The summation of $P_{jj'}$ over all states of the system of interest, that is, over all j, yields the probability given in words on the lefthand side of equation (4.7.1). We should require, as a constraint on the probabilities $P_{jj'}$, that such a summation yield precisely the righthand side of equation (4.7.1). Thus the constraint takes the form

CONSTRAINT I:

$$\sum_j P_{jj'} = \frac{e^{-\beta E'_{j'}}}{\left(\sum_{i'} e^{-\beta E'_{i'}}\right)} \quad \text{with} \quad \beta = \frac{1}{kT}. \qquad (4.7.2)$$

Beyond imposing a condition on the probability distribution for the joint system, this relation leads us to consider in more detail the dependence of $P_{jj'}$ on the energies.

The contact between the gas thermometer and the system of interest, which leads to the *mutual* equilibrium, ensures that the two systems can exchange energy and hence do share a common energy. To be sure, we do not know what the common total energy actually is. Still, it seems natural that if there are different states of the joint system with the same total energy, those states should be assigned the same probability; states of the joint system with different total energies could quite reasonably have different probabilities.

Criterion II of Section 3.7 enables us to give the "natural" conclusion a firm basis. Suppose two different states of the joint system do have equal total energies. We ask, should they then have equal probabilities? To assign unequal probabilities would be unwarranted. Indeed, if we were initially to assign unequal probabilities, we could always increase the missing information of the probability distribution by making those two probabilities more nearly equal. To maximize the missing information we should assign the same probability to all states of the joint system that have the same total energy.

The firm conclusion carries a strong implication: the dependence of the probability $P_{jj'}$ on the energy should be a dependence on only the *sum* of the energies of the system of interest and the thermometer. Only such a dependence can ensure that two different states of the joint system with the same total energy will be assigned the same probability.* With this implication we have a second constraint on the probability distribution for the joint system. If we introduce a presently unknown function $f(E_j + E'_{j'})$ of the sum of the energies, we may state the constraint as the relation

CONSTRAINT II:

$$P_{jj'} = \frac{f(E_j + E'_{j'})}{\sum_i \sum_{i'} f(E_i + E'_{i'})}. \tag{4.7.3}$$

The sum in the denominator ensures the correct normalization for the probabilities.

We should note that it is through Constraint II that the physical coupling of the gas thermometer and the system of interest—the *mutual* equilibrium—

---

* Alternatively, only if we wanted to assign different numerical probabilities to different states of the joint system with the same total energy would we need to write $P_{jj'}$ as a function of E_j and $E'_{j'}$ separately, rather than as a function of their sum.

is inserted into the mathematics. It is through that constraint that we introduce the possibility of the gas thermometer's telling us something about the
system of interest.

There is no difficulty in writing down a general expression for the probability
$P_{jj'}$ that satisfies both constraints and the normalization condition:

$$P_{jj'} = \frac{e^{-\beta(E_j + E'_{j'})}}{\sum_i \sum_{i'} e^{-\beta(E_i + E'_{i'})}}.$$

From this distribution we may derive the probability distribution for the
system of interest by summing over the states of the thermometer. Let us
denote by P_j the probability of the inference, "The system of interest is
appropriately described by its energy eigenstate ψ_j," given the background
hypothesis. Then we compute P_j as

$$P_j = \sum_{j'} P_{jj'} = \frac{\sum_{j'} e^{-\beta(E_j + E'_{j'})}}{\sum_i \sum_{i'} e^{-\beta(E_i + E'_{i'})}}$$

$$= \frac{e^{-\beta E_j} \left(\sum_{j'} e^{-\beta E'_{j'}} \right)}{\left(\sum_i e^{-\beta E_i} \right) \left(\sum_{i'} e^{-\beta E'_{i'}} \right)}.$$

The summations directly referring to the thermometer may be canceled, top
and bottom. The upshot of the analysis is that we find a universal form for
the "best" probability distribution for the system of interest when the
temperature is known:

$$P_j = \frac{e^{-\beta E_j}}{\left(\sum_i e^{-\beta E_i} \right)} \quad \text{with} \quad \beta = \frac{1}{kT}. \tag{4.7.4}$$

To derive this probability distribution, we had to use a characteristic
property of the temperature measurement process: the system of interest and
the gas thermometer are in contact and achieve a mutual equilibrium. The
distribution must apply, however, even after contact between thermometer
and system of interest is broken, for the system remains in equilibrium and
our information does not change in any significant fashion. The probability
distribution is quite generally applicable to a system in equilibrium whose
environment and temperature one knows.

Let us be clear on the matter of uniqueness: the expression for the probability P_j in equation (4.7.4) is the only expression permitted by the two constraints. A proof is a matter of two sentences. Constraint II implies that the form of the probability $P_{jj'}$ is symmetric in E_j and $E'_{j'}$. Since Constraint I requires that a summation of $P_{jj'}$ over j yield the exponential form in equation (4.7.2), the symmetry implies that a summation of $P_{jj'}$ over j' must yield a similar exponential form, in particular, that of equation (4.7.4). Indeed, this argument could have been used to avoid a number of steps in the derivation, though perhaps at the cost of clarity and conviction. The conclusion, though, is what matters: the expression in equation (4.7.4) for the probability P_j follows uniquely from the constraints.

Lest there be misunderstanding, one more point should be made here. The "system of interest" may be a dense gas or a solid or any one of a host of physical systems. When one estimates its energy by calculating $\langle E \rangle = \Sigma E_j P_j$, with P_j as given in equation (4.7.4), the temperature T will appear via β, but there is no reason to expect $\langle E \rangle$ to be proportional to T (as it is for a dilute gas). Intermolecular potential energies, or an interaction with a magnetic field, or just peculiar quantum-mechanical effects, may make the connection between $\langle E \rangle$ and T highly nonlinear, and dependent on the external parameters as well. Numerous applications in the chapters to come will bear out this claim.

To get a better appreciation of the logic underlying the derivation, we should retrace our steps, noting the crucial landmarks. The first is the gas-thermometer definition of temperature. Adoption of the gas-thermometer definition brings the benefit that, although temperature is a macroscopic, nonmechanical concept, we may readily analyze the thermometer itself in microscopic, mechanical terms. In that analysis we established a connection between the statistical theory and observation by imposing on the statistical theory the requirement that $\langle p \rangle$ be numerically equal to p_{obs}. This is a modest requirement but an essential one. The energy-temperature relation for the dilute-gas thermometer follows from it.

When we know that a physical system is in equilibrium, know the environment, and know the molecular constitution, we are still very far from knowing the quantum-mechanical state that should be used to describe the system. We may conclude that the state should be one of the energy eigenstates, but which of many is still an open question. A statistical description is the best we can hope to achieve. If we supplement our knowledge with information enabling us to force an expectation value calculation of the energy to yield a specified numerical value, we are certainly better off. Nonetheless, this addition still leaves us without a determination of the probability distribution.

To choose a probability distribution without the introduction of bias, we turn to the measure of missing information. Out of all the probability distributions compatible with what we know about the system, we choose, as "best," that distribution with the largest missing information. This is a quantitative procedure for the avoidance of unwarranted preference or neglect among the inferences. Given the severe limitations on the information available to us, a description in terms of probabilities is unavoidable; the maximum missing information criterion ensures that we choose the probability distribution in the most bias-free manner. This is the heart of the statistical approach.

The determination of the best probability distribution, when we specify the expectation value of the energy, is a necessary preliminary calculation. That calculation enables us to analyze further the gas thermometer itself and to work out an explicit probability distribution for it. It is here that we find the relation $\beta = 1/kT$.

To formulate the probability distribution for a general system whose temperature is "known," we must examine the temperature-measurement process itself. When we do this, we find that 'knowing the temperature" provides us with relatively little information about the system, and that that information is macroscopic. There is a mutual equilibrium between system of interest and gas thermometer, and there are observations on the thermometer itself. Progress can come only if we make use of the "mutual equilibrium" property and look for a probability distribution for the joint system of thermometer and system of interest. On such a probability distribution we may place two strong constraints. The bases for these are consistency with the previously derived probability distribution for the gas thermometer itself and the maximum missing information criterion. The analysis leads, for the statistical description of a system whose temperature is known, to the probability distribution in equation (4.7.4), with the parameter β being $1/kT$.

There are many different routes to the end-result of this section. The reader should be forewarned that in other books he may find different derivations of the same relation. There is something to be learned from each of these. A procedure like ours is reasonable if one takes seriously the proposition that a probability distribution depends on what one *knows*, on the specific hypothesis.

In the analysis thus far we have depended quite strongly on an operational definition of temperature with a dilute-gas thermometer. Later considerations, in Chapter 7, will much relax this dependence.

J. Willard Gibbs was the first to suggest the exponential form for the general case of a system in equilibrium at a known temperature. In the preface to his *Elementary Principles in Statistical Mechanics,* published posthumously in 1902, Gibbs referred to this probability distribution and introduced the name now commonly used: "This distribution, on account of its unique importance in the theory of statistical equilibrium, I have ventured to call *canonical.*" Let us follow Gibbs and refer to the distribution in equation (4.7.4) as the *canonical probability distribution.*

One last point warrants discussion before we go on to an extensive application. The canonical probability distribution associates a non-zero probability with each energy eigenstate of the physical system (provided the energy eigenvalue is not infinite, a limiting case). Isn't this absurd, for doesn't the system have some one definite energy? No, it is not absurd: even if the system does have some one definite energy, we don't know what that energy is; with what we do know, we can do no more than estimate the energy. The canonical probability distribution is based on the realistic hypothesis that we know the atomic constituents of the system, their interactions, the fixed environment, and the temperature of the system. We are not provided with a knowledge of the energy. So the limited scope of our knowledge compels us to consider a variety of possibilities for the system's energy; in the specified context the canonical distribution associates with each energy eigenstate a probability in the "best," most bias-free manner. The range of possible energies is inherent in the limitations to our knowledge about the system.

The test for a good probability distribution is twofold. (1) Does the probability distribution correctly represent our knowledge? (2) Does it admit the limitations to our knowledge in a bias-free manner? Given the specified physical context, the canonical probability distribution passes both aspects of the test.

REFERENCES for Chapter 4

The work of Edwin T. Jaynes on statistical mechanics lies at the very foundation of the present development. His 1962 Brandeis lectures—in *Statistical Physics*, Brandeis Theoretical Physics Lectures, Vol. 3, 1962 (New York: Benjamin, 1963)—provide what is perhaps the best single exposition of his views. The more formal basis is laid in two papers: *Physical Review*, **106**, 620 (1957), and **108**, 171 (1957). An article, *American Journal of Physics*, **33**, 291 (1965), is particularly germane to Chapter 11 of this book and to Appendix D, whereas the article, *IEEE Transactions on Systems Science and Cybernetics*, **4**, 227 (1968), outlines a general approach to a still-unresolved difficulty in probability theory, and thereby contributes to the foundations of classical statistical mechanics and of statistical inference in general. Some aspects of

Jaynes' views did, of course, have forerunners, for example, Walter M. Elsasser, *Physical Review*, **52**, 987 (1937).

Amnon Katz has written an elegant, sophisticated book, *Principles of Statistical Mechanics; The Information Theory Approach* (San Francisco: Freeman, 1967). in which classical and quantum statistical mechanicals are developed in parallel on the basis of information theory (effectively, on a basis identical to our Criteria I and II for the assignment of probabilities). The proofs are all there, though this implies that the reader should (for some sections) know some graduate-school-level mechanics of both the classical and quantum varieties. Chapter 9 deals with "Nonequilibrium Equations" in a clear and compelling manner. There is one weakness: Katz does not make clear what he means by "probability." On the whole, the book is superb; besides rigor, it provides penetrating insights into the meaning of information in statistical mechanics.

The approach taken by Myron Tribus in his *Thermostatics and Thermodynamics* (Princeton, N.J.: Van Nostrand, 1961), is in some respects identical to the approach we have taken: the notion of probability is much the same, and the criteria for the assignment of probabilities follow Jaynes' idea of using information theory. The book is valuable for the parallel development it presents on these aspects. (A warning: the "proof" on p. 58 in connection with the derivation of *MI* is mathematically untenable.) Thereafter Tribus's book and this book diverge at a continually increasing rate, for Tribus is much concerned with thermodynamics. Since his book was designed primarily as a thermodynamics text for engineering students, the concern is understandable, but it does make the book quite complicated after the first few chapters.

Richard C. Tolman's *Principles of Statistical Mechanics*, (London: Oxford University Press, 1938), is the classic of a (now) conventional approach to statistical mechanics. He was indeed concerned with the principles and enunciated his postulates clearly. Tolman's views are always worth considering; the reader of this book should at some time look carefully at Tolman's approach. Even if one disagrees with Tolman, one must commend him for having had the reader firmly in mind while he wrote. The reader should be forewarned that Tolman's view of probability differed from that adopted in the present book, his being the more conventional kind; the "translation" can, however be made.

J. Willard Gibb's little book, *Elementary Principles in Statistical Mechanics*, reprinted by Dover Publications (New York, 1960), is a landmark in the development of statistical mechanics. It presents a rather axiomatic approach in an attempt to bypass some of the difficulties that had—and still do—beset the subject. Einstein advanced some of the same ideas, independently and somewhat more physically, in two little-known papers: *Annalen der Physik*, **9**, 417 (1902), and **11**, 170 (1902).

PROBLEMS

4.1. Suppose that the dilute gas in a constant-volume gas thermometer has a particle density of $N/V = 5 \times 10^{17}$ molecules per cubic centimeter. If the observed gas pressure is 0.025 times atmospheric pressure, what is the temperature of the gas? (Take atmospheric pressure to be 10^6 dynes per square centimeter.)

4.2. Maximum for MI. The objective is a demonstration that the exponential solution in Section 4.5 does provide an actual maximum for MI, subject to the normalization and energy constraints. Let us call that solution $\tilde{P}_j$. The objective will be achieved if each "neighboring" probability distribution, $P_j = \tilde{P}_j + \delta P_j$, which also satisfies the two constraints, yields a smaller value of MI. The term δP_j is taken to be small relative to $\tilde{P}_j$. Insert this form for P_j into the expression for MI and expand through second order in δP_j, using a Taylor's series for the logarithms. You should be able to show that the summations linear in δP_j vanish because the "neighboring" probability distribution is required to satisfy the constraint equations. Then you should find that the summations quadratic in δP_j contribute negatively, indicating a maximum.

4.3. The integral approximation in Section 4.6 was based on $\beta\gamma$ being small relative to unity. Is the ultimate relation, $\beta = 1/kT$, consistent with this?

4.4. The length of the tail. For a gas, the canonical distribution extends with non-zero P_j to indefinitely high energy. Surely the probability of an energy greater than 1,000 $\tilde{E}$, say, ought to be strictly zero, for if the gas had such an energy, the gas would be highly ionized, and we know it isn't. The canonical distribution doesn't cut off at a finite energy because we never inserted the information that we know the energy is below some high (and ill-defined) upper limit. What form emerges for the "best" probability distribution if one imposes

$$P_j = 0 \qquad \text{when} \qquad E_j > 1{,}000 \; \tilde{E}$$

in addition to equations (4.5.1,2)? Is there likely to be a discernible difference in statistical estimates from the two probability distributions? What is the implication for the length of the tail?

4.5. The microcanonical probability distribution. Suppose we know that a system is in equilibrium, know the constituents of the system and the quantum-mechanical energy operator, but do not know the temperature (for it has not been measured). Instead of information about a temperature, we are given the information that the total energy of the system lies between E^* and $E^* + \Delta E^*$, endpoints included (if it matters). The interval ΔE^* may be small by macroscopic standards, but is much larger than the typical separation between energy levels for the system. This is all we know. If we wish to discuss the properties of the system, we will have to estimate them (for we certainly do not have the information requisite for the determination of a unique wave function for the system.)

a. Present an argument indicating why we should consider, as possibly appropriate states, the energy eigenstates of the entire system.

b. What is the "best" probability distribution for those states? By what criterion?

The criteria developed in Chapter 3 lead to

$$P_J = \begin{cases} \text{zero} & \text{if } E_J \text{ does } not \text{ lie within the range } E^* \text{ to } E^* + \Delta E^*, \\ \text{a fixed} \\ \text{constant} & \text{if } E_J \text{ does lie within the range } E^* \text{ to } E^* + \Delta E^*. \end{cases}$$

The "fixed constant" is to be determined by the normalization conditions. Indicate verbally its value. The probability distribution just derived is called the *microcanonical probability distribution*. The root of the difference between it and the canonical distribution lies in the difference in the information provided about the system.

5

PARAMAGNETISM AND STATISTICAL ESTIMATES

Comparison of these numbers shows ... how certain are
such theorems that theoretically are merely probability laws,
but in practice have the same significance as laws of nature.

Ludwig Boltzmann
" Reply to Zermelo's Remarks on the Theory of Heat "

Since an application of the probability distribution to which the preceding
chapters have led is certainly in order, this chapter is devoted to a physically
significant yet mathematically tractable application: the statistical description
of a macroscopic paramagnetic sample at equilibrium in an external magnetic
field. We can analyze this problem in considerable detail and can begin to
understand why statistical mechanics is generally so successful in estimating
the macroscopic properties of macroscopic systems in equilibrium. The key
to such understanding is the recognition that many states of a macroscopic
system, though differing in microscopic properties, show closely comparable
macroscopic properties. When we ask a question about a macroscopic
property, we can set up a probability distribution for the occurrence of that
property itself. The analysis begins with a very broad probability distribution
for individual states—the canonical distribution. But after a grouping of

states with comparable macroscopic properties, the analysis leads (typically) to a very sharp probability distribution for the occurrence of the macroscopic property itself. In consequence the statistical mechanical estimates are almost certain to be correct.

5.1. THE EXPECTATION VALUE OF THE TOTAL MAGNETIC MOMENT

We set out to estimate the total magnetic moment of a paramagnetic sample when the system is in equilibrium at a known temperature and is located in an external magnetic field. Since the dependence of the total moment on both temperature and field is experimentally accessible, there are excellent means of comparing statistical estimates with observation.

To avoid unnecessary algebra at this point, we consider a sample in which the individual paramagnetic particles have a total angular momentum, arising from a net spin, of $\frac{1}{2}\hbar$. (Examples of experimental interest are provided by the protons in water, by the nuclei of the flourine isotope F^{19} in a lithium fluoride crystal, and by triply ionized titanium as found in some salts.) The magnetic moment associated with such a paramagnetic particle we write as a vector $\boldsymbol{\mu}$ with magnitude μ. Its direction is determined by that of the angular momentum. Whether $\boldsymbol{\mu}$ is parallel, or antiparallel, to the angular momentum depends on the kind of particle. For definiteness let us take $\boldsymbol{\mu}$ to be parallel to the angular momentum. (For the proton and the fluorine isotope they are parallel.) Further, we specify that the paramagnetic particles are located in a crystal lattice. This means that the particles are spatially fixed in a more or less rigid fashion, though there is no initial restriction on the orientation of the individual moments. The number of such particles will be denoted by N. The sample is located in a spatially uniform, time-independent, external magnetic field $\mathbf{H}$, which, again for the sake of definiteness, we take to be directed along the z-axis: $\mathbf{H} = H\hat{\mathbf{z}}$.

Some interaction among the individual paramagnetic particles is inevitable. (The interaction is in some ways like the classical interaction of a host of tiny bar magnets.) The mutual magnetic interactions, together with forces arising from the small vibrations of the particles about their sites in the lattice structure, can produce changes in the orientations of the magnetic moments; hence they permit changes in the total magnetic moment in response to changes in the external field and in the temperature. Often, however, the mutual interactions are weak relative to the interaction of the individual moments with the external magnetic field. In such a case the non-zero mutual interactions serve primarily to ensure that the system may respond to an altered environment. They have little other effect on the total magnetic moment. In calculating probabilities and estimates of the total magnetic moment, one may then neglect the mutual interactions and work solely with the energy of interaction with the strong external field. Often one may also neglect the vibrational motion of the paramagnetic particles and consider them to be spatially fixed in a rigid fashion. (If other, nonparamagnetic particles are present in the crystal, they, too, may be regarded as spatially fixed and may be neglected in the sequel.) A considerable number of paramagnetic systems met in the laboratory may adequately be approximated in this fashion.

A good example of such a paramagnetic system is provided by cesium titanium alum, $CsTi(SO_4)_2 \cdot 12H_2O$. The effective magnetic moment of this compound comes from the single Ti^{+++} ion, which exhibits a net angular momentum of $\frac{1}{2}\hbar$. The interaction between titanium ions is relatively weak because they are so far separated from one another in the crystal structure.

For the sake of simplicity, we will suppose that our sample is one for which the foregoing approximation scheme is valid. The major advantage of the approximation scheme is that an analysis in detail of the energy eigenstates of the system becomes feasible. Since the estimation problem is still not trivial, we will work toward the solution of the realistic N-particle problem by examining first the somewhat artificial one-particle case.

Let us single out some *one* paramagnetic particle and treat it as a (microscopic) system in equilibrium in the external magnetic field. The specification that the mutual interactions are weak relative to the interaction with the external field provides the property needed to justify this preliminary approach. To employ the canonical distribution in an analysis of this "system" of one particle, we need its energy eigenstates and their eigenvalues. The energy of interaction between a single magnetic moment μ and an external magnetic field $\mathbf{H}$ is given by a scalar product:

$$- \mu \cdot \mathbf{H}.$$

In classical terms this is the potential energy associated with the torque on the individual moment exerted by the external field. Unless the moment is precisely parallel or antiparallel to $\mathbf{H}$, the torque tends to align μ parallel to $\mathbf{H}$. Hence the parallel position is that of minimum energy.

Since we neglect the mutual interactions, the term just written is the sole term in the energy expression for a single paramagnetic particle. Under these circumstances a particle of spin $\frac{1}{2}\hbar$ has only two energy eigenstates: the spin may be either parallel to the field $\mathbf{H}$ or antiparallel to it. Calling these "up" and "down," respectively (and remembering that we have taken $\boldsymbol{\mu}$ parallel to the angular momentum), we have

$$\text{energy for spin up}\ \ \ = -\mu H,$$

$$\text{energy for spin down} = +\mu H.$$

With the two energy eigenstates described and their energies known, we can readily write out the probabilities of spin up and of spin down that follow from the canonical distribution. Since that probability distribution has the general form

$$P_j = \frac{e^{-\beta E_j}}{\sum_i e^{-\beta E_i}}$$

for energy eigenstates labeled with the symbol j, we have, in the two-state case,

$$P_{\text{up}} = \frac{e^{-\beta(-\mu H)}}{e^{-\beta(-\mu H)} + e^{-\beta(+\mu H)}} = \frac{e^{+\beta\mu H}}{e^{+\beta\mu H} + e^{-\beta\mu H}},$$

$$P_{\text{down}} = \frac{e^{-\beta(+\mu H)}}{e^{-\beta(-\mu H)} + e^{-\beta(+\mu H)}} = \frac{e^{-\beta\mu H}}{e^{+\beta\mu H} + e^{-\beta\mu H}}.$$

With $\beta\mu H = \mu H/kT > 0$, the spin up state has the greater probability. This agrees with the classical idea of a torque that tends to align the magnetic moment parallel to the field.

Now we compute the expectation value $\langle\boldsymbol{\mu}\rangle$ of the magnetic moment of this single paramagnetic particle. First we work out the component parallel to the external field, that is, the component $\langle\mu_z\rangle$. In the up state μ_z has the value $+\mu$; in the down state, the value $-\mu$. Hence we have

$$\langle\mu_z\rangle = (+\mu)P_{\text{up}} + (-\mu)P_{\text{down}}$$

$$= \mu\,\frac{e^{\beta\mu H} - e^{-\beta\mu H}}{e^{\beta\mu H} + e^{-\beta\mu H}}.$$

For $\beta\mu H > 0$, this is positive, as anticipated from a classical argument with a torque tending to align $\boldsymbol{\mu}$ parallel to $\mathbf{H}$.

For the remaining two components, $\langle\mu_x\rangle$ and $\langle\mu_y\rangle$, we can make use of a symmetry argument to show that they are zero. In the plane perpendicular to the external field $\mathbf{H}$, there is no preferred direction. The spin has "no reason" to prefer pointing in the $+x$ direction over pointing in the $-x$

direction. Hence the expectation value of the x-component, $\langle \mu_x \rangle$, must be zero.* The same argument and result apply to $\langle \mu_y \rangle$.

This completes the determination of $\langle \boldsymbol{\mu} \rangle$ when we may single out some one paramagnetic particle. Before we analyze the temperature dependence, let us work out the expectation value of the total magnetic moment $\mathbf{M}$. We do so by setting up the canonical distribution for the entire system.

Our first task is the determination of the energy eigenstates and associated energies for the system of N paramagnetic particles in the external field. Let us first establish the possible values of the energy. Under the assumption that we may neglect the mutual interactions of the paramagnetic particles relative to their interaction with the external field, we may directly write down the energy expression for the entire system:

$$\begin{pmatrix} \text{the energy expression} \\ \text{for the system of } N \\ \text{paramagnetic particles} \\ \text{in the field } \mathbf{H} \end{pmatrix} = \sum_{i=1}^{N} -\boldsymbol{\mu}_i \cdot \mathbf{H}$$

$$= -\left(\sum_{i=1}^{N} \boldsymbol{\mu}_i\right) \cdot \mathbf{H}. \qquad (5.1.1)$$

The individual paramagnets have been labeled with an index i. Any one particle may have its spin either up or down. An energy eigenstate of the entire system is described when we specify, for each of the N particles, that the spin is up or is down. Nonetheless, the energy itself depends merely on how many spins are up and how many down, not on which are up and which down. Let us denote by n the number of spins up, parallel to the external field. An up spin contributes $-\mu H$ to the total energy; a down spin $+\mu H$. For a state with n spins up, the total energy is

$$n(-\mu H) + (N - n)(+\mu H) = -(2n - N)\mu H.$$

---

* We have resorted to this symmetry argument because a direct calculation of $\langle \mu_x \rangle$ requires some additional quantum-mechanical analysis. In the spin up state, i.e., spin parallel to $\hat{z}$, the x-component of spin does not have unique value. (Upon measurement, it is actually as likely to be $-\frac{1}{2}\hbar$ as $+\frac{1}{2}\hbar$ and will be one of those values.) The same holds for the spin down state. To use directly the expression given in equation (2.8.1) for an expectation value calculation, we have to know the probabilities of inferences that a variable, here μ_x, takes on its definite possible values. Here that would entail a transformation (only) of the canonical probability distribution into one giving probabilities of $\mu_x = +\mu$ and $\mu_x = -\mu$. With some additional quantum-mechanical analysis the transformation can be made, but that is unnecessary for us, since the new distribution leads merely to the same zero value for $\langle \mu_x \rangle$ that we have more easily determined by symmetry.

Since the number n may range from 0 to N in steps of one unit, the energy eigenvalues have the range

$$-N\mu H \leqslant (\text{total energy}) \leqslant N\mu H.$$

Having determined the energy eigenvalues, we turn to a further analysis of the states themselves, and we do so by counting. We count, for each energy value, the number of distinct states with that energy. The extremes are easy enough. There is only one state with energy $N\mu H$; that is the state with no spins up, all spins down. Similarly, there is only one state with energy $-N\mu H$; that is the state with all spins up, none down. the energy eigenvalues with $n \neq 0$ or N present the real problem. For $n = 1$, the single up spin may be any one of the N spins. (More picturesquely, the single up spin may be at any one of the N sites in the crystal.) So there are N different energy eigenstates with $n = 1$.

We could go on and analyze $n = 2, 3, \ldots$ directly, but we have already performed this kind of counting operation in Section 2.10. If we regard the up and down "arrows" of that section as corresponding to up and down spins, we may immediately extract the general answer to the counting problem. The number of distinct energy eigenstates with n spins up is precisely the same as the number of different observational sequences in which property α appears n times out of N observations. Taking over the result given in equation (2.10.8), we may directly write down the answer to the present counting problem:

$$\begin{pmatrix} \text{the number of distinct} \\ \text{energy eigenstates with} \\ n \text{ spins up} \end{pmatrix} = \frac{N!}{n!(N-n)!}. \tag{5.1.2}$$

Since each spin may be either up or down, there will be a total of $2 \times 2 \times \ldots \times 2 = 2^N$ distinct states. A summation over all n of the expression given here does properly yield 2^N.

The preliminaries are almost at a close. For an energy eigenstate ψ_j we write the value of the total magnetic moment along the external field as $m_j\mu$. The value of the integer m_j will lie between the limits $+N$ (for the single state with all spins up) and $-N$ (for the state with all spins down). The relation with the energy eigenvalue E_j is, by virtue of equation (5.1.1),

$$E_j = -m_j\mu H. \tag{5.1.3}$$

The index j will run from $j = 1$ to $j = 2^N$.

Let us note here the connection between m_j and the number of spins up in the particular state ψ_j. If there are n spins up, the net magnetic moment along the external field is

$$n(\mu) + (N - n)(-\mu) = (2n - N)\mu.$$

So the connection between m_j and n will be

$$\left(\begin{array}{l}\text{the numerical value of } m_j \\ \text{in state } \psi_j \text{ with } n \text{ spins} \\ \text{up}\end{array}\right) = (2n - N). \qquad (5.1.4)$$

Now we may calculate the expectation value $\langle \mathbf{M} \rangle$. The component along the field direction is the most relevant component, and so we begin with M_z. The data specified in the initial paragraphs of this section (equilibrium, known temperature, and so on) imply that the canonical distribution is the appropriate probability distribution. We therefore estimate M_z by taking the value $m_j\mu$ of M_z in state ψ_j, weighting that value with the probability P_j that the state provides the appropriate description of the system, and then summing over all states. Thus we write

$$\langle M_z \rangle = \sum_{j=1}^{2^N} m_j \mu P_j = \frac{\displaystyle\sum_{j=1}^{2^N} m_j \mu e^{-\beta E_j}}{\displaystyle\sum_{i=1}^{2^N} e^{-\beta E_i}}. \qquad (5.1.5)$$

This is an awkward expression, but it is amenable to simplification and, ultimately, exact evaluation. We can put it into better shape by using our knowledge of the number of distinct states with any given value of the number of spins up. When the summation goes over states having the same number n of spins up, the numerical value of m_j is the same for all, namely $m_j = (2n - N)$. Moreover, the value of the exponential remains constant. So we may first sum over all states within each such set ($m_j = 2n - N$ for all states in the set) and then finish with a summation over all such sets.

Stating in words the intermediate result of this procedure for the numerator on the righthand side of equation (5.1.5) may help clarify the meaning of the process:

$$\sum_{j=1}^{2^N} m_j \mu e^{-\beta E_j}$$

becomes

$$\sum_{\substack{\text{all different} \\ \text{values of } n}} \left(\begin{array}{l}\text{the number of} \\ \text{distinct states} \\ \text{with } n \text{ spins up}\end{array}\right) \left(\begin{array}{l}\text{the value of} \\ \text{the magnetic} \\ \text{moment of a} \\ \text{state with } n \\ \text{spins up}\end{array}\right) \exp -\beta \left(\begin{array}{l}\text{the energy of} \\ \text{a state with} \\ n \text{ spins up}\end{array}\right).$$

With the insertion of mathematical expressions in place of the words for both numerator and denominator, the procedure leads to some progress in computing $\langle M_z \rangle$:

$$\langle M_z \rangle = \frac{\sum_{n=0}^{N} \frac{N!}{n!(N-n)!}(2n-N)\mu e^{+(2n-N)\beta\mu H}}{\sum_{n'=0}^{N} \frac{N!}{n'!(N-n')!} e^{+(2n'-N)\beta\mu H}}. \tag{5.1.6}$$

The remaining summations are still formidable in appearance, but with the binomial theorem and another trick we may perform them.

The summation in the denominator may be directly handled with that theorem:

$$\sum_{n'=0}^{N} \frac{N!}{n'!(N-n')!} e^{+(2n'-N)\beta\mu H}$$

$$= e^{-N\beta\mu H} \sum_{n'=0}^{N} \frac{N!}{n'!(N-n')!} e^{n'2\beta\mu H}$$

$$= e^{-N\beta\mu H} \sum_{n'=0}^{N} \frac{N!}{n'!(N-n')!} (e^{2\beta\mu H})^{n'}(1)^{N-n'}$$

$$= e^{-N\beta\mu H}(e^{2\beta\mu H} + 1)^N$$

$$= (e^{\beta\mu H} + e^{-\beta\mu H})^N. \tag{5.1.7}$$

The intermediate forms are intended to facilitate comparison with the binomial theorem, given in equation (2.10.11). The last step makes the result a bit tidier. Thus we have the denominator in closed form.

A trick is required to perform the summation in the numerator. Let us note that the summation is similar to that which we have performed, the sole difference being a factor of $(2n-N)\mu$ in the summand. That factor we may relate to a partial derivative of the exponential with respect to βH:

$$(2n-N)\mu e^{(2n-N)\mu\beta H} = \frac{\partial}{\partial(\beta H)}[e^{(2n-N)\mu\beta H}]$$

With this noted the summation follows relatively easily:

$$\sum_{n=0}^{N} \frac{N!}{n!(N-n)!}(2n-N)\mu e^{(2n-N)\mu\beta H}$$

$$= \sum_{n=0}^{N} \frac{N!}{n!(N-n)!} \frac{\partial}{\partial(\beta H)}[e^{(2n-N)\mu\beta H}]$$

$$= \frac{\partial}{\partial(\beta H)} \left[\sum_{n=0}^{N} \frac{N!}{n!(N-n)!} e^{(2n-N)\mu\beta H} \right]$$

$$= \frac{\partial}{\partial(\beta H)} \left[(e^{\mu\beta H} + e^{-\mu\beta H})^{N} \right]$$

$$= N\mu(e^{\mu\beta H} - e^{-\mu\beta H})(e^{\mu\beta H} + e^{-\mu\beta H})^{N-1}. \tag{5.1.8}$$

By combining this form with the result in equation (5.1.7) we achieve an explicit expression for $\langle M_z \rangle$:

$$\langle M_z \rangle = N\mu \frac{e^{\beta\mu H} - e^{-\beta\mu H}}{e^{\beta\mu H} + e^{-\beta\mu H}}.$$

As we might have anticipated, the expectation value of the total moment along the field direction is N times the earlier result for $\langle \mu_z \rangle$. But now we have an explicit demonstration that when we analyze the system of N particles *as a whole, as we properly should*, a comfortingly reasonable result emerges.

For the other two components, $\langle M_x \rangle$ and $\langle M_y \rangle$, we may argue as before. There is no preferred direction in the plane perpendicular to the external magnetic field. Hence those expectation values must, by symmetry, be zero.

To summarize, we have determined the expectation value of the total magnetic moment of N spin $\frac{1}{2}\hbar$ paramagnetic particles at thermal equilibrium in an external magnetic field as follows:

$$\langle M_z \rangle = \sum_{j=1}^{2^N} m_j \mu P_j = N\mu \frac{e^{\beta\mu H} - e^{-\beta\mu H}}{e^{\beta\mu H} + e^{-\beta\mu H}}, \tag{5.1.9}$$

$$\langle M_x \rangle = \langle M_y \rangle = 0.$$

We now have a general expression for $\langle \mathbf{M} \rangle$ as a function of temperature, with $\beta\mu H = \mu H/kT$ being the essential parameter. Whenever an expression is somewhat complicated, as it is here, it is worthwhile to look first at the limiting cases. Here they are "low" and "high" temperature, where the qualitative designations are made quantitative by saying that "low" means $kT \ll \mu H$ and "high" means $kT \gg \mu H$.

For low temperature the essential parameter is very large, and $\exp(-\mu H/kT)$ is negligible relative to the positive exponential. Thus we find that the *low-temperature limit* is

$$\langle M_z \rangle \simeq N\mu.$$

It is as though all the spins were aligned parallel to the external field.

In the high temperature extreme the essential parameter is very small. To assess $\langle M_z \rangle$ in this limit we can use the Taylor's series expansion of an exponential (around zero value for the exponent):

$$e^x = 1 + x + \tfrac{1}{2}x^2 + \dots.$$

This expansion converges quite rapidly for $|x| < 1$. In the combination of exponentials in equation (5.1.9), the leading contribution of the numerator is

$$e^{\beta\mu H} - e^{-\beta\mu H} = (1 + \beta\mu H + \ldots) - (1 - \beta\mu H + \ldots) \simeq 2\beta\mu H,$$

while that of the denominator is

$$e^{\beta\mu H} + e^{-\beta\mu H} = (1 + \beta\mu H + \ldots) + (1 - \beta\mu H + \ldots) \simeq 2.$$

So the leading term in the expansion of $\langle M_z \rangle$, which gives us the *high-temperature limit*, is

$$\langle M_z \rangle \simeq N\mu \frac{2\beta\mu H}{2} = \frac{N\mu^2 H}{kT}.$$

This provides the first term in a general expansion in powers of $1/T$. The next term is of order $1/T^3$. The proportionality with H, as well as the $1/T$ dependence in this high-temperature limit, are well-confirmed experimentally. The limiting expression is known as *Curie's law*.*

The entire run of the function $\langle M_z \rangle$ is given in figure 5.1.1, where $\langle M_z \rangle / N\mu$ is plotted against the parameter $\mu H/kT$. The "low" temperature behavior

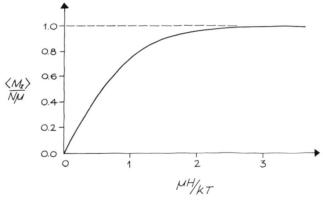

FIGURE 5.1.1
The behavior of $\langle M_z \rangle$ as a function of the parameter $\mu H/kT$.

---

* One can generate a "mechanistic" picture of why $\langle M_z \rangle$ decreases with increasing temperature. High temperature implies appreciable vibration of the paramagnetic particles about their sites in the lattice. That leads to appreciable electromagnetic forces on the magnetic moments, forces in competition with the external magnetic field. Less perfect alignment along the field direction is a natural result. Though we have not explicitly included such forces in the calculation, they appear implicitly through the temperature T. This microscopic "mechanistic" picture is explored in more detail in Section 11.4.

(to the right in the diagram) is approached quite rapidly because of the exponential dependence.

An estimate of the value of $\mu H/kT$ under typical laboratory conditions is in order. For the external magnetic field H, a value of 10^4 gauss is readily attainable. That is about 2×10^4 times as strong as the Earth's field or comparable to that in a cyclotron. For a characteristic value of μ, we have two broad choices: electronic (or atomic) size or nuclear, the latter being about 2,000 times smaller. Let us choose the former, and take for μ a value close in magnitude to the magnetic moment of the electron:*

$$\mu = 10^{-20} \text{ ergs/gauss.}$$

For the product μH this gives

$$\mu H = \begin{cases} 10^{-16} \text{ ergs,} \\ 6.24 \times 10^{-5} \text{ electron volts.} \end{cases}$$

This is quite small; so one might guess that for most commonplace temperatures one would have $kT \gg \mu H$. For paramagnetic studies a typical temperature would be $T = 4$ K, for that is readily attained by cooling with liquid helium. This choice of temperature leads to

$$kT = \begin{cases} 5.52 \times 10^{-16} \text{ ergs,} \\ 3.45 \times 10^{-4} \text{ electron volts.} \end{cases}$$

For the essential parameter one has then

$$\frac{\mu H}{kT} = \frac{10^{-16}}{5.52 \times 10^{-16}} = 0.181.$$

This is sufficiently small so that the high-temperature limit should give a good approximation to $\langle M_z \rangle$. To reach the low-temperature limit, either higher magnetic fields ($\simeq 5 \times 10^4$ gauss) or lower temperatures ($\simeq 1$ K or less) are required. For nuclear magnetic moments the problem of attaining the low-temperature limit is about 2,000 times more difficult.

---

* A classical picture of a spinning, spherical electron (spin $\frac{1}{2}\hbar$) leads directly to the idea of a magnetic moment associated with the spinning charge distribution. To be sure, in this case the property that the electron is negatively charged implies μ antiparallel to the spin. But we want just an order-of-magnitude estimate for an electronic or atomic permanent magnetic moment, so we need not worry about the "wrong" orientation. The value of the electron's magnetic moment is very nearly $e\hbar/2mc$. Both experiment and quantum electrodynamics indicate a correction of about 0.1 per cent (toward larger absolute value). The appearance of the particle mass in the denominator indicates that nuclear moments are likely to be roughly 2,000 times smaller.

Now for the expectation value of the total magnetic moment. Substitution of $\beta\mu H = 0.181$ in the exact expression, equation (5.1.9), yields

$$\langle M_z \rangle = 0.179 N\mu.$$

This may be compared with the high-temperature approximation:

$$\langle M_z \rangle \simeq \left(\frac{\mu H}{kT}\right)N\mu = 0.181N\mu.$$

In a field like 10^4 gauss, a temperature of 4 K is already high temperature for the paramagnetic sample.

The calculation for a system of paramagnetic particles with a higher value for the individual angular momentum, such as $\frac{3}{2}\hbar$ or $2\hbar$, is analogous. The final expression is a bit more complicated (because there are more than two states for a single particle), but is again a combination of exponentials with the same limiting behavior. The derivation will be given in Chapter 7, after we have developed more powerful techniques for handling the summations.

The extent of the agreement between the expectation value estimates and the experimentally observed total moments is nothing short of astonishing. A superb example is displayed in figure 5.1.2. This prompts one to ask whether there is any reason to anticipate such good agreement. After all, we work with the canonical distribution. Essential in its derivation was the notion that one should choose the most broad and noncommittal probability distribution compatible with the information one possesses. Offhand, one would anticipate significant deviations between estimates and observations. To analyze the success of the statistical approach, we need to examine more closely the full set of 2^N probabilities. To this we turn in the next section.

5.2. THE ANALYSIS OF ANTICIPATED DEVIATIONS

We will take two approaches to the question, why is the agreement between statistical estimate and experiment so good? First we will calculate the root mean square estimate of the deviations of M_z from $\langle M_z \rangle$ and briefly discuss the result. This is a rather blind approach, for we never really "see" the probability distribution; we merely manipulate the probabilities and out pops a relevant result. The prime reason for doing the root mean square calculation at all is that it will give us confidence in later results based on an approximation.

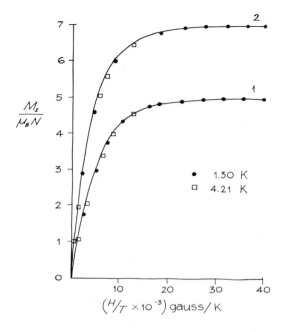

FIGURE 5.1.2
The confrontation between experiment and statistical theory for two paramagnetic materials and a wide range of the magnetic-field-to-temperature ratio H/T. The vertical scale gives the magnitude of the total magnetic moment along the magnetic field direction after division by $\mu_B N$. The symbol μ_B denotes one *Bohr magneton*: $|e|\hbar/2mc$. The data points labeled (1) are for Fe^{+++} (with a net spin of $\frac{5}{2}\hbar$) in iron ammonium alum; those labeled (2) are for Gd^{+++} (with a net spin of $\frac{7}{2}\hbar$) in gadolinium sulfate octahydrate. The smooth curves are expectation value calculations for the two cases. The samples were in the form of solid spheres, about two centimeters in diameter, and hence definitely of macroscopic size. Magnetic fields as high as 50,000 gauss and temperatures as low as 1.3 K were used to achieve the "low temperature" limits (large H/T). The experimental procedure, together with some general conclusions, are presented in the very readable paper by W. E. Henry, *Physical Review*, **88**, 559 (1952), from which the data points and theoretical curves have been taken.

In the second approach we will work quite explicitly with the relevant probability distributions. With the aid of some graphs we will be able to "see" why the deviations from the estimated behavior are likely to be exceedingly small when the number N of paramagnetic particles is very large, as it is for a macroscopic system. In this approach we can hope for a real beginning in understanding why statistical mechanics is so successful in estimating the macroscopic properties of macroscopic systems. But first the blind calculation.

As an estimate of the deviations to be anticipated, we may calculate the root mean square estimate:

$$\Delta M_z \equiv \langle (M_z - \langle M_z \rangle)^2 \rangle^{1/2}. \tag{5.2.1}$$

To facilitate the calculation we first square the expression for ΔM_z and then expand the internal square:

$$(\Delta M_z)^2 = \langle (M_z - \langle M_z \rangle)^2 \rangle = \langle M_z^2 \rangle - \langle M_z \rangle^2. \qquad (5.2.2)$$

The seond term on the right we already know; only the first needs to be determined. So we write

$$\langle M_z^2 \rangle = \sum_{j=1}^{2^N} (m_j \mu)^2 P_j = \frac{\sum_j (m_j \mu)^2 e^{-\beta E_j}}{\sum_i e^{\beta E_i}}$$

Again we may split the full summation into sums within sets of states having the same numerical value for m_j (that is, the same number of spins up) and then finish with a summation over all such sets. The first kind of summation leads to

$$\langle M_z^2 \rangle = \frac{\displaystyle\sum_{n=0}^{N} \frac{N!}{n!(N-n)!}[(2n-N)\mu]^2 e^{+(2n-N)\beta \mu H}}{\displaystyle\sum_{n'=0}^{N} \frac{N!}{n'!(N-n')!} e^{+(2n'-N)\beta \mu H}}.$$

The sum appearing in the denominator was calculated in equation (5.1.7). The sum in the numerator is similar to the sum we evaluated in equation (5.1.8). If we differentiate both sides of that equation with respect to βH, then on the left we get the desired summation and on the right the desired answer. Upon combining numerator and denominator and doing some canceling, we arrive at

$$\langle M_z^2 \rangle = \frac{N\mu^2(e^{\mu \beta H} + e^{-\mu \beta H})^2 + N(N-1)\mu^2(e^{\mu \beta H} - e^{-\mu \beta H})^2}{(e^{\mu \beta H} + e^{-\mu \beta H})^2}.$$

This expression is to be inserted into equation (5.2.2), along with an explicit expression for $\langle M_z \rangle^2$ from equation (5.1.9). Tidying up the difference and taking a square root leads to a concise result:

$$\Delta M_z = N^{1/2} \mu \left(\frac{2}{e^{\beta \mu H} + e^{-\beta \mu H}} \right). \qquad (5.2.3)$$

We now have the root mean square estimate of the deviations in M_z from $\langle M_z \rangle$ that are to be anticipated for our system of N paramagnetic particles.

Of first concern to us is the size of ΔM_z relative to the maximum possible deviations from $\langle M_z \rangle$. For a temperature $T \geqslant 0$ the expectation value $\langle M_z \rangle$ lies somewhere between zero and $N\mu$. Since the observed total magnetic moment may range from $-N\mu$ to $N\mu$, the maximum possible deviations from $\langle M_z \rangle$ are of order $N\mu$; let us write this as $O(N\mu)$.

In the expression for ΔM_z the factor in parentheses, though dependent on temperature and magnetic field, is never greater than unity, and so ΔM_z is of order $N^{1/2}\mu$ or less. Therefore the ratio of ΔM_z to the maximum possible deviation is

$$\frac{\Delta M_z}{O(N\mu)} \leqslant \frac{N^{1/2}\mu}{O(N\mu)} = O\left(\frac{1}{N^{1/2}}\right). \tag{5.2.4}$$

For large N this is an exceedingly small ratio. A typical macroscopic sample would have $N \simeq 10^{20}$, with the consequence that the ratio would be of order 10^{-10}.

Furthermore, except for exceedingly high temperature or negligible magnetic field, the value of $\langle M_z \rangle$ will be of order $N\mu$. So the ratio of ΔM_z to $\langle M_z \rangle$ will be of order $N^{-1/2}$.

Thus the ratio of ΔM_z to either the maximum possible deviation or to $\langle M_z \rangle$ will go as $1/N^{1/2}$, and so will be exceedingly small for the macroscopic case of very large N. Though large deviations are possible, the implication is that they are highly improbable when N is large.

Although the preceding calculation has indicated that deviations from $\langle M_z \rangle$ will be quite small, it has failed to provide any insight into why this is so. To gain an adequate understanding we must work much more explicitly with the probabilities themselves. To this second approach we now turn.

First we note a rather obvious point: we have been estimating, and asking questions about, a *macroscopic* property of the physical system. The component of the *total* magnetic moment along the external field direction is surely a macroscopic quantity. Details about the system, such as which spins are up and which are down, though part of the intermediate analysis, are of no concern in the final estimates.

Since we are concerned here with only the total magnetic moment, we can usefully classify and group the energy eigenstates according to their total moment. Let us designate the component of the total moment along the field direction by $m\mu$, where m is an integer that must run from $m = -N$ to $m = +N$. To relate the integer m to the number n of spins up, we need only refer to equation (5.1.4). That equation implies

$$m = (2n - N)$$

and hence

$$n = \frac{N + m}{2}. \tag{5.2.5}$$

Thus the number of states with component $m\mu$ is the same as the number of states with $n = (N + m)/2$ spins up, and so we may use equation (5.1.2) to write

$$\begin{pmatrix} \text{the number of distinct} \\ \text{states with total moment} \\ m\mu \text{ along the external} \\ \text{field} \end{pmatrix} = \frac{N!}{\left(\dfrac{N + m}{2}\right)!\left(\dfrac{N - m}{2}\right)!}.$$

This expression exhibits more clearly the symmetry between positive and negative values of the moment. There are equally many states with $m = |m|$ and $m = -|m|$.

For the relatively low N case of $N = 10$, one can readily work out the number of states for each m. They are exhibited in figure 5.2.1.

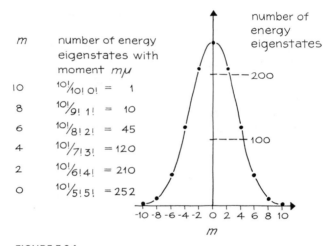

m	number of energy eigenstates with moment $m\mu$	
10	$\dfrac{10!}{10!\,0!}$	$= 1$
8	$\dfrac{10!}{9!\,1!}$	$= 10$
6	$\dfrac{10!}{8!\,2!}$	$= 45$
4	$\dfrac{10!}{7!\,3!}$	$= 120$
2	$\dfrac{10!}{6!\,4!}$	$= 210$
0	$\dfrac{10!}{5!\,5!}$	$= 252$

FIGURE 5.2.1
A partial tabulation and complete plot of the number of energy eigenstates with total moment $m\mu$ when $N = 10$.

Two points are worthy of note. The first is that m always changes by two units. When an additional spin is flipped up, one loses a negative contribution to the moment and gains a positive contribution. The two-unit change in m follows also in an algebraic fashion from equation (5.2.5) or the equation preceding that one.

The second point is that the number of states with given m rises rather sharply as one approaches $m = 0$ from either side. This is noticeable already here with merely $N = 10$. When N is of order 10^{20}, the peak is extraordinarily sharp. There are generally many states that have the same value for m, though they differ in which spins are up and which down. Though microscopically different, the states have the same macroscopic magnetic properties. We will find in this part of the reason why statistical mechanics can make confident predictions about a macroscopic property like the total magnetic moment.

For the sake of brevity let us introduce a special symbol for the cumbersome combination of factorials and write

$$(\mathcal{N}m) \equiv \frac{N!}{\left(\dfrac{N+m}{2}\right)!\left(\dfrac{N-m}{2}\right)!} = \begin{pmatrix} \text{the number of energy} \\ \text{eigenstates with} \\ \text{total moment } m\mu \\ \text{along the field} \end{pmatrix}.$$

With these preliminaries accomplished, we may return to the starting point of the calculation of $\langle M_z \rangle$, the relation given in equation (5.1.5). For convenience it is reproduced here:

$$\langle M_z \rangle = \sum_{j=1}^{2^N} m_j \mu P_j = \frac{\displaystyle\sum_{j=1}^{2^N} m_j \mu e^{-\beta E_j}}{\displaystyle\sum_{i=1}^{2^N} e^{-\beta E_i}}. \tag{5.2.6}$$

Previously we noted that this is an awkward yet tractable expression. Again we may make progress by splitting up the summation process. When the summation goes over states having the same value of m ($m_j = m$ for all of them), the exponential remains constant, as well as the factor $m_j \mu$. So we may first sum over all states within such a set ($m_j = m$ for all those states) and then finish with a summation over all such sets. The summations of the first kind lead to

$$\langle M_z \rangle = \frac{\displaystyle\sum_{m=-N}^{N} m\mu \mathcal{N}(m) e^{+m\beta\mu H}}{\displaystyle\sum_{m'=-N}^{N} \mathcal{N}(m') e^{+m'\beta\mu H}}. \tag{5.2.7}$$

The remaining summations go merely over all values of m, and we must remember that when m changes, it changes by two units.

A little thought indicates that what we really have now is a summation over the probabilities that the system has a particular moment $m\mu$, regardless of how that value of m is achieved by flipping this or that spin. We can readily confirm this.

Let us designate by $P(m)$ the probability that the system has a moment $m\mu$, regardless of the precise way in which the individual spins are oriented —up or down—to produce that m. To compute $P(m)$ we need to consider two elements:

    (1) the number of distinct states with moment $m\mu$;

    (2) the probability of each of those states.

Because all states with given m have the same energy, the canonical distribution implies that they are equally probable. So, for given m, $P(m)$ is the sum of $\mathcal{N}(m)$ equal probabilities. This implies the explicit relation

$$P(m) = \mathcal{N}(m) \times \frac{e^{-\beta(-m\mu H)}}{\sum\limits_{t=1}^{2^N} e^{-\beta E_t}}$$

$$= \mathcal{N}(m) \times \frac{e^{+m\beta\mu H}}{\sum\limits_{m'=-N}^{N} \mathcal{N}(m')e^{+m'\beta\mu H}} \tag{5.2.8}$$

The transcription of the denominator parallels that in the step from equation (5.2.6) to equation (5.2.7). A glance back at the expression for $\langle M_z \rangle$ in equation (5.2.7) shows that our interpretation is correct, and that we may write

$$\langle M_z \rangle = \sum\limits_{m=-N}^{N} m\mu P(m). \tag{5.2.9}$$

To compute $\langle M_z \rangle$ we need only $P(m)$, the probability that the system has moment $m\mu$, and that probability we acquire by a judicious combination of probabilities from the canonical distribution.

We have arrived at the conclusion just stated in a rather backward fashion. The notion of an expectation value estimate, as defined in equation (2.8.1), really requires that a calculation of $\langle M_z \rangle$ take the form of equation (5.2.9) or be reducible to that form. In general, expectation value calculations may be written with the full set of probabilities of the canonical distribution, for such a form will be reducible (by grouping) to the form of equation (2.8.1). That was the way in which we started the calculation of $\langle M_z \rangle$ back in equation (5.1.5).

For relatively low N, say, our previous example with $N = 10$, one can compute explicit numerical values of $P(m)$ with a modest expenditure of arithmetic effort. We do, however, first need a value for the parameter $\beta\mu H$. From the preceding section we may take $\beta\mu H = 0.2$ as a typical laboratory value. Figure 5.2.2 gives a plot of $P(m)$ against m for $N = 10$. (Time required: about an hour with a desk calculator and a table of exponentials.) The expectation value of the moment comes out as

$$\langle M_z \rangle = \sum_{m=-10}^{10} m\mu P(m) = 1.97\mu,$$

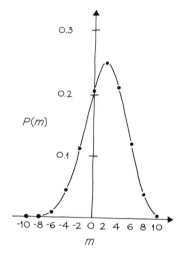

0.3

0.2

$P(m)$

0.1

-10 -8 -6 -4 -2 0 2 4 6 8 10

m

FIGURE 5.2.2
A plot of $P(m)$, the probability that the system has a moment $m\mu$, versus m, for $N = 10$ and $\beta\mu H = 0.2$.

rather close to μ times the value of m for which the peak in $P(m)$ occurs. (The expectation value given here can, of course, be computed directly from the closed-form expression for $\langle M_z \rangle$ derived in the preceding section.) A comparison of the diagram for $P(m)$ with that for the number of states with given m, provided in figure 5.2.1, shows that the diagrams are quite similar. To be sure, the plot for $P(m)$ is no longer fully symmetric about its peak. This should not be surprising, for the $P(m)$ plot is effectively the preceding plot with each point in it weighted by a probability proportional to $\exp(+m\beta\mu H)$. The asymmetric weighting—compare $m = +|m|$ with $m = -|m|$—both displaces the peak toward positive m and makes the wings asymmetric.

For the N typical of a macroscopic paramagnetic sample, the desk-calculator approach to $P(m)$ is unrealistic. We will see the physics most clearly if we turn "large N" from a liability into an asset and use the condition $N \gg 1$ to justify an approximation procedure. We look for an approximate expression for $\mathcal{N}(m)$, the cumbersome combination of factorials in $P(m)$.

To the practiced eye the plot in figure 5.2.1 suggests a Gaussian curve, an m-dependence going as $\exp(-\text{const } m^2)$. This form is symmetric about $m = 0$ and falls off rapidly with increasing $|m|$. If this might be what an approximate expression will lead to, the suggestion is at hand that the logarithm of $\mathcal{N}(m)$ might be the quantity to work with. The leading terms in an expansion of $\ln \mathcal{N}(m)$ might then go as low powers of m, in particular, as m^2. (A general comment: approximating the logarithm of a rapidly varying function typically gives a better approximation than does directly approximating the function. The reason lies in the slower variation of the logarithm.)

Upon taking the natural logarithm of $\mathcal{N}(m)$, we have

$$\ln \mathcal{N}(m) = \ln N! - \ln\left(\frac{N+m}{2}\right)! - \ln\left(\frac{N-m}{2}\right)!.$$

With $N \gg 1$ we may certainly use Stirling's approximation for the logarithm of a factorial to simplify the first term. For the other two terms, the justification for a comparable procedure is not evident, for $N \pm m$ may become small and even zero. Figure 5.2.1 showed, however, that the number of states with m close to $\pm N$ is quite small, and when N is large, these few states are quite unimportant. Moreover, even for 2! Stirling's approximation is good to about 10 per cent. So we may use Stirling's approximation for all three terms, though bearing in mind that we should not expect high accuracy for $m \simeq \pm N$.

For the generic variable x, Stirling's approximation for $\ln x!$ is

$$\ln x! = \frac{1}{2}\ln 2\pi + \frac{1}{2}\ln x + x \ln x - x + O\left(\frac{1}{10x}\right).$$

Using this logarithmic version of Stirling's approximation, and, in anticipation of a further approximation, doing a bit of judicious rearranging, one may write the approximate expression for $\mathcal{N}(m)$ as

$\ln \mathcal{N}(m)$

$$\simeq \ln\left(2^N \sqrt{\frac{2}{\pi N}}\right) - \frac{(N+m+1)}{2}\ln\left(1+\frac{m}{N}\right) - \frac{(N-m+1)}{2}\ln\left(1-\frac{m}{N}\right).$$

All terms prior to $O(1/10x)$ have been retained.

The second and third terms in the intermediate result can be further simplified. Let us consider the second. The full range of m is $-N \leqslant m \leqslant N$. For the central part of that range, the argument of the logarithm, $(1 + m/N)$, will be quite close to unity. We can expand $\ln(1 + m/N)$ about the value one for the argument of the logarithm. This will give us an excellent approximation for the part of the range of m that really concerns us.* The same analysis holds for the third term in the central part of the range of m.

To approximate the logarithms when $(1 \pm m/N)$ is close to unity, we use the Taylor's series expansion,

$$\ln\left(1 \pm \frac{m}{N}\right) = \pm \frac{m}{N} - \frac{1}{2}\left(\frac{m}{N}\right)^2 + \cdots,$$

valid for $|m/N| < 1$. When this approximation is inserted and the expression again rearranged, this time in powers of m, one finds

$$\ln \mathcal{N}(m) \simeq \ln\left(2^N \sqrt{\frac{2}{\pi N}}\right) - \frac{m^2}{2N}\left(1 - \frac{1}{N}\right),$$

correct to order m^2. (The next terms must be of order m^4 because of the symmetry in $m = \pm |m|$). We drop the $1/N$ in $(1 - 1/N)$ as surely negligible for large N and solve for $\mathcal{N}(m)$:

$$\mathcal{N}(m) = e^{\ln \mathcal{N}(m)} \simeq 2^N \sqrt{\frac{2}{\pi N}}\, e^{-m^2/2N}. \tag{5.2.10}$$

To this order of approximation, the function $\mathcal{N}(m)$ does have a Gaussian shape.

The steps in the derivation rigorously require $N \gg 1$ and $|m| \ll N$ for their validity, but let us examine this approximation to $\mathcal{N}(m)$ for the relatively low N for which we know the exact values, namely $N = 10$. Table 5.2.1 provides three comparisons: center, about halfway out, and extreme end-point. The approximation is far better than we had any right to expect.

TABLE 5.2.1

Comparison of the exact and approximate expressions for $\mathcal{N}(m)$ when N has the relatively low value of $N = 10$.

Value of m	Exact $\mathcal{N}$(m)	Approximate $\mathcal{N}$(m)
0	252	259
4	120	116
10	1	1.7

* As m approaches the end point $-N$ and $(1 + m/N)$ approaches zero, the logarithm $\ln(1 + m/N)$, grows large and negative. At the same time, however, the outside factor of $(N + m + 1)/2$ approaches zero (actually, only 1/2). This keeps the second term well within bounds except for $m = -N$, when the term becomes infinite. But for $m = -N$ the use of Stirling's approximation was not valid to begin with, for it fails badly for $\ln(0!)$. The approximation we are about to make will, remarkably, take care of this problem and keep $\mathcal{N}(m)$ well-behaved at $m = -N$, though it is not specially designed to do so.

Indeed, the *points* in figure 5.2.1 are the exact integers, but the suggestive *curve* closely connecting them was drawn from the approximate expression just derived, with m taken as a continuous variable. To the eye the fit is unimpeachable. With increasing N the approximation becomes better and better. For $N = 50$ and $m = 10$ (for which exact values are still available in tables), the error is already less than 1 per cent. For the extremely large N typical of a macroscopic system, we may confidently use the Gaussian approximation to $\mathcal{N}(m)$.

Now we return to the probability $P(m)$, first written in equation (5.2.8). Of primary interest is the dependence of $P(m)$ on m for large N. If we use the Gaussian approximation for $\mathcal{N}(m)$, we may concisely express $P(m)$ as

$$P(m) \simeq Ce^{-m^2/2N}e^{+m\beta\mu H}. \tag{5.2.11}$$

Here C is a normalization factor, dependent on N and $\beta\mu H$ but not on m. The product form exhibits again the dependence of $P(m)$ on its two essential constituents: (1) the probability of an individual state with moment $m\mu$, proportional to $\exp(+m\beta\mu H)$; and (2) the number of individual states with given m, proportional to $\exp(-m^2/2N)$. The former is rather broad as a function of m. It is the sharpness of the latter that leads to the sharp peak in $P(m)$ which we saw appearing already for $N = 10$ in figure 5.2.2.

The existence of a peak in $P(m)$ can be made more evident by completing the square in the exponent and thereby giving to $P(m)$ an explicit Gaussian-like appearance. This procedure starts with

$$-\frac{m^2}{2N} + m\beta\mu H = -\frac{(m - N\beta\mu H)^2}{2N} + \frac{N(\beta\mu H)^2}{2}$$

and leads to

$$P(m) \simeq (Ce^{N(\beta\mu H)^2/2})e^{-(m-N\beta\mu H)^2/2N}. \tag{5.2.12}$$

The diagrams in figure 5.2.3 show $P(m)$ for three different values of $\beta\mu H$. The Gaussian approximation to $P(m)$ does yield values for $\langle M_z \rangle$ in good agreement with the exact evaluation of the expectation value. Moreover, both yield results in good agreement with an estimate of M_z based on the peak value of $P(m)$. At larger values of N the agreement among the three will be even better.

As $\beta\mu H$ increases, the $P(m)$ distribution becomes noticeably asymmetric. Indeed, for $\beta\mu H > 1$, which may be achieved in the laboratory with exceedingly high fields and low temperatures, $P(m)$ rises continuously to a sharp maximum at $m = N$. All that appears is the "lefthand wing" of the Gaussian.

This asymmetry has little effect on estimates of the deviations in M_z from the expectation value $\langle M_z \rangle$. The widths of the three curves at half-maximum closely approximate that for the spread in the exact points. Moreover, all

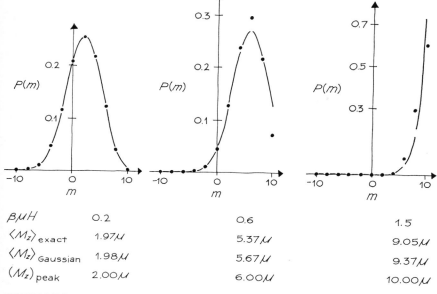

$\beta\mu H$	0.2	0.6	1.5
$\langle M_z \rangle_{\text{exact}}$	1.97μ	5.37μ	9.05μ
$\langle M_z \rangle_{\text{Gaussian}}$	1.98μ	5.67μ	9.37μ
$(M_z)_{\text{peak}}$	2.00μ	6.00μ	10.00μ

FIGURE 5.2.3

These diagams give $P(m)$ as a function of m for $N = 10$ and three different values of the parameter $\beta\mu H$. The points give the exact values; an interpolation of the gaps in the curve gives the Gaussian approximation.

three are comparable, though the low-temperature case (the third) is somewhat narrower. To get a reliable estimate of the width as a function of N, we need merely return to the Gaussian approximation for $P(m)$. We consider a situation with $\beta\mu H < 1$, so that the peak in the $P(m)$ distribution comes at $m = N\beta\mu H < N$. We ask, how far must m deviate from its value at the peak before $P(m)$ falls to one-half its peak value? Calling Δm the deviation in m from the peak at $m = N\beta\mu H$, we solve for Δm from

$$P(N\beta\mu H + \Delta m) = \tfrac{1}{2}P(N\beta\mu H).$$

Shorn of constant factors, this means

$$e^{-(\Delta m)^2/2N} = \tfrac{1}{2}e^{-(\text{zero})}$$

and leads to

$$\Delta m = \pm\sqrt{2N\ln 2} \simeq \pm N^{1/2}. \tag{5.2.13}$$

The deviation estimate Δm is to be compared with the full range of m, which is $2N$. The ratio of the width of $P(m)$ at half-maximum to the full range of m is thus of order $1/N^{1/2}$ and hence exceedingly small for the large N of a macroscopic sample. An observation of M_z is almost certain to give a value very close to the expectation value $\langle M_z \rangle$.

We should try to develop an acute appreciation for the sharpness of the peak in the probability distribution $P(m)$. A value for N of 10^{20} is a modest laboratory value. The ratio of the width of the peak in $P(m)$ to the full range of m is then of order 10^{-10}. If $P(m)$ were plotted as a typical graph, in which the full range of m occupied 10 cm, the peak would have a width of about 1/10 the diameter of a hydrogen atom. This is so extreme that one cannot visualize it. Let's turn things the other way around. We specify a graph in which the peak in $P(m)$ has a width equal to the width of the letter i printed here. For the scale of the full range of m, look across the room and out the window, and try to imagine a distance of about one-twentieth the circumference of the Earth.

In the sharpness of a probability distribution like $P(m)$ lies a reason why statistical mechanics can usually make *confident* predictions of the properties and behavior of a macroscopic system. Application of the theory starts with a broad probability distribution for individual states of the system. Here it is the canonical distribution, found by looking quantitatively for the distribution with the maximum missing information when temperature and environment are the given information. When one asks a question about a macroscopic property, one can group individual states that have closely comparable values of the macroscopic property (though the states necessarily differ microscopically). From the grouping and from the probabilities of individual states given by the canonical distribution, one can then construct a probability distribution for the occurrence of the specific macroscopic property. It is the resultant distribution that almost invariably exhibits a sharp peak or, stated differently, a width that is narrow relative to the range of the macroscopic property. That sharp peak justifies confidence in expectation value estimates of the macroscopic property. Using such estimates for predictions, and for comparison with results from experiment, makes good sense.

A few words should be said to avoid a possible misconception. In the paramagnetic case, the grouping already gives a peak, as evinced by $\mathcal{N}(m)$, and that is merely transferred to $P(m)$. For a gas and a question about total energy, the grouping of individual states with closely comparable energy leads to an " $\mathcal{N}(E)$ " that rises continuously, approaching infinity as E does. It has no peak. Nonetheless, the product of that *rapidly rising* function of E with the probability for an individual state, $\propto e^{-\beta E}$, which *falls* with E, does produce a peak. For a macroscopic number of gas molecules, the peak is again exceedingly sharp and leads to confident predictions about the total energy and pressure of the gas, both macroscopic properties. In brief, the details of the origin of the peak may vary.

In closing this section, let us note another implication of the sharp peak in $P(m)$. The canonical probability distribution is based on exceedingly little information (as hypothesis). Yet it leads, ultimately, to confident predictions of the macroscopic properties of macroscopic systems. The implication is that little information substantially determines the observable macroscopic properties. Roughly but suggestively, it must be that most of the (micro-scopically different) quantum-mechanical states that are compatible with the known macroscopic data (and to which appreciable probability is assigned) have, for their other macroscopic properties, quite comparable values. This is a conjectured property of the quantum-mechanical states of macroscopic systems, a start toward sharpening the notion that the fantastic diversity of a system's microscopic configurations is lost, washed out, when one looks only at its macroscopic properties. Herein lies at least a clue to the success of statistical mechanics (and also of thermodynamics).

PROBLEMS

5.1. Suppose the external magnetic field applied to a sample of paramagnets (spin $\frac{1}{2}\hbar$) at low temperature is such that $\mu H/kT = 0.6$. How small must N be in order that the uncertainty ΔM_z in one's estimate (based on the canonical distribu-tion) of the total magnetic moment be as large as 1 per cent of $\langle M_z \rangle$? After you have estimated N, estimate the mass of such a sample in grams and the volume in cubic centimeters (both to within a factor of ten or so). Conclusions?

5.2. We consider a nucleus with spin $1\hbar$ (such as a nitrogen nucelus or a deuteron) that possesses an electric quadruple moment. It is one nucleus among many in a sample in equilibrium at a temperature T in an external electric field. We may imagine the nucleus to be spatially fixed, but free to take on various orientations. Provided we neglect the mutual interactions, there are three energy eigenstates (as given in table P5.2), each with a definite value for the projection of the spin along the field direction $\hat{z}$.

TABLE P5.2
Eigenstates.

Angular momentum state		Associated energy
spin up	$(J_z = +1\hbar)$	$+\varepsilon$
spin "sideways"	$(J_z = 0)$	0
spin down	$(J_z = -1\hbar)$	$+\varepsilon$

a. What is the probability of finding the nucleus with spin up? In what limit would this be 1/3?

b. Calculate an expression for the expectation value of the energy (in terms of $\varepsilon, T, \ldots$).

c. What is the value of $\langle M_z \rangle$?

6

THE CLASSICAL LIMIT

For the purposes of calculation and of visualization, one often wants to treat a problem in classical, rather than quantum-mechanical, terms. One has a better-developed intuition for classical positions and momenta than for quantum-mechanical states. This chapter is devoted to adapting the preceding results to the classical way of describing a physical system.

First we will discuss the conditions under which a classical description ought to be useful and permissible. The conclusions drawn here will be supported by detailed calculations in Chapter 9. Then we will adapt the results of Chapter 4 to a classical statistical description of equilibrium. Some qualitative arguments will suffice. The analysis could be made more rigorous by repeating the steps in Chapter 4 in classical terms, but doing so seems not worth the effort.

The chapter closes with some applications, specifically, the Maxwell velocity distribution and a situation in which one can see a classical treatment fail.

6.1. THE DOMAIN IN WHICH A CLASSICAL ANALYSIS IS ADMISSIBLE

We seek now the conditions under which a transition from a quantum-mechanical treatment to a classical treatment may be permissible. Such a transition is most relevant for a gas. For a solid, an analysis with classical physics is all but useless; only with the advent of quantum mechanics could theory and experiment be brought into agreement in more than a few instances. A solid is essentially a quantum-mechanical system. Liquids form an intermediate kind of system, often, though certainly not always, amenable to a classical analysis. For the purpose of this discussion, we can subsume liquids under the heading of gases.

Our basic problem is to determine the circumstances under which we may consider a gas to be a classical system of small bodies in chaotic motion. To each particle would be ascribed a well-defined position and momentum at each instant of time, even if we do not happen to know them. In quantum-mechanical language, this would be a situation in which the particles could be described by individual wave packets moving rather independently, except for "collisions" occurring when two well-localized wave packets overlap. In proposing to ascribe a well-defined position and momentum to a particle, one must be circumspect: the Heisenberg uncertainty principle sets limits to the precision with which the ascription may legitimately be imagined to be done. If we focus attention on the x-axis coordinate of a gas molecule and on the x-component of the momentum, p_x, the uncertainty principle assures us that the uncertainties, Δx and Δp_x, in those two quantities have a mutually determined lower bound:

$$\Delta x \, \Delta p_x \gtrsim \hbar.$$

Although *either* the x-coordinate *or* the momentum component p_x may be specified arbitrarily sharply, the product of the uncertainties cannot be smaller than order $\hbar$. Viewed as a limitation on measurement, the foregoing means that we can have simultaneous experimental knowledge of position and momentum only within the bounds of the inequality. We can employ the Heisenberg relation to establish a criterion that must be satisfied if a classical picture is to be permissible and useful.

Before accepting a classical description as adequate, we would surely want the (quantum-mechanical) uncertainty about position to be much less than the typical separation between molecules. Any picture of little bodies in motion implicitly assumes as much.* If we denote the typical molecular separation by $\bar{R}$, we impose the strong inequality

$$\bar{R} \gg \Delta x \qquad (6.1.1)$$

(and similarly for Δy and Δz) as a prerequisite for a classical description.

An analogous statement can be made about the momentum. We would surely want the (quantum-mechanical) uncertainty about momentum to be much less than the typical momentum that can be ascribed to the erratic thermal motion. For the x-component we may express this requirement as the inequality

$$\bar{p}_x \gg \Delta p_x, \qquad (6.1.2)$$

where $\bar{p}_x$ stands for the magnitude of a typical thermal momentum component.

For a classical picture to be valid, these two strong inequalities must hold simultaneously. Upon forming their product, we arrive at the relation

$$\bar{R}\bar{p}_x \gg \Delta x \, \Delta p_x \gtrsim \hbar. \qquad (6.1.3)$$

This inequality frames a stringent restriction on the physical conditions under which a classical picture is useful or permissible. (The product was formed so that the individual uncertainties, Δx and Δp_x, could be eliminated. It is only their *mutual* bound that is prescribed by the Heisenberg principle.)

Another way of expressing the restriction follows naturally if we move $\bar{p}_x$ to the righthand side:

$$\bar{R} \gg \frac{\hbar}{\bar{p}_x}. \qquad (6.1.4)$$

The quantity on the new righthand side is (roughly) the de Broglie wavelength of the waves making up the wave packet associated with a typical molecule. (A range of wavelengths will be necessary for the construction of the wave packet, but $\hbar/\bar{p}_x$ will give the order of magnitude of the dominant wavelenths.) We might call the quantity $\hbar/\bar{p}_x$ the *thermal de Broglie wavelength*. The statement of the restriction becomes this: classical reasoning can be valid only if the typical molecular separation is large relative to the thermal de Broglie wavelength.

---

* The uncertainty about position, Δx, is roughly the diameter of the wave packet that we associate with a molecule. Unorthodox though it may be, I think of scattered, puffy white clouds drifting in an otherwise blue sky. When the clouds spread out and overlap most of the time, then we're in for trouble.

We can make the restriction still more informative by introducing some estimates for $\bar{R}$ and $\bar{p}_x$ in terms of the temperature and density of the gas, together with the molecular mass. The volume of space, free from neighbors, around a typical molecule is of order $\bar{R}^3$. For N molecules in a volume V, this leads to

$$N\bar{R}^3 \simeq V$$

and hence

$$\bar{R} \simeq (V/N)^{1/3}. \tag{6.1.5}$$

For an estimate of $\bar{p}_x$ we use our results from the statistical mechanics of a single gas molecule. Since the expectation value of the kinetic energy is $\frac{3}{2}kT$, we may write

$$\tfrac{1}{2}m(\bar{v}_x)^2 = \frac{(\bar{p}_x)^2}{2m} \simeq \tfrac{1}{3}(\tfrac{3}{2}kT)$$

and hence

$$\bar{p}_x \simeq \sqrt{mkT}. \tag{6.1.6}$$

Upon inserting these estimates into the inequality (6.1.4), we arrive at

$$\left(\frac{V}{N}\right)^{1/3} \gg \frac{\hbar}{\sqrt{mkT}} \tag{6.1.7}$$

as an inequality that must be satisfied if classical reasoning is to be applicable. The term on the left is a measure of the typical intermolecular separation; that on the right, a measure of the thermal de Broglie wavelength. Qualitatively, the restriction declares that the density of particles N/V may not be too large, the temperature too low, or the particle mass too small.

A few numerical estimates are now in order. Let us first take air at typical room temperature and pressure. For data we have:

temperature, $T \simeq 300$ K;

pressure, $p_{\text{obs}} \simeq 10^6$ dynes/cm^2 (atmospheric pressure);

particle mass, $m = 2\,m_{\text{nitrogen}} \simeq 4.6 \times 10^{-23}$ gm.

The estimate for the particle mass is based on the facts that diatomic nitrogen is the dominant constituent of air and that diatomic oxygen, the other significant constituent, has a comparable mass.

The gas law gives an estimate for the particle density:

$$\frac{N}{V} = \frac{p_{\text{obs}}}{kT} \simeq 2.4 \times 10^{19} \text{ molecules/cm}^3.$$

This density implies a typical molecular separation of order

$$\bar{R} \simeq \left(\frac{V}{N}\right)^{1/3} \simeq 3.5 \times 10^{-7} \text{ cm}$$

(or about 35 Ångstroms). This value for the typical separation is to be compared with an estimate of the thermal de Broglie wavelength:

$$\frac{\hbar}{\sqrt{mkT}} \simeq 8 \times 10^{-10} \text{ cm.}$$

There is a favorable inequality of better than two orders of magnitude. A classical analysis of air under typical room conditions ought to be valid.

The situation with liquid air is less certain. When boiling into open air, the liquid has a temperature of 83 K and a density rather like that of liquid water. The typical molecular separation is reduced from that in ordinary gaseous air by about a factor of ten:

$$\bar{R} \simeq 3.8 \times 10^{-8} \text{ cm.}$$

This separation is about the " size " of a small molecule; the molecules in a liquid are virtually in constant contact. The typical separation is again to be compared with an estimate of the thermal wavelength:

$$\frac{\hbar}{\sqrt{mkT}} \simeq 1.5 \times 10^{-9} \text{ cm.}$$

The difference is only a factor of 20 or so. In view of the approximate nature of our restriction, this is approaching the edge. For liquid helium, with both a smaller mass and a much lower temperature ($\simeq 4$ K), the two terms are comparable. Classical reasoning is inadmissible, and indeed only a quantum-mechanical analysis has begun to elucidate the peculiar properties of liquid helium. There are, however, gases with molecular masses greater than that of air and with higher condensation temperatures; for their condensed state, a classical analysis should certainly be permissible. No general statement can be made about liquids.

As the final example, we might look at the conduction electrons in a good metallic conductor. As a satisfactory first approximation, one may neglect the electromagnetic interactions and treat the conduction electrons as a gas of noninteracting particles. With an estimate of one conduction electron per atom and an atomic spacing of about 3×10^{-8} cm, we have a typical separation of

$$\bar{R} \simeq 3 \times 10^{-8} \text{ cm}$$

for the electrons, roughly like that of a typical liquid. The particle mass, however, is now very small, about 1/7000th that of helium. For room temperature the value to be compared with $\bar{R}$ is

$$\frac{\hbar}{\sqrt{mkT}} \simeq 17 \times 10^{-8} \text{ cm.}$$

The inequality is the reverse of that required by the classical restriction. There is good reason to believe that for conduction electrons a classical treatment is inadmissible and will be sure to fall if pushed. The confrontation of theory with experiment amply bears out this judgement.

From these numerical estimates, we may tentatively conclude that: for gases at temperatures and densities like those of typical room conditions, a classical treatment should be satisfactory; for liquids the situation varies considerably, but is definitely against a classical treatment for liquid helium; and for conduction electrons in a solid, a quantum-mechanical treatment is mandatory.

We must note that the analysis thus far has neglected both the structure of molecules and the inevitable intermolecular forces. Some modification of the tentative conclusions will be required, and this will further restrict the classical domain.

A classical treatment of atomic or molecular structure is hopelessly inadequate. It is only the motion of the center of mass of an atom or molecule that classical statistical mechanics may be able to handle correctly. Where the tentative conclusions spoke favorably for a classical analysis, they must be understood to refer to statistical descriptions of the center-of-mass motions.

There is still the question of intermolecular forces. Quantum mechanics informs us that the position of a molecule is "spread out" over a region whose dimensions are of the order of the dominant wavelengths in the wave packet associated with the molecule. For us this means dimensions of the order of the thermal de Broglie wavelength. For a molecule to respond to intermolecular forces in the manner of a classical particle, the thermal de Broglie wavelength must be small relative to the distance in which the classical force changes appreciably. Otherwise the spread-out molecule will "sample" widely differing values of the force and hence not respond like a classical particle, which sees only one value of the force at a time.

The foregoing is a good start. Now we should note that equilibrium statistical mechanics is generally concerned with energies, rather than the forces themselves, in both the quantum and classical versions. This suggests that we consider the variation in the potential energy which a molecule samples. We should compare that variation with kT, the natural scale for energy in equilibrium statistical mechanics. A plausible second restriction on the applicability of a classical analysis would be this: the variation of the potential energy, in a distance equal to the thermal de Broglie wavelength, must be small relative to kT. Often it is sufficient that the expectation value of the variation be small relative to kT, for a relatively large variation with only a small probability of occurrence would be tolerable. The expectation value would be calculated classically, for one may make a provisional assumption that classical statistical mechanics is adequate, and then check for consistency.

Although a bit vague, and only made plausible here, the foregoing is the essence of a second condition for the adequacy of a classical treatment. One may, for example, try to analyze the vibrational motion of the atoms in a solid with classical statistical mechanics, but one will fail, and that failure can be traced back to a violation of the second condition.

Before we can derive the classical version of statistical mechanics by "taking a limit" and thereby passing from quantum physics to classical physics, we will need to extend somewhat our development of probability theory. To this we turn in the next section.

6.2. CONTINUOUS PROBABILITY DISTRIBUTIONS

Thus far in our development of statistical mechanics—a quantum-mechanical development—we have dealt with inferences about energy eigenstates of a finite physical system. If (momentarily) we look on those states as the possible values that a variable may take on, we note that they form a discrete set: we may enumerate them as ψ_1, ψ_2, and so on. (Whether the total number of states is finite or infinite is irrelevant here.) In classical statistical mechanics, the concern is not with states but rather with the position and momentum of each of many particles. The values that these may take on vary *continuously* over an infinite range. We therefore need to develop a mathematical apparatus for dealing with inferences and probabilities associated with continuous variables. The logic underlying our theory of probability remains unaltered.

In the classical limit of statistical mechanics, we will need to assign a probability to the following *type* of inference: the position of a particle labeled so-and-so has an x-coordinate in the infinitesimal interval Δx around the value x. The new element is that we must phrase the inference in terms of a small interval, rather than a precise point. The nature of quantum mechanics and the limiting process of deriving classical statistical mechanics will provide us with a physical reason for using an interval, and we will see that reason in the next section. Here we are concerned with the mathematical reason, which is readily illustrated.

Suppose one knew that the particle were somewhere between $x = 0$ and $x = 1$, endpoints included. Further, suppose that the particle were as likely to be found at any one point on that line segment as at any other. The probability of finding the particle somewhere in the segment would be unity. The probability of finding it in the first half of the segment would be $1/2$; in the first quarter of the segment, $1/4$; in the first eighth, $1/8$; and so on. As one reduces the interval of concern by a factor of one-half, the probability decreases proportionately. If one were to imagine shrinking the interval indefinitely—an attempt to specify the probability of finding the particle at some given *point* along the line—one would find the probability tending toward zero, and this would be true for any given point. Yet, by hypothesis, the particle is at some point along the line segment. For any finite interval, no matter how small though not zero, there is a meaningful non-zero probability. An attempt to pass to the point limit generates nothing but an apparent absurdity.

With the preceding as an introduction, we turn now to a reasonably careful development of a mathematical apparatus for handling probability distributions for continuous variables. Let us work with a variable that may take on any value between a and b, with $b > a$ and the endpoints of the interval being included. The variable might be a polar angle running continuously from zero to π. If the variable were a position or momentum component of a particle, we would need to let a go to $-\infty$ and b to $+\infty$.

The introductory remarks have indicated that we cannot profitably phrase inferences and probabilities in terms of this or that precise value of the variable. The use of an interval or a range is called for. Let us begin modestly with an inference whose probability will surely make sense, the inference stated as follows: "The value of the variable is greater than or equal to a and is less than or equal to x, with $x \geqslant a$." For x greater than a, this inference encompasses a finite range of possible values and so should have associated with it a meaningful probability. Let us denote the inference by $A(x)$.

On a hypothesis, which we denote by B, about the behavior of the variable, we may write the probability of the inference as $P(A(x)\,|\,B)$. For the sake of clarity let us write this out as follows:

$$P(A(x)\,|\,B) \equiv \left(\begin{array}{l} \text{the probability, on hypothesis } B, \\ \text{that the following relation holds:} \\ a \leqslant (\text{value of variable}) \leqslant x \end{array} \right). \qquad (6.2.1)$$

The probability will increase with increasing x, for then the range considered in the inference will increase. If the point x is taken to be the point b, the probability will be unity, for the value of the variable must, by hypothesis, lie in the interval a to b. A possible behavior of the probability as a function of x is indicated in figure 6.2.1.

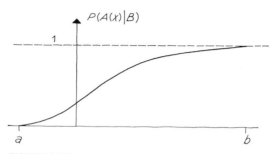

FIGURE 6.2.1
A sketch of how the probability $P(A(x)|B)$ might behave as a function of x.

Typically our interest will focus on a more detailed probability, the probability that the value of the variable lies in a small interval Δx around a value x. The foregoing does enable us to state such a probability as a difference:

$$\begin{pmatrix} \text{the probability, on hypothesis } B, \\ \text{that the following relation holds:} \\ x < (\text{value of variable}) \leqslant x + \Delta x \end{pmatrix} = P(A(x + \Delta x)\,|\,B) - P(A(x)\,|\,B).$$

$$(6.2.2)$$

Although the difference on the righthand side does provide a formal expression for the probability presented in words on the left, none of this is of much use unless we can find a way of making it analytically tractable.

In the situations we will encounter in classical statistical mechanics, a probability like $P(A(x)\,|\,B)$ will change smoothly with x. Then it makes sense to take the derivative of that probability with respect to x, and to relate the difference in equation (6.2.2) to the first derivative and the size Δx of the interval. On the assumption that the derivative does exist and will be useful, let us introduce an abbreviation for it:

$$\mathcal{P}(x) \equiv \frac{d}{dx} P(A(x)\,|\,B) \geqslant 0. \tag{6.2.3}$$

Because the probability $P(A(x)\,|\,B)$ certainly cannot decrease with increasing x, the function $\mathcal{P}(x)$ is nonnegative. An explicit display of hypothesis and inference has been dropped in order to keep the size of subsequent mathematical expressions within reasonable bounds. Given $\mathcal{P}(x)$, we can regain the probability by integration:

$$P(A(x)\,|\,B) = \int_{x'=a}^{x'=x} \mathcal{P}(x')\,dx'. \qquad (6.2.4)$$

There is nothing profound in this, but we may now derive a genuinely useful expression for the probability formally expressed in equation (6.2.2). The difference on the righthand side we write as a difference of two integrals, one from a to $x + \Delta x$, the other from a to x. Then we combine the two integrals as a single integral going merely from x to $x + \Delta x$. Writing out the relevant probability once again, we find

$$\left(\begin{array}{l} \text{the probability, on hypothesis } B, \\ \text{that the following relation holds:} \\ x < (\text{value of variable}) \leqslant x + \Delta x \end{array} \right) = \int_{x'=x}^{x'=x+\Delta x} \mathcal{P}(x')\,dx'$$

$$= \mathcal{P}(x)\,\Delta x + \left(\begin{array}{l} \text{terms of higher} \\ \text{order in } \Delta x \end{array} \right). \qquad (6.2.5)$$

The "terms of higher order in Δx" arise from the variation in $\mathcal{P}(x)$ over the range x to $x + \Delta x$. When the variation of $\mathcal{P}(x)$ is not extreme and when Δx is relatively small, the terms of higher order in Δx may be neglected. Then the numerical value of the probability stated on the left is proportional to the size Δx of the interval and depends on the position of the interval through $\mathcal{P}(x)$ only. Generally we will work with the relation just derived, with the terms of higher order dropped in the limit of infinitesimal Δx. We will refer to the function $\mathcal{P}(x)$ as a *probability density*. The relations are illustrated in figure 6.2.2.

We have gone the long way round to a simple, almost "intuitively obvious" result. There are two reasons for doing this.

1. A probability proportional to Δx, provided Δx is infinitesimal, emerges without recourse to plausibility arguments.

2. The approach makes clear from the outset that $\mathcal{P}(x)$, the probability density, is *not* itself a probability but the derivative of a probability. As defined, $\mathcal{P}(x)$ is nonnegative but may be greater than unity (as an example will shortly show).

There is a minor technical matter that we should discuss. The product $\mathcal{P}(x)\Delta x$ gives the probability for the interval Δx *beyond* x when the higher-order terms may be neglected. Often one likes to consider the probability for an interval Δx *around* x, that is, for the range $x - \frac{1}{2}\Delta x$ to $x + \frac{1}{2}\Delta x$. If the

158

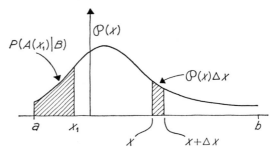

FIGURE 6.2.2
A sketch of the probability density $\mathcal{P}(x)$, calculated
by differentiation from the curve for the probability
$P(A(x)|B)$ given in figure 6.2.1. The shaded area on
the left indicates how one may regain the
probability by integration over the probability
density. The area of the small shaded column on the
right indicates the probability that the value of the
variable will lie in the interval Δx beyond x.

probability density $\mathcal{P}(x)$ is continuous and Δx is small, that probability can
also be written as $\mathcal{P}(x)\Delta x$, again provided one neglects the higher-order
terms.

The preceding analysis is sufficiently abstract that we should now turn to
an illustration.

The Gaussian Probability Distribution

There can be little doubt that the Gaussian probability distribution is the
most famous of all continuous probability distributions. The endpoints of the
range of the variable, corresponding to our a and b, are $\pm \infty$. The probability
density,

$$\mathcal{P}(x) = \frac{1}{\sqrt{2\pi}\,\sigma}\, e^{-(x-x_0)^2/2\sigma^2},$$

depends on two parameters, x_0 and σ. Figure 6.2.3 presents a graph of this
probability density. From the sketch we can see that the peak in $\mathcal{P}(x)$ occurs
when x is equal to x_0, that the probability density is symmetric about x_0,
and that the parameter σ provides a measure of the width of the peak in the
distribution.

By hypothesis the value of the variable must lie somewhere in the range
$-\infty$ to $+\infty$. If we sum the probability associated with each interval in a set

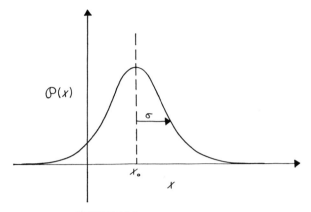

FIGURE 6.2.3
A sketch of the probability density for
the Gaussian distribution.

of nonoverlapping intervals that, together, cover the full range, the result must yield unit probability. So let us first confirm that the quoted probability density satisfies this normalization condition. The indicated summation is, for this continuous variable, performed by integration:

$$\left(\begin{array}{l}\text{the limit of a sum of probabilities,}\\ \text{each of the form}\\ \qquad \mathcal{P}(x_i)\,\Delta x_i,\\ \text{associated with a set of intervals}\\ \text{that span the range without}\\ \text{overlapping}\end{array}\right) = \int_{-\infty}^{\infty} \mathcal{P}(x)\,dx$$

$$= \frac{1}{\sqrt{2\pi}\,\sigma} \int_{-\infty}^{\infty} e^{-(x-x_0)^2/2\sigma^2}\,dx$$

$$= \frac{1}{\sqrt{2\pi}\,\sigma} \int_{-\infty}^{\infty} e^{-y^2/2\sigma^2}\,dy$$

$$= \frac{1}{\sqrt{2\pi}\,\sigma} \sqrt{\pi(2\sigma^2)^{1/2}} = 1.$$

The step to the third line is merely a change of variable to $y = x - x_0$. For finite x_0 this leaves the limits of integration effectively unchanged at $\pm\infty$. The $y =$ integral itself is evaluated in Appendix A. And in the end we find that the factor preceding the exponential in $\mathcal{P}(x)$ provides the correct normalization.

The symmetry of $\mathcal{P}(x)$ about the point x_0 suggests that the expectation value of the variable will be x_0. We may readily confirm this tentative conclusion. First, however, we must transcribe the definition of an expectation value, given in equation (2.8.1) for a discrete variable, so that the definition is applicable to a continuous variable. We do this for a general function $f(x)$, writing

$$\langle f(x) \rangle = \begin{pmatrix} \text{the limit of a sum, each} \\ \text{term being of the form} \\ f(x_i)\mathcal{P}(x_i)\,\Delta x_i \end{pmatrix}$$

$$= \int_{-\infty}^{\infty} f(x)\mathcal{P}(x)\,dx.$$

Now we take the function $f(x)$ to be merely x, insert the present form for the probability density, and manipulate the integrals a bit:

$$\langle x \rangle = \int_{-\infty}^{\infty} x \frac{1}{\sqrt{2\pi}\,\sigma} e^{-(x-x_0)^2/2\sigma^2}\,dx$$

$$= \frac{1}{\sqrt{2\pi}\,\sigma} \left[\int_{-\infty}^{\infty} y e^{-y^2/2\sigma^2}\,dy + x_0 \int_{-\infty}^{\infty} e^{-y^2/2\sigma^2}\,dy \right]$$

$$= \frac{1}{\sqrt{2\pi}\,\sigma} [0 + x_0\sqrt{2\pi}\,\sigma] = x_0$$

The step to the second line uses the same change of variable as before. The first integral in the second line has a net value of zero because the integrand is an odd function of y. The expectation value of the variable is x_0, as the symmetry argument suggested.

The value of the variable will not necessarily be close to x_0. The spread of reasonably probable values will depend on the size of the second parameter, σ. To find a precise relation let us calculate the root mean square estimate of deviations from the expectation value. First we calculate the expectation value of the square of the deviations, bearing in mind that here $\langle x \rangle$ is simply x_0:

$$\langle (x - \langle x \rangle)^2 \rangle = \int_{-\infty}^{\infty} (x - x_0)^2 \mathcal{P}(x)\,dx$$

$$= \frac{1}{\sqrt{2\pi}\,\sigma} \int_{-\infty}^{\infty} y^2 e^{-y^2/2\sigma^2}\,dy$$

$$= \frac{1}{\sqrt{2\pi}\,\sigma} \frac{\sqrt{\pi}}{2} (2\sigma^2)^{3/2} = \sigma^2.$$

The definite integral is among those computed in Appendix A. Upon taking a square root, we find

$$\langle (x - \langle x \rangle)^2 \rangle^{1/2} = \sigma.$$

The root mean square estimate of deviations from $\langle x \rangle$ is given directly by the parameter σ. Incidentally, the factorization, as $2\sigma^2$, in the exponent of $\mathcal{P}(x)$ is made specifically so that this equality between root mean square deviation and width parameter will hold.

The smaller σ is, the sharper the peak will be. Let us note that for σ smaller than $1/\sqrt{2\pi}$, the peak value of the probability density will be greater than unity:

$$\mathcal{P}(x_0) = \frac{1}{\sqrt{2\pi}\,\sigma} > 1 \quad \text{if } \sigma < \frac{1}{\sqrt{2\pi}}.$$

There is no inconsistency here, for neither $\mathcal{P}(x_0)$ nor, more generally, $\mathcal{P}(x)$ is itself a probability. Each is merely a probability density. Only the *product* of $\mathcal{P}(x)$ and a sufficiently small interval Δx is a probability.

The Gaussian distribution provides adequate illustration of how one deals with a continuous probability distribution for a single variable. The extension to two or more continuous variables entails nothing more than a proliferation of infinitesimal intervals and of the variables on which the probability density depends. We may turn now to the application of this mathematical apparatus in classical statistical mechanics.

6.3. THE CANONICAL DISTRIBUTION IN THE CLASSICAL LIMIT

Knowing that a classical treatment is sometimes admissible, we can consider how a system of particles should be described in the classical limit of quantum mechanics. Let us begin modestly, with a single particle constrained to one-dimensional motion. Once we have established the scheme for handling this case, we will find the generalization to three dimensions and then to many particles not difficult.

In the classical limit we may, in principle, ascribe to the single particle a rather well-defined position (x-coordinate, say) and momentum (correspondingly, p_x) at each instant of time. The specification of the values of x and p_x constitutes the classical specification of the state of the one-particle system. To have a picture of this, we may use an auxiliary two-dimensional space, with axes labeled x-axis and p_x-axis. Formally, this is the *phase space* for the system. The classical state of the system is specified when we specify a particular point in the auxiliary space. Figure 6.3.1 indicates such a two-dimensional phase space and a point specifying the system classically.

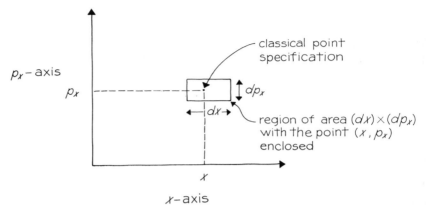

FIGURE 6.3.1
The two-dimensional phase space for a single particle constrained to one-dimensional motion. The point indicates an uncompromisingly classical specification of the state of the system: a precise position (x) and a precise momentum (p_x).

Now we consider inferences about the particle and the corresponding probabilities. A typical question would be: what is the probability that the particle is to be found with position variable in some small range dx around x and with momentum variable in some small range dp_x around p_x? In the phase-space language, the question asks for the probability of finding the point specifying the particle's state in a small rectangle of area $(dx) \times (dp_x)$ with the point (x, p_x) inside.

There are two reasons for stating the question in terms of small ranges rather than of precise position and momentum specifications. We have already noted the mathematical reason for such a procedure. The physical reason is grounded in the tenets of quantum mechanics and the Heisenberg uncertainty principle. However appealing a classical "point" specification in the phase space may be, we may not totally forget about quantum mechanics. From the Heisenberg principle we know that unavoidable experimental uncertainties arise when one tries to achieve simultaneous knowledge of position and momentum. The best one can really do is limited by the inequality

$$\Delta x \, \Delta p_x \gtrsim \hbar.$$

Moreover, no matter how hard one tries to construct a wave packet that will spatially localize a particle and give it a definite momentum, one can do no better than that limit.* There is no need to construct a probability for an area $(dx) \times (dp_x)$ that is significantly smaller than $\hbar$, for our knowledge of the "point" specification is always limited by an uncertainty region of order $\hbar$.

For the reasons outlined here and in the preceding section, we may conclude that all meaningful probabilities concerning position and momentum can be constructed with small but non-zero ranges dx and dp_x. Let us set up the general product form for the relevant probability:

$$\left(\begin{array}{l}\text{the probability that the}\\ \text{particle is to be found with}\\ \text{position variable in } dx\\ \text{around } x \text{ and momentum}\\ \text{variable in } dp_x \text{ around } p_x\end{array}\right) = \mathcal{P}(x,\, p_x)dx\, dp_x. \qquad (6.3.1)$$

The function $\mathcal{P}(x,\, p_x)$ is the probability density for the two variables; the ranges dx, dp_x, are taken to be infinitesimals. The probability density will generally vary with x and p_x, indicating that regions of equal area (equal magnitude for the product $dx\, dp_x$) in the two-dimensional phase space will have different probabilities that the "point" classically specifying the system is to be found in the corresponding region. We must bear in mind that the factorization of the probability into a probability density times a region in phase space is admissible only when dx and dp_x are individually small (really, infinitesimal). Only then may we neglect the "higher-order terms" and satisfactorily take the probability to be proportional to the individual ranges and hence to the size of the (infinitesimal) region in phase space.

Of primary concern to us is the relation between equation (6.3.1) and the canonical distribution. According to the latter, the probability that a system is appropriately described by an energy eigenstate ψ_j is proportional to $\exp(-\beta E_j)$. The change from energy eigenvalues to a classical energy expression is not likely to be troublesome. The real question is the connection between a region $dx\, dp_x$ in phase space and quantum-mechanical states. The connection can be established by argument from the Heisenberg uncertainty principle. The most precise specification of x and p_x allowable by quantum mechanics leaves the particle's position and momentum "spread out" over a region in phase space whose area is of order $\hbar$. This strongly suggests that

---

* For Δx, Δp_x, taken as root mean square estimates of uncertainties, the precise lower limit on wave-function construction is $\frac{1}{2}\hbar$, a result derivable within the framework of quantum mechanics.

a region of roughly that extent should be regarded as the analog of a quantum-mechanical state. Given a general small rectangle, of area dx by dp_x, we may say that the region contains

$$\frac{dx \, dp_x}{O(\hbar)}$$

distinct quantum-mechanical states in the classical limit. At this time we can justifiably write only $O(\hbar)$—of order $\hbar$—in the denominator. A precise determination must wait for some time.

> Later we will find "$O(\hbar)$" to be exactly Planck's constant h. The number of "distinct quantum-mechanical states" is to be understood as the number of mutually orthogonal, or linearly independent, quantum-mechanical states. The crucial point is that we may take the number of quantum-mechanical states associated with a region of area $dx \, dp_x$ to be the same for all regions of equal area in the two-dimensional phase space, regardless of where the regions are located.

Now we are ready to pass to the classical limit. Given that a quantum-mechanical treatment of the statistical situation would lead to the quantum-mechanical form of the canonical distribution, we may say that the probability $\mathcal{P}(x, p_x)dx \, dp_x$ is proportional to the product of two factors.

The first factor is the probability, from the canonical distribution, of any specific state in the region around x, p_x. That probability is proportional to

$$e^{-\beta E(x, \, p_x)}.$$

The classical energy expression $E(x, p_x)$ replaces the energy eigenvalue of the quantum-mechanical treatment.

The second factor is the number of distinct quantum-mechanical states in the region $dx \, dp_x$ of the two-dimensional phase space. That number is given by

$$\frac{dx \, dp_x}{O(\hbar)}.$$

Upon combining the two factors, we may write

$$\mathcal{P}(x, p_x) \, dx \, dp_x \propto e^{-\beta E(x, \, p_x)} \times \frac{dx \, dp_x}{O(\hbar)}. \qquad (6.3.2)$$

To establish an equality, we note that the particle, upon inspection, must be found with position and momentum placing it in some area of the two-dimensional phase space. A set of inferences about various small regions, provided the regions do not overlap and do, when summed, cover the entire phase space, is a set of mutually exclusive and exhaustive inferences on the hypothesis that the particle must be somewhere with some momentum. Hence a form of the normalization condition must hold. The sole difference from previous considerations is that the sum over inferences is a sum over small regions $dx\, dp_x$, most naturally performed as an integration. The probabilities must, therefore, satisfy a normalization condition of the form

$$\int_{-\infty}^{\infty} \int_{-\infty}^{\infty} \mathcal{P}(x, p_x)\, dx\, dp_x = 1. \tag{6.3.3}$$

Upon returning to equation (6.3.2) and introducing a proportionality constant C, we write

$$\mathcal{P}(x, p_x)\, dx\, dp_x = \frac{Ce^{-\beta E(x,\, p_x)}}{O(\hbar)}\, dx\, dp_x \tag{6.3.4}$$

and determine the constant C from the condition

$$\iint \frac{Ce^{-\beta E(x,\, p_x)}}{O(\hbar)}\, dx\, dp_x = 1,$$

with the solution

$$C = \left[\iint \frac{e^{-\beta E(x,\, p_x)}}{O(\hbar)}\, dx\, dp_x\right]^{-1}.$$

The expression for C may be inserted in equation (6.3.4) to yield the final result:

$$\mathcal{P}(x, p_x)\, dx\, dp_x = \frac{e^{-\beta E(x,\, p_x)}}{\displaystyle\int_{\text{phase space}} e^{-\beta E(x',\, p'_x)}\, dx'\, dp'_x}\, dx\, dp_x. \tag{6.3.5}$$

The point most worthy of note is that Planck's constant has disappeared from the final expression for the probability, as it must in the classical limit. The change in the notation for the normalization factor in the denominator is merely a matter of convenience. Only an explicit form for the classical energy $E(x, p_x)$, which will depend on the specific physical problem, is required to complete the determination of the probability in the classical limit.

If we extend the preceding ideas to a single particle free to move in three dimensions, we must consider a phase space of six dimensions, three position-component axes and three momentum-component axes. A single point in the six-dimensional space provides the classical specification of the particle's position and momentum. Needless to say, this severely taxes one's power of visualization, but it is the natural generalization. For small spatial ranges dx, dy, dz, and small momentum ranges dp_x, dp_y, dp_z, there is a "volume" of magnitude

$$dx \, dy \, dz \, dp_x \, dp_y \, dp_z .$$

A counting of quantum-mechanical states in the volume proceeds as before. With each pair of the form $dx \, dp_x$, we associate one (inverse) factor of $O(\hbar)$. We may say that the small volume of six-dimensional phase space contains

$$\frac{dx \, dp_x}{O(\hbar)} \cdot \frac{dy \, dp_y}{O(\hbar)} \cdot \frac{dz \, dp_z}{O(\hbar)} = \frac{d^3x \, d^3p}{O(\hbar)^3}$$

distinct quantum-mechanical states in the classical limit. For the expression on the right the convenient abbreviations

$$d^3x \equiv dx \, dy \, dz, \qquad d^3p \equiv dp_x \, dp_y \, dp_z ,$$

have been introduced.

Now we may frame and answer the following question. Given that the canonical distribution is appropriate for the quantum-mechanical statistical description of a single particle free to move in three dimensions, what is the probability of finding the particle with position variable in a small (three-dimensional) volume d^3x around $\mathbf{x}$ and momentum variable in a small (three-dimensional) volume d^3p around $\mathbf{p}$? Proceeding by analogy with the one-dimensional analysis, we write the probability as $\mathcal{P}(\mathbf{x}, \mathbf{p}) \, d^3x \, d^3p$, and relate it to the classical energy$(\mathbf{x}, \mathbf{p})$ by

$$\mathcal{P}(\mathbf{x}, \mathbf{p}) \, d^3x \, d^3p = \frac{e^{-\beta E(\mathbf{x}, \mathbf{p})}}{\int e^{-\beta E(\mathbf{x'}, \mathbf{p'})} \, d^3x' \, d^3p'} \, d^3x \, d^3p. \qquad (6.3.6)$$

The extension to two particles proceeds in much the same fashion. We consider a phase space of 12 dimensions, three position-component axes and three momentum-component axes for each of the two particles. A probability refers now to a small "volume" $d^3x_1 \, d^3p_1 \, d^3x_2 \, d^3p_2$ in the full 12-dimensional phase space. Here the two particles have been labeled as "particle 1" and "particle 2." The classical energy is now a function of the variables for both particles:

$$E = E(\mathbf{x}_1, \mathbf{p}_1, \mathbf{x}_2, \mathbf{p}_2).$$

The prose statement of a question about probabilities becomes much more cumbersome, but it would go like this: what is the probability, in the classical limit, of finding particle 1 with position variable in d^3x_1 around $\mathbf{x}_1$ and momentum variable in d^3p_1 around $\mathbf{p}_1$ and, simultaneously, of finding particle 2 with position variable in d^3x_2 around $\mathbf{x}_2$ and momentum variable in d^3p_2 around $\mathbf{p}_2$? With an evident interpretation for the symbols, the reply would take the form:

$$\mathcal{P}(\mathbf{x}_1, \mathbf{p}_1, \mathbf{x}_2, \mathbf{p}_2)\, d^3x_1\, d^3p_1\, d^3x_2\, d^3p_2$$

$$= \frac{e^{-\beta E(\mathbf{x}_1, \mathbf{p}_1', \mathbf{x}_2', \mathbf{p}_2)}}{\displaystyle\int_{\substack{\text{all} \\ \text{phase} \\ \text{space}}} e^{-\beta E}\, d^3x_1' \cdots}\, d^3x_1\, d^3p_1\, d^3x_2\, d^3p_2.$$

The assumption that one may meaningfully distinguish particle 1 from particle 2 is implicit in both question and answer. If the two particles are a helium atom and a neon atom, say, the assumption is certainly justified by the difference in mass (among other differences). If the two particles were two helium atoms, meaningful distinguishability would be, at best, questionable. This introduces a subtle point in the association of regions of phase space with quantum-mechanical states in the classical limit. In later chapters we will need to discuss this point in some detail. For now we can rest with merely raising the issue, for nothing in the remainder of this chapter depends on it. We would do well to see how the classical calculations go in simple cases before we delve into subtleties.

The extension to N particles follows in an analogous manner, by the enlargement of the phase space to $2 \times 3 \times N = 6N$ dimensions. The probability density depends on $6N$ variables because the classical energy in the exponential does, and the probability itself contains a factor with $6N$ infinitesimal ranges.

We should, at this point, examine another question of principle. In classical statistical mechanics one often talks about the probability distribution for a single particle in equilibrium at some specified temperature. In Section 4.3 we noted that one cannot ascribe a temperature to a particle in the way that one can ascribe a position or an energy, for the concept of temperature is primarily a macroscopic notion. A single particle can be "in equilibrium at some specified temperature" only if it is part of a macroscopic system with those properties. This observation implies that the logical starting point for a calculation of probabilities is with the probability distribution for the entire system of N particles. The derivation of the distribution for the single particle then proceeds by the reduction process of Section 2.11. Starting with the probability distribution for the N particles, one fixes the variables referring to the chosen single particle and integrates over the variables referring to all others. The result will be a probability distribution for the chosen particle.

The logical necessity for such a conceptual procedure is well worth bearing in mind: we can justify a use of the canonical distribution for a single particle *only if* that particle is an integral element of some macroscopic system and *only if* reduction from the probability distribution for the entire system does lead to the single-particle version of the canonical distribution. Moreover, when we say that the particle is in equilibrium, the statement can mean only that the particle is part of a system that is in equilibrium. We should not be surprised to find, for example, that a diatomic molecule "in equilibrium" will typically exhibit vibration of its constituent atoms and over-all rotational motion. These are not characteristics of equilibrium in the strict mechanical sense, but they are typical of what one may call "thermal equilibrium."

Significantly, these remarks indicate that in the development of this chapter we should really have worked from the N-particle probability distribution down to that for one particle constrained to one-dimensional motion. The adopted course is, however, far easier to follow.

Before we apply the classical theory, there is another point worthy of mention. We must take into account the possibility that a particle has an intrinsic angular momentum: a spin. This is a purely quantum-mechanical phenomenon and finds no real analog in classical physics. If a particle has a spin $S\hbar$ (where S will necessarily be zero, a positive integer or one-half a positive integer), there are $2S + 1$ orientations possible for the spin relative to some fixed axis. Thus, for each state of motion of the particle, there are $2S + 1$ different spin states, and we must include this multiplicity. In specifying a probability, we must specify the orientation of the spin, and in the normalizing factor that appears in the denominator, we must introduce an explicit summation over the spin states of each particle, bearing in mind that the "classical" energy may have a part dependent on the spin orientation (relative to an external magnetic field, say). This is, however, not an issue of pressing concern.

6.4. THE MAXWELL VELOCITY DISTRIBUTION

Let us focus our attention on a single molecule in a box containing many molecules (identical or not) in equilibrium at a temperature T. If we specify that the gas is quite dilute, we may avoid the problem of dealing explicitly, at this stage, with the inevitable intermolecular forces. We confine ourselves to an analysis of the motion and position of the molecule's center of mass. Thus we can omit from consideration the internal structure of the molecule as well as the spin (if any). For such a molecule we expect that the canonical distribution in its classical form will provide a valid basis for assessing probabilities.

The primary single-particle probability is the probability that the molecule's position is within an infinitesimal region d^3x around the position $\mathbf{x}$ and that its momentum is within a region d^3p around the value $\mathbf{p}$. We carry over the canonical form for this probability from equation (6.3.6):

$$\mathcal{P}(\mathbf{x}, \mathbf{p})\, d^3x\, d^3p = \frac{e^{-\beta E(\mathbf{x}, \mathbf{p})}}{\int e^{-\beta E(\mathbf{x}', \mathbf{p}')}\, d^3x'\, d^3p'}\, d^3x\, d^3p. \qquad (6.4.1)$$

There remains the question of the classical energy associated with the molecule's position and center-of-mass motion. For a molecule of mass m confined to a box, the energy is the sum of a kinetic and a potential part:

$$E(\mathbf{x}, \mathbf{p}) = \frac{p^2}{2m} + U_{box}(x, y, z). \qquad (6.4.2)$$

The function U_{box} represents the potential energy associated with the walls. It is zero when the molecule's center of mass is inside the box and infinite when that center is considered to be outside. The infinitely sharp rise at the walls represents the force that the walls exert to confine the molecule to the box. (Problem 6.9 offers an assist in investigating the consequences of including a gravitational potential energy.)

To tidy up the expression for the probability, we should calculate the normalization integral in the denominator. Despite its being a six-dimensional integral, it is not difficult to evaluate. First we hold the momentum variables fixed and integrate over all "ordinary" space. The integrand is then constant within the box (with $U_{box} = 0$) and vanishes outside, because U_{box} takes on a positive infinite value there. The integration produces a factor of V, the volume of the container:

$$\int_{\substack{\text{all phase}\\ \text{space}}} \exp\left[-\beta\left(\frac{p^2}{2m} + U_{box}\right)\right] d^3x\, d^3p = \iint_{box} \exp\left[-\beta\left(\frac{p^2}{2m} + 0\right)\right] d^3x\, d^3p$$

$$= V \int_{\substack{\text{momentum}\\ \text{space}}} e^{-\beta p^2/2m}\, d^3p.$$

The new integrand can be factored, and the three-dimensional integral over all momentum space split into the product of three identical one-dimensional integrals:

$$V \int e^{-\beta p^2/2m} \, d^3p = V \int e^{-\beta(p_x^2 + p_y^2 + p_z^2)/2m} \, dp_x \, dp_y \, dp_z$$

$$= V \left(\int_{-\infty}^{\infty} e^{-\beta p_x^2/2m} \, dp_x \right) \left(\int_{-\infty}^{\infty} e^{-\beta p_y^2/2m} \, dp_y \right) \left(\int_{-\infty}^{\infty} e^{-\beta p_z^2/2m} \, dp_z \right)$$

$$= V \left(\sqrt{\frac{2\pi m}{\beta}} \right)^3 .$$

The last step follows with the aid of the table of integrals in Appendix A. We can insert this result into equation (6.4.1) and arrive at a tidy form for the probability:

$$\mathcal{P}(\mathbf{x}, \mathbf{p}) \, d^3x \, d^3p = \frac{1}{V} \left(\frac{\beta}{2\pi m} \right)^{3/2} \exp \left[-\beta \left(\frac{p^2}{2m} + U_{box} \right) \right] d^3x \, d^3p. \quad (6.43)$$

As we might have expected, the spatial dependence asserts that regions of equal volume within the box are equally probable; those outside are assigned zero probability. The momentum distribution has much more to offer. We can facilitate discussion of it by integrating over all possible positions to arrive at a probability distribution purely for the momentum:

$$\mathcal{P}(\mathbf{p}) \, d^3p = \int_{\substack{\text{ordinary} \\ \text{space}}} \mathcal{P}(\mathbf{x}, \mathbf{p}) \, d^3x \, d^3p = \left(\frac{\beta}{2\pi m} \right)^{3/2} e^{-\beta p^2/2m} \, d^3p.$$

Let us calculate, with the momentum probability distribution, the expectation value of the kinetic energy, though the outcome can hardly be in doubt. Some algebra can be avoided by noting that the symmetry of the probability distribution implies the equalities

$$\langle p_x^2 \rangle = \langle p_y^2 \rangle = \langle p_z^2 \rangle$$

and hence

$$\left\langle \frac{p^2}{2m} \right\rangle = 3 \left\langle \frac{p_x^2}{2m} \right\rangle .$$

So the problem is reducible to that of calculating $\langle p_x^2/2m \rangle$:

$$\left\langle \frac{p_x^2}{2m} \right\rangle = \int \frac{p_x^2}{2m} \mathcal{P}(\mathbf{p}) \, d^3 p$$

$$= \left(\frac{\beta}{2\pi m} \right)^{3/2} \left(\int_{-\infty}^{\infty} \frac{p_x^2}{2m} e^{-\beta p_x^2/2m} \, dp_x \right) \left(\int_{-\infty}^{\infty} e^{-\beta p_y^2/2m} \, dp_y \right)$$

$$\times \left(\int_{-\infty}^{\infty} e^{-\beta p_z^2/2m} \, dp_z \right)$$

$$= \left(\frac{\beta}{2\pi m} \right)^{3/2} \left(\frac{\pi m}{2\beta^3} \right)^{1/2} \left(\frac{2\pi m}{\beta} \right)^{1/2} \left(\frac{2\pi m}{\beta} \right)^{1/2} = \frac{1}{2} \frac{1}{\beta} = \frac{1}{2} kT,$$

again with the aid of Appendix A. We arrive at the anticipated result:

$$\left\langle \frac{p^2}{2m} \right\rangle = \frac{3}{2} kT.$$

Though expected, the result is nonetheless a favorable outcome of what we might regard as a test of the classical phase-space scheme for doing statistical problems. To be sure, in the quantum-mechanical calculation of Section 4.6 we did use the equation $\langle E \rangle = \frac{3}{2}kT$ to determine the value of the parameter β. That, however, was a quantum-mechanical calculation, even though to complete it we were forced to replace—justifiably in that case—a summation by an integration. In Section 6.1 we examined the reasons why we would expect a classical calculation to give valid results for a dilute gas. Now we have support for the correctness of the classical phase-space scheme (as a sometimes appropriate limit of quantum mechanics).

We tend to feel more at home when talking about velocities than about momenta. Let us transform the momentum probability distribution into a velocity distribution. For the probability that the molecule has velocity in a region $d^3 v$ around $\mathbf{v}$ we write $f(\mathbf{v})d^3 v$. Since momentum and velocity are related by the equation $\mathbf{p} = m\mathbf{v}$, the connection between the probability densities $\mathcal{P}(\mathbf{p})$ and $f(\mathbf{v})$ should be rather simple. Except for the normalization, we could reliably guess the connection, and then correct the normalization, if necessary. But there is a systematic procedure.

To each region $d^3 v$ in velocity space, there corresponds a unique region $d^3 p$ in momentum space. The sizes are related as

$$d^3 p = dp_x \, dp_y \, dp_z = (m dv_x)(m dv_y)(m dv_z) = m^3 d^3 v. \qquad (6.4.4)$$

The "volumes" of the two regions are related by a scale factor of m^3. The probability that the molecule is to be found with velocity in the region $d^3 v$ must be the same as the probability that it is to be found with momentum in the uniquely corresponding region $d^3 p$. This implies the equality

$$f(\mathbf{v})d^3 v = \mathcal{P}(\mathbf{p})d^3 p$$

for *corresponding* regions d^3v and d^3p, around $\mathbf{v}$ and $\mathbf{p}$, respectively. If we substitute for d^3p from the relation (6.4.4), we find

$$(\mathbf{v})\, d^3v = \mathcal{P}(m\mathbf{v})m^3\, d^3v$$

$$= \left(\frac{\beta m}{2\pi}\right)^{3/2} e^{-\beta mv^2/2}\, d^3v. \tag{6.4.5}$$

This is the justly celebrated Maxwell probability distribution for the molecular velocity. James Clerk Maxwell, even better known as the man who put electromagnetism on a solid mathematical foundation, derived a result of this form, when multiplied by the number N of molecules, in 1859.

> In his paper,* Maxwell proposed to analyze the state of a gas after the particles had been colliding for "a certain time." He noted that then "the average number of particles whose velocity lies between certain limits [is] ascertainable, though the velocity of each particle changes at every collision," and he set himself the task of finding "the average number of particles whose velocities lie between given limits, after a great number of collisions among a great number of equal particles." By "average" Maxwell meant—it means clear—some kind of time average, and his actual calculation was based on clever symmetry assumptions, one of which even he was dubious about. Thus both Maxwell's approach and the precise meaning of his result are different from ours. An exploration of the differences and connections is initiated in problems 6.13 through 6.15.

Since the probability density of the Maxwell velocity distribution depends on merely the magnitude v of $\mathbf{v}$, that is, on the speed, we can readily compute a probability distribution for speed alone. The general idea is once again a use of the reduction procedure developed in Section 2.11. To compute the speed distribution we need only integrate over all directions for a fixed magnitude of the velocity. For the speed range v to $v + dv$, this means an integration over a thin spherical shell in velocity space of radius v, thickness dv, and hence volume $4\pi v^2 dv$. This is illustrated in figure 6.4.1. Invoking the reduction procedure, we write

$$\begin{pmatrix} \text{the probability that} \\ \text{the molecule has a} \\ \text{speed in the range} \\ v \text{ to } v + dv \end{pmatrix} \equiv \tilde{f}(v)\, dv = \int_{\substack{\text{thin} \\ \text{shell}}} f(\mathbf{v})\, d^3v$$

$$= 4\pi\left(\frac{\beta m}{2\pi}\right)^{3/2} e^{-\beta mv^2/2} v^2\, dv. \tag{6.4.6}$$

---

* J. C. Maxwell, *Philosophical Magazine*, **19**, 19 (1860).

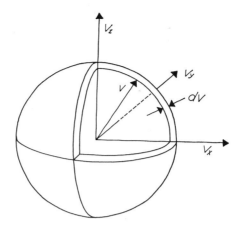

FIGURE 6.4.1
A thin shell in the velocity space.

The introduction of a symbol $\tilde{f}(v)$ for the speed probability density is a matter of convenience.

Over the range $0 \leqslant v \leqslant \infty$, the explicit expression for the speed probability distribution is correctly normalized:

$$\int_0^\infty \tilde{f}(v)\, dv = 1.$$

A word or two about the upper limit is in order. Special relativity theory emphatically denies that a material particle may have a speed equal to that of light, and it is generally assumed that speeds greater than c, the speed of light, are also forbidden. Properly, therefore, the probability of a speed $v \geqslant c$ should be zero for a particle with non-zero rest mass. That it isn't zero here is a result of our using nonrelativistic, as well as classical, expressions for the energy. (The canonical distribution holds for the relativistic domain as well as the nonrelativistic. One must merely use the correct relativistic expressions for the energy eigenvalues.) For gases in other than astrophysical situations, the error is seldom a worry, for the exponential ensures that the "probability" of a speed greater than c is extremely small.

With a little effort we can estimate the "probability" of a speed greater than c for an air molecule at room temperature. The calculation entails an integration from c to infinity:

$$\begin{pmatrix} \text{the "probability"} \\ \text{of } v \geqslant c \end{pmatrix} = \int_c^\infty \tilde{f}(v)\, dv = 4\pi \left(\frac{\beta m}{2\pi}\right)^{3/2} \int_c^\infty e^{-\beta m v^2/2} v^2\, dv.$$

For $v = c$, the numerical value of the exponent is this:

$$-\beta m c^2/2 = -\frac{mc}{2kT} \simeq -5.6 \times 10^{11}.$$

The exponent is so large—and negative—that the integrand at $v = c$ is practically zero and diminishes extremely rapidly despite a multiplicative factor of v^2. The integral cannot be evaluated in closed form, but we can afford to be quite casual in approximating it. To integrate the exponential, one would like $v\,dv$, but $v^2\,dv$ is troublesome. So let us set one of the factors of v equal to c and integrate with the other:

$$\int_c^\infty e^{-\beta m v^2/2} v^2 \, dv \simeq c \int_c^\infty e^{-\beta m v^2/2} v \, dv = \frac{c}{\beta m} e^{-\beta m c^2/2}.$$

Upon using this approximation and then inserting numerical values, we find

$$\int_c^\infty \tilde{f}(v) \, dv \simeq 4\pi \left(\frac{\beta m}{2\pi}\right)^{3/2} \frac{c}{\beta m} e^{-\beta m c^2/2}$$

$$\simeq 10^5 \times e^{-5.6 \times 10^{11}} \simeq 10^{-(10^{11})}.$$

By almost anyone's standards, this "probability" is negligibly small.

The speed probability density is shown in figure 6.4.2. It exhibits a gentle rise from zero at $v = 0$ to a maximum and then declines exponentially,

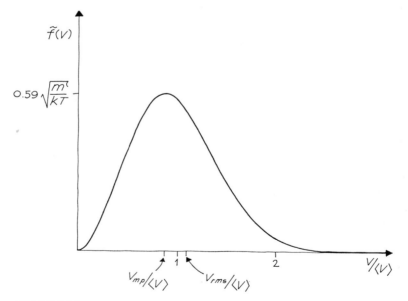

FIGURE 6.4.2
The Maxwell speed distribution. A natural scale, $\langle v \rangle = (8kT/\pi m)^{1/2}$, has been used for the speed axis.

approaching zero only asymptotically. The source of the initial rise is the rapid increase in the number of quantum states with speed in the range v to $v + dv$, as v itself grows.

The value "most probable" speed is the speed for which $\tilde{f}(v)$ reaches its maximum. To determine it, we write

$$\frac{d}{dv}\,\tilde{f}(v) = 4\pi \left(\frac{\beta m}{2\pi}\right)^{3/2} [2v + (-\beta mv)v^2]e^{-\beta mv^2/2} = 0.$$

Let us denote the most probable speed by the symbol v_{mp}. We find for that speed the explicit expression

$$v_{mp} = \sqrt{\frac{2}{\beta m}} = \sqrt{\frac{2kT}{m}}.$$

We can extract another characteristic speed, this time from the expectation value of the kinetic energy. Calling that the root mean square speed and denoting it by v_{rms}, we find

$$v_{rms} \equiv \langle v^2 \rangle^{1/2} = \left(\frac{2}{m}\,\langle \tfrac{1}{2}mv^2 \rangle\right)^{1/2} = \sqrt{\frac{3kT}{m}}.$$

(We could compute v_{rms} directly from the speed probability distribution. Using the already known expectation value of the kinetic energy avoids the repetition of some integrations.) The two speeds are related by the ratio

$$\frac{v_{rms}}{v_{mp}} = \sqrt{\frac{3}{2}} \simeq 1.224.$$

We have no reason to expect the same numerical value for the two speeds, for we have asked of the probability distribution different questions, though we would understandably be disturbed if the values were radically different.

Substitution of numerical values for air at room temperature yields

$$v_{rms} \simeq \begin{cases} 5 \times 10^4 \text{ cm/sec,} \\ 10^3 \text{ miles/hour.} \end{cases}$$

This characteristic speed compares favorably with the observed speed of sound in air, as indeed it should (to within a factor of 2 or so).

6.5.   EXPERIMENTAL EVIDENCE FOR
THE MAXWELL DISTRIBUTION

Probably A. A. Michelson provided the first direct experimental support for the Maxwell velocity distribution. In his search "for a radiation of sufficient homogeneity to serve as an ultimate standard of length,"* Michelson examined the spread in wavelength of spectral lines from some half-dozen chemical elements with his justly famous interferometric techniques. The light from a given atomic transition showed an approximately Gaussian spread in wavelength around a central wavelength. The central wavelength may be looked upon as characteristic of the transition. For the spread in the light from his gaseous samples at high temperature, Michelson suggested two reasons: Doppler broadening and collision broadening.

Since the emitting atoms were certainly not standing still, their motion along the line of sight gave rise to Doppler shifts of the frequency (and hence wavelength) of the observed light. In relativistic terms, the atom emits with its characteristic frequency in its rest frame, but the light has a different frequency when observed in the laboratory frame, with respect to which the atom is moving. Analysis with the Maxwell velocity distribution (outlined in problem 6.5) leads to a Gaussian spread around the characteristic wavelength. The width of the spread will depend on the temperature and particle mass.

Collision broadening is much more directly connected with the emission process. The collision of one atom with another, while the first is in the process of emitting a photon, can interrupt the emission process. A typical time during which an atom emits is of order 10^{-8} seconds. This can be a long time relative to the time between successive collisions. (For air at room temperature and pressure, the latter is a few times 10^{-10} seconds.) A result of the interruptions by collision is a statistical spread in the energy of the emitted photons and hence in the wavelengths.

Michelson observed the red hydrogen line (6563 Ångstroms) at successively lower pressures, thus reducing the density and hence the contribution to the spread due to collision broadening. The extrapolated (to zero pressure) value of the spread he attributed primarily to Doppler broadening. He examined spectral lines from a host of other gases as well.

---

* This and the subsequent quotation are from A. A. Michelson, *Philosophical Magazine*, **34**, 280 (1892).

A quotation from the 1892 paper of this master of optical experiments gives Michelson's conclusions:

> The accordance between the measured widths of eighteen lines shows, further, that this broadening of lines in a rare gas can be fully accounted for by the application of Döppler's principle to the motion of the vibrating atoms in the line of sight, and indeed furnishes what may be considered one of the most direct proofs of the kinetic energy of gases.
>
> The form of the ultimate components of all the groups of spectral lines thus far examined is found to agree fairly well with an exponential curve [The equation given at this point indicates that a Gaussian curve is meant.] which shows that the distribution of velocities cannot vary widely from that demanded by Maxwell's theory.

In the following years, other, still more direct tests of the theory were made, most notably that of Otto Stern in 1920 with a beam of silver atoms. The classic experiment is perhaps that performed by I. F. Zartman at the University of California, Berkeley, and published in 1931 as "A Direct Measurement of Molecular Velocities."*

The essence of Zartman's experiment is indicated schematically in figure 6.5.1. Zartman heated metallic bismuth in a small oven. This produced a gas

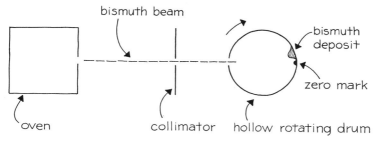

FIGURE 6.5.1

A schematic drawing of Zartman's apparatus for determining the probability distribution for speeds in a bismuth beam. The region through which the beam passed was highly evacuated. The drum is shown in the momentary position in which a bunch of particles is admitted.

of bismuth at a controlled and monitored temperature. Some of the bismuth gas emerged from a small slit (0.05 mm in width by 10 mm in length) in the oven wall. The emergent gas was formed into a bismuth beam by a second, collimating slit. The primary element of the detection apparatus consisted of

---

* I. F. Zartman, *Physical Review*, **37**, 383 (1931).

a rapidly rotating hollow drum with another knife-edge slit. A bunch of particles from the beam was admitted to the drum once every revolution, as the drum slit swept across the collimated beam from the oven. As the bunch moved across the diameter of the revolving drum, it spread out, the faster particles reaching the far side earlier than the slower ones. Thus the particles in the bunch were distributed over the glass-plate receiver on the far side in accordance with their speeds, the faster ones being closer to the zero mark (infinite speed) than the slower ones. The zero mark was established by holding the drum fixed with slits aligned at the start of each experimental run and allowing some particles to pass. The density of deposited bismuth on the glass plate, as a function of distance along the plate, could then be compared with a theoretical prediction from the Maxwell velocity distribution. In Zartman's words: "The maximum of the deposit first becomes visible in from three to six hours depending upon the temperature and speed of the run [rate of revolution of the drum]. The run is continued for periods ranging from eight to twenty-two hours."

Because of complications arising from the finite slit widths, the theoretical calculation is not trivial, but the speed distribution within the beam is readily calculated. Within the *oven* the theoretical speed distribution for a single particle is that of equation (6.4.6). To determine the distribution of speeds within the *beam* itself, we consider the region within the oven in which a particle could be and still join the emergent beam within unit time. This is most easily done by "looking back" from just before the drum slit through the two preceding slits into the oven. If a particle is to emerge and make it through all the slits, it must be in a wafer-like region in the oven. If the particle is to emerge within unit time, the length of the wafer-like region in which it may be is proportional to its speed. This enhances the probability of finding high-speed particles in the beam over that for low-speed particles, and does so by a factor of v. For the probability $\mathcal{P}_b(v)dv$ that a particle in the *beam* has a speed between v and $v + dv$, we have

$$\mathcal{P}_b(v)\,dv = \tfrac{1}{2}(m\beta)^2 e^{-\beta m v^2/2} v^3\,dv.$$

Because v^3 has replaced v^2, new constant factors were needed for normalization to unity.

The steps from here to the theoretical distribution on the glass receiving plate are of two kinds: (1) a determination of where a particle with given speed hits the receiver—the distance from the zero mark will be inversely proportional to the speed—and (2) corrections for the finite width of the slits.

Figure 6.5.2 presents a reproduction from Zartman's paper of data for bismuth at a temperature of 1124 K and at two different rates of revolution of the drum. Note that the distance from the zero mark increases to the right, while speed *decreases* to the right. Zartman found that he was unable to fit a theoretical curve, calculated from the Maxwell distribution, to the data if

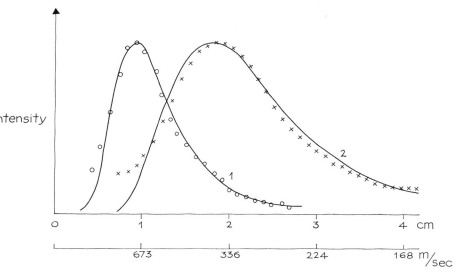

FIGURE 6.5.2

A figure from Zartman's paper (*Physical Review*, **37**, 383, 1931) presenting theoretical and experimental intensity distributions on the receiving plate. The theoretical distributions are based on the Maxwell velocity distribution and the assumption of a vapor composition of 40% Bi and 60% Bi_2. Zartman gives the temperature of the oven as 851°C (about 1124 K). For curve (1) the drum revolution rate was 120.7 revolutions per second; for curve (2), 241.4 rps. The horizontal scale gives distance from the zero mark and also the molecular speed appropriate to curve (2).

he assumed that the beam consisted of bismuth atoms, Bi. A large admixture of bismuth "molecules," Bi_2, had to be assumed. (With their larger mass, the Bi_2 particles will have a smaller expectation value for the speed and so will tend to hit at larger distances from the zero mark than will bismuth atoms.) The theoretical curves are for a mixture of 40 per cent Bi and 60 per cent Bi_2 in the bismuth beam. This amounts to a fit to the data with (only) two adjustable parameters: the Bi_2/Bi ratio and the overall intensity. Again in Zartman's words:

> In all cases the experimental results would not agree with the theoretical distribution for Bi or Bi_2 as given by Eq. (8) [Zartman's theoretical calculation from the Maxwell velocity distribution]. The work of Leu on the magnetic

deflection of bismuth suggests that the beam is composed of Bi and Bi_2. On the basis of this suggestion various percentages of Bi and Bi_2 were assumed and using Eq. (8) the resultant intensity distributions plotted. These distributions were then compared with the experimental results. The theoretical distribution which best fitted the experimental results is taken as the composition of the vapor stream.

In [figure 6.4.2] the solid lines represent the theoretical distributions for the conditions of operation given in [Zartman's] Table I. The circles represent the experimental results obtained from Run 1, and the crosses those from Run 2. The experimental points are in close agreement (within the experimental error) with the theoretically derived curve except for a few points on the high velocity side. These points disagree by an amount greater than the experimental error. The disagreement is probably due to the lack of a sharply defined molecular beam during the entire course of the experiment. It may also be due to a "slipping" of the molecules forming the zero mark. The possibility is being investigated.

Though giving good support to the Maxwell distribution, Zartman's work does leave one with the desire for a more clearcut examination. Experiments by R. C. Miller and P. Kusch, reported in a paper entitled "Velocity Distributions in Potassium and Thallium Atomic Beams,"[*] provide a major step in that direction. Again a beam was produced by an oven, but the scheme for determining the velocity distribution was quite different. A solid rotating cylinder was placed in the path of the beam, with its rotational axis parallel to the beam. Had it not been for some helical slots cut along the curved sides of the cylinder, the beam would have been completely stopped. For a given angular velocity of the cylinder, most particles entering a slot and proceeding straight ahead collided with the slot walls and were lost from the beam. Particles with speed in a narrow range determined by the geometry and angular velocity of the cylinder were, however, able to pass through. For them the slot continually rotated to an "open" position as they proceeded in a straight line toward the detector. A measurement of the beam intensity at the detector as a function of the angular velocity of the cylinder was tantamount to a measurement of beam intensity as a function of the speed of the particles in the beam.

Miller and Kusch tested separately (by magnetic resonance methods) for the presence of "molecules" such as had plagued Zartman's experiment. The independent experiment indicated less than 1 per cent in the beam (and probably less even than that).

_____

[*] R. C. Miller and P. Kusch, *Physical Review*, **99**, 1314 (1955).

The location of the peak in the curve of experimental intensity versus speed agreed to within 0.5 per cent with the Maxwellian calculation for the measured oven temperature. For further comparison, the data and theory must be reduced to a common intensity scale, by imposing equal heights for the peaks, say. The speeds considered were in the range 0.3 to 2.5 times the most probable speed in the oven. For this entire range, the intensities from the experimental points and those from the theoretical curve did not differ by more than 1 per cent of the maximum intensity.

6.6. THE EQUIPARTITION THEOREM

The equipartition theorem is much overrated, but it does belong in every physicist's store of knowledge. The reason for my denigrating attitude is not that the theorem doesn't follow from the assumptions—it does follow—but rather than the assumptions are only seldom fulfilled in physical situations of interest.

The set of assumptions has three elements:

1. A classical treatment of the physical system, at equilibrium at a temperature T, is permissible and adequate. The description is in terms of some set of position variables $x_1, \ldots, x_{3N}$ and associated momentum variables $p_1, \ldots, p_{3N}$.

2. The classical expression for the total energy splits additively into two parts, with one part depending on only a single variable (p_i, say) and the other part being independent of that variable:

$$E = E'(p_i) + E'' \text{ (all the other variables)}.$$

3. The term $E'(p_i)$ depends *quadratically* on p_i:

$$E'(p_i) = ap_i^2.$$

The numerical coefficient a must be positive (to ensure convergence of the normalizing integral) but may otherwise remain unspecified.

The Equipartition Theorem: Provided the three assumptions just listed are met, the expectation value of $E'(p_i)$ is always $\frac{1}{2}kT$, regardless of the numerical value of the constant a and of the rest of the system:

$$\langle E' \rangle = \tfrac{1}{2}kT. \tag{6.6.1}$$

The proof takes only a few lines. With assumption 1 and the canonical distribution, we may write

$$\langle E' \rangle = \frac{\int E' e^{-\beta E} \, dp_i \, d(\text{others})}{\int e^{-\beta E} \, dp_i \, d(\text{others})} ,$$

where the integrals are over all phase space. Assumption 2 permits us to write E in the exponents as the special sum and then to factor both numerator and denominator:

$$\langle E' \rangle = \frac{\int E' e^{-\beta E'} \, dp_i \int e^{-\beta E''} \, d(\text{others})}{\int e^{-\beta E'} \, dp_i \int e^{-\beta E''} \, d(\text{others})} .$$

The factors on the right, top and bottom, are identical, and so we may cancel them. With assumption 3 about the form of E' invoked, the problem is reduced to that of calculating the ratio

$$\langle E' \rangle = \frac{\displaystyle\int_{-\infty}^{\infty} a p_i^2 e^{-\beta a p_i^2} \, dp_i}{\displaystyle\int_{-\infty}^{\infty} e^{-\beta a p_i^2} \, dp_i} .$$

Use of the integral table in Appendix A leads to

$$\langle E' \rangle = \frac{a\tfrac{1}{2}\sqrt{\pi}(\beta a)^{-3/2}}{\sqrt{\pi}(\beta a)^{-1/2}} = \frac{1}{2}\frac{1}{\beta} = \frac{1}{2} kT. \qquad \text{Q.E.D.}$$

The theorem does follow from the assumptions, and it can be an exceedingly helpful theorem. In Section 6.4 we expended considerable effort to compute $\langle p_x^2/2m \rangle$, eventually finding it equal to $\tfrac{1}{2}kT$. The equipartition theorem, once established with relatively little effort, gives the result immediately upon inspection of the problem.

Unfortunately, the theorem's domain of applicability is decidedly restricted. The crux of the matter is assumption 1, that a classical description is permissible and adequate. The one-dimensional harmonic oscillator provides a good model for an examination of the typical limitations. Let us denote the mass by m and the spring constant by κ. The classical expression for the energy is

$$E(x, p_x) = \frac{p_x^2}{2m} + \tfrac{1}{2}\kappa x^2 \qquad (6.6.2)$$

and fits nicely into the framework of the equipartition theorem.

In a diatomic molecule, vibrational motion of the two atoms about their equilibrium position has an associated potential energy rather like that of the one-dimensional harmonic oscillator. The vibration of atoms in a solid is rather like that of a three-dimensional harmonic oscillator, which can be split into three separate one-dimensional oscillators, one along each of three orthogonal axes. We have here a model with considerable physical interest. In estimating the energy of any such harmonic oscillator, the real question is whether a classical analysis is or is not valid.

With the equipartition theorem, the classical treatment is now simplicity itself:

$$\langle E \rangle_{\text{classical}} = \left\langle \frac{p_x^2}{2m} \right\rangle + \left\langle \tfrac{1}{2}\kappa x^2 \right\rangle$$

$$= \tfrac{1}{2}kT + \tfrac{1}{2}kT = kT. \tag{6.6.3}$$

For the quantum-mechanical treatment, one needs some knowledge of the energy eigenstates and associated energies of the one-dimensional harmonic oscillator. There are infinitely many energy eigenstates, each with a different energy. If, following convention, we arrange the states in order of increasing energy and label each with an index n, the associated energy eigenvalues are:

$$E_n = (n + \tfrac{1}{2})\, hv \quad n = 0, 1, 2, \ldots, \infty. \tag{6.6.4}$$

The symbol v denotes the classical frequency of vibration:

$$v = \frac{1}{2\pi}\sqrt{\frac{\kappa}{m}}. \tag{6.6.5}$$

With these data we can write down, and then evaluate, the quantum-mechanical expression for the expectation value of the energy:

$$\langle E \rangle_{\text{QM}} = \frac{\sum\limits_{n=0}^{\infty} E_n e^{-\beta E_n}}{\sum\limits_{n=0}^{\infty} e^{-\beta E_n}}. \tag{6.6.6}$$

The sum in the denominator can be calculated in closed form. First a convenient factorization:

$$\sum_{n=0}^{\infty} e^{-\beta E_n} = \sum_{n=0}^{\infty} e^{-\beta(n+1/2)hv} = e^{-\beta hv/2} \sum_{n=0}^{\infty} e^{-n\beta hv}.$$

The last summation is an infinite geometric series:

$$\sum_{n=0}^{\infty} e^{-n\beta hv} = 1 + e^{-\beta hv} + (e^{-\beta hv})^2 + \ldots = \frac{1}{1 - e^{-\beta hv}}.$$

The summation converges because $\exp(-\beta h\nu)$ is less than unity. Thus for the denominator we arrive at the expression

$$\sum_{n=0}^{\infty} e^{-\beta E_n} = \frac{e^{-\beta h\nu/2}}{1 - e^{-\beta h\nu}}.$$

The trick for evaluating the numerator is the observation that the numerator may be written as the partial derivative with respect to β of a summation like that just evaluated:

$$\sum_{n=0}^{\infty} E_n e^{-\beta E_n} = -\frac{\partial}{\partial \beta} \sum_{n=0}^{\infty} e^{-\beta E_n}.$$

When the algebraic smoke has cleared, the exact result emerges as

$$\langle E \rangle_{QM} = \tfrac{1}{2}h\nu + \frac{h\nu}{(e^{h\nu/kT} - 1)}. \tag{6.6.7}$$

This bears scant resemblance to the classical result.

Only in the limit

$$\frac{h\nu}{kT} \ll 1 \tag{6.6.8}$$

can one retrieve the classical value from the quantum-mechanical expression. When this inequality is satisfied, the denominator of the second term in equation (6.6.7) may be expanded as

$$e^{h\nu/kT} - 1 = \left[1 + \left(\frac{h\nu}{kT}\right) + \cdots \right] - 1 \simeq \frac{h\nu}{kT}$$

to the lowest nonvanishing order. The full expression becomes

$$\langle E \rangle_{QM} \Big|_{h\nu/kT \ll 1} \simeq \tfrac{1}{2}h\nu + \frac{h\nu}{(h\nu/kT)}$$

$$\simeq \tfrac{1}{2}h\nu + kT \simeq kT. \tag{6.6.9}$$

The neglect of $\tfrac{1}{2}h\nu$ in the last step is justified by the assumed inequality. The explicit dependence on ν (and hence on m and κ) disappears. In an appropriate *limit*, the classical result does emerge from the exact quantum-mechanical result.

For the harmonic oscillator, the inequality (6.6.8) provides the criterion for the attainment of the classical limit. When the inequality is satisfied, the

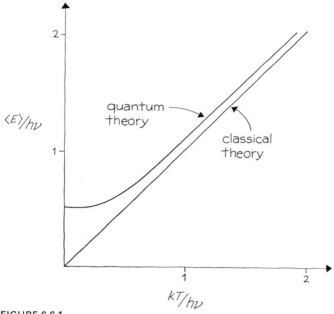

FIGURE 6.6.1

Classical versus quantum theory for the one-dimensional harmonic oscillator. The striking difference at (relatively) low temperatures is readily probed by experiment.

energy step hv between adjacent energy eigenvalues is small in comparison with a "typical" thermal energy kT. Then the discreteness of the energy levels, which go as $(n + \frac{1}{2})hv$, is of relatively little significance, and a classical calculation, which explicitly assumes a continuous range of energies smoothly dependent on x and p_x, is an adequate approximation.

Is the classical limit likely to apply when the harmonic oscillator is of physical interest? Spectroscopic data on the diatomic nitrogen molecule give a vibrational frequency of 7×10^{13} cycles per second as typical. This implies

$$hv \simeq \begin{cases} 4.6 \times 10^{-13} \text{ ergs,} \\ 0.3 \text{ electron volts.} \end{cases}$$

The comparison with kT at room temperature leads to

$$\frac{hv}{kT} \simeq 10.$$

The inequality needed to justify the classical result is distinctly not satisfied. For a nitrogen molecule at room temperature, the equipartition theorem fails to give a correct answer for the expectation value of the vibrational energy and for its temperature dependence because one of the basic premises does not hold, namely, a classical description is not valid.

The case with a solid is more subtle, for a host of frequencies in the range from essentially zero to the order of 10^{13} cycles per second are involved. For the lower frequencies the equipartition theorem will hold at room temperature, but its validity will at best be marginal for the higher frequencies. At low temperature the theorem is sure to lead to disaster. Only when Einstein (in 1907) and Debye (in 1912) applied quantum mechanics to the vibrational energy of a solid could any agreement between experiment and theory be established for low temperatures.

In summary, the equipartition theorem is a remarkable result of *classical* statistical mechanics, but its domain of applicability is severely restricted.

REFERENCES for Chapter 6

The discussion of classical statistical mechanics by Tolman in Chapter III of his *Principles* (cited in the references to Chapter 4) is, as always with him, both readable and profound. The required mathematical background is reviewed earlier in his book.

A wealth of reprinted papers on kinetic theory and heat, including Maxwell's paper, are to be found in S. G. Brush, ed., *Kinetic Theory* (New York: Pergamon, 1965), Vol. I. Brush has, in addition, provided a historical introduction and a summary of each paper.

PROBLEMS

6.1. Suppose the probability density for the speed s of a car on a highway is given by

$$\mathcal{P}(s) = Ase^{-s/s_0}, \quad 0 \leqslant s \leqslant \infty,$$

where A and s_0 are positive constants. More explicitly, $\mathcal{P}(s)\,ds$ gives the probability that a car has a speed between s and $s + ds$.

a. Determine A in terms of s_0.

b. What is the expectation value of the speed?

c. What is the "most probable" speed, that is, the speed for which the probability density has its maximum? Compare this with the expectation value.

d. What is the probability that a car has a speed more than three times as large as the expectation value of the speed?

6.2. a. Use the Maxwell speed probability distribution to compute the quantities $\langle v \rangle$ and $\langle v^2 \rangle^{1/2}$ in terms of kT, m, and pure numbers.

b. Each of the three quantities—$\langle v \rangle$, v_{rms}, and v_{mp}—may be considered a characteristic speed. Is there any reason for preferring one over the other two?

6.3. Estimate the numerical value of the most probable speed of an " air molecule " under typical room conditions. Would you expect a molecule to be able to travel about that far across a room in one second?

6.4. From the Maxwell velocity distribution, determine the probability distribution for the velocity component v_x. The values of v_y and v_z are to be considered irrelevant. Sketch the ensuing probability density. Do you recognize the profile? Then calculate $\langle v_x \rangle$ and $\langle v_x^2 \rangle$. Why is a zero value for the former quite reasonable?

6.5. A gas of atoms, each of mass m, is maintained at a temperature T inside an enclosure. The atoms emit light, some of which passes through a window of the enclosure, down a long collimating tube, and is then observed as a spectral line in a spectroscope. Let us take the direction of the collimating tube as the x-axis. A stationary atom would emit light at the sharply defined frequency v_0. Because of the Doppler effect, the frequency of the light observed from an atom having an x-component of velocity v_x is not simply equal to the frequency v_0 but is approximately given by

$$v = v_0 \left(1 + \frac{v_x}{c} \right),$$

where c is the speed of light. As a result, not all of the light arriving at the spectroscope is at the frequency v_0. Instead, the light is characterized by an intensity distribution $I(v) \, dv$ specifying the fraction of light intensity lying in the frequency range between v and $v + dv$.

a. Calculate the expectation value $\langle v \rangle$ of the frequency of the light observed in the spectroscope.

b. Calculate the root mean square frequency shift $(\Delta v)_{rms} = \langle (v - \langle v \rangle)^2 \rangle^{1/2}$ of the light observed in the spectroscope.

c. Determine the intensity distribution $I(v) \, dv$ of the light observed in the spectroscope. Does $I(v)$ have a Gaussian shape? How does the width depend on particle mass and temperature? Is the behavior of $I(v)$ for the very low and very high frequency extremes reasonable? Accurate?

6.6. The typical *relative* speed of two molecules, 1 and 2, in a dilute gas may well differ from the typical speed of each relative to the container. Make a root mean square kind of estimate of the relative speed by computing the quantity $\langle (\mathbf{v_1} - \mathbf{v_2})^2 \rangle^{1/2}$. Mentally constructing a velocity probability distribution for the *pair* of molecules, and noting that (given the canonical distribution for the entire system) the distribution for the pair factors nicely, should enable you to perform the calculation without doing any integrals. Compare your answer with v_{rms}.

6.7. a. Starting with the Maxwell speed probability distribution and the relation $\frac{1}{2}mv^2 = \varepsilon$, where ε is the energy of a single molecule, derive a probability distribution for the molecular energy. The quantity $\mathcal{P}(\varepsilon)\,d\varepsilon$ is to give the probability that a molecule has an energy between ε and $\varepsilon + d\varepsilon$.

b. Check your result in part (a) by using it to calculate $\langle\varepsilon\rangle$, for which you know the correct answer.

c. What is the "most probable" kinetic energy? Is that equal to $\frac{1}{2}m(v_{mp})^2$? Need it be?

6.8. Compute the expectation value of the kinetic energy of a particle in Zartman's *beam*. Why is this greater than $\frac{3}{2}kT$?

6.9. Particle in a gravitational field. The physical situation is the same as in Section 6.4, except that we include a constant gravitational force mg acting in the negative z direction. Take the box to have a height L and a cross-sectional area A.

a. Determine explicitly the new form of $\mathcal{P}(\mathbf{x}, \mathbf{p})$. (To start, you will need to add mgz to the previous energy expression.)

b. Check that your new form reduces to the old when you carefully take the limit $g \to 0$.

c. Calculate a probability distribution for height alone (by a reduction procedure), that is, for height regardless of the momentum or the "horizontal" position.

d. What kind of heights are needed for significant effects in the height probability distribution? (Centimeters? Kilometers?)

e. What is the expectation value of the gravitational potential energy? Is there any reason to expect it to be $\frac{1}{2}kT$? Why? What might one expect? Do you find it to be so, approximately, when the inequality $\beta mgL \ll 1$ holds?

6.10. Let us analyze a system of paramagnetic particles, each having a magnetic moment $\boldsymbol{\mu}$, with the classical version of the canonical probability distribution. If we neglect the mutual interactions we may satisfactorily focus on a single paramagnet. The magnetic moment $\boldsymbol{\mu}$ can, in classical theory, make any arbitrary angle θ with respect to a given direction, which we take to be the z-axis. In the absence of a magnetic field the probability that the orientation of the moment lies between θ and $\theta + d\theta$ is proportional to the solid angle $2\pi \sin\theta\,d\theta$ enclosed in this range, with some *constant* of proportionality. In the presence of a magnetic field H in the z direction, this probability must further be proportional to the factor $\exp[-\beta(-\mu H \cos\theta)]$, where $-\mu H \cos\theta$ is the classical magnetic energy of the moment making an angle θ with respect to the field.

a. Construct the properly normalized probability distribution $\mathcal{P}(\theta)\, d\theta$ giving the probability that the orientation of $\boldsymbol{\mu}$ lies between θ and $\theta + d\theta$ for the magnetic case.

b. Then compute, classically, the expectation value of the component of magnetic moment along the z-axis (the field direction).

c. Examine the high and low temperature limits of your results in part (b). In the high temperature case, determine the leading nonvanishing term in an expansion in powers of $\mu H/kT$.

6.11. Guaranteed failure of the equipartition theorem. Even without a detailed knowledge of the energy eigenvalues of the one-dimensional harmonic oscillator, one can show that the equipartition theorem must fail to give a correct result for the expectation value of the energy at low temperature. Show, by constructing an argument based on the Heisenberg uncertainty principle and on the classical energy expression displayed in equation (6.6.2), that the minimum energy of the oscillator must be of order $h\nu$. Then show that, for a low enough temperature, the equipartition theorem of *classical* statistical mechanics leads to a contradiction.

6.12. This problem employs the one-dimensional harmonic oscillator to illustrate the second restriction on the applicability of classical statistical mechanics (which was discussed in Section 6.1). For the "expectation value of the variation of the potential energy," let us take the square root of the classical expectation value of the square of $\partial(\frac{1}{2}\kappa x^2)/\partial x$. Here κ is the spring constant of Section 6.6. The squaring ensures that positive and negative variations count equally.

a. Show that the quantity that emerges is $(\kappa kT)^{1/2}$. Thus a classical analysis may be permissible if the thermal de Broglie wavelength times $(\kappa kT)^{1/2}$ is much less than kT.

b. By rearrangement, show that when the latter condition is satisfied, then the "classical limit" of Section 6.6 has indeed been attained.

6.13. Maxwell's way. This problem invites you to derive our results in equation (6.4.5), which is a *probability distribution* for the velocity of a chosen molecule, along the lines Maxwell used to derive his *average number* of molecules with velocity in d^3v around $\mathbf{v}$.

Part one of Maxwell's argument starts with the proposition that "the existence of the velocity [component v_x] does not in any way affect that of the [velocity components v_y] or [v_z], since these are all at right angles to each other and independent." (Here one might ask, really?) Therefore we must be able to write

$$f(\mathbf{v})\, d^3v = [f_1(v_x)\, dv_x] \times [f_1(v_y)\, dv_y] \times [f_1(v_z)\, dv_z]$$
$$= f_1(v_x) f_1(v_y) f_1(v_z)\, d^3v, \tag{1}$$

where f_1 is some presently unknown probability density for a single velocity component.

Part two is much simpler. Since all direction in space are equivalent, the velocity probability density should not depend on the direction of the velocity but only on the magnitude. Using the square of the magnitude for analytic convenience, we may write

$$f(\mathbf{v}) \, d^3v = f_2(v_x^2 + v_y^2 + v_z^2) \, d^3v, \tag{2}$$

where f_2 is a second unknown probability density.

Enough has now been asserted so that one may derive the form of $f(\mathbf{v})$. Here is an outline of the derivation.

a. Equate the righthand sides of equations (1) and (2). Differentiate both sides with respect to v_x, bearing in mind that differentiation of f_2 requires the use of the chain rule. Then start afresh and repeat with v_y. Form the ratio of your two equations and rearrange to arrive at

$$\frac{1}{v_x \, f_1(v_x)} \frac{df_1(v_x)}{dv_x} = \frac{1}{v_y \, f_1(v_y)} \frac{df_1(v_y)}{dv_y}.$$

b. Support the following argument. Since the two sides of the above equation depend on *different independent* variables, each side must be merely a constant and, of course, the same constant.

$$\frac{1}{v_x \, f_1(v_x)} \frac{d f_1(v_x)}{dv_x} = \text{constant},$$

and the same for $f_1(v_y)$.

c. Integrate this equation and find that $f_1(v_x)$ has an exponential dependence on v_x^2. There will be two constants, one of which has already appeared. Now you have the form of $f(\mathbf{v})$.

d. Determine the values of the two constants by imposing the normalization condition and by requiring that the relation $\langle \frac{1}{2}mv^2 \rangle = \frac{3}{2}kT$ hold. The expression for $f(\mathbf{v})$ derived in the text should now be in your hands.

6.14. On the *average* distribution of molecular velocities. The canonical probability distribution for the N-particle dilute gas enables us to compute an *average* distribution of molecular velocities, with " average " taken to mean an expectation value. This statistical average is *not* necessarily the same thing as Maxwell's time average, but the calculation provides a fruitful comparison.

Let us denote the *actual* number of particles with velocity in d^3v around $\mathbf{v}$ by $f^*(\mathbf{v}) \, d^3v$. We do not know what that number is, but we may ask, what is the probability that $f^*(\mathbf{v}) \, d^3v$ has the integral value n? Let us call that probability $P(n)$ and work it out.

a. Show that the probability density for the velocities of all N particles is

$$\mathcal{P}(\mathbf{v}_1, \mathbf{v}_2, \ldots, \mathbf{v}_N) = \left(\frac{\beta m}{2\pi}\right)^{\frac{3}{2}N} \exp\left(-\frac{1}{2}\beta m \sum_{i=1}^{N} v_i^2\right).$$

b. Next we need the probability that particles 1 through n have their velocity in d^3v around $\mathbf{v}$ and that the other particles have their velocity elsewhere. Show, from (a), that this probability is equal to

$$\left[\prod_{i=1}^{n} \zeta_i \right] \times \left[\prod_{i=n+1}^{N} (1 - \zeta_i) \right],$$

where

$$\zeta_i \equiv \left(\frac{\beta m}{2\pi} \right)^{\frac{3}{2}} e^{-\frac{1}{2}\beta m v_i^2} \, d^3 v$$

and each v_i is numerically equal to v.

c. Now we note that we do not care which n particles, out of the N particles, are in the chosen volume. Use the counting scheme developed in Section 2.10, together with the result in (b), to arrive at

$$P(n) = \frac{N!}{n!(N-n)!} \, \zeta^n (1-\zeta)^{N-n},$$

where ζ is just ζ_i without the subscript. Lo and behold, the binomial distribution!

d. Finally, compute the expectation value of $f^*(v) \, d^3 v$ and find

$$\langle f^*(v) \, d^3 v \rangle = N \left(\frac{\beta m}{2\pi} \right)^{\frac{3}{2}} e^{-\frac{1}{2}\beta m v^2} \, d^3 v.$$

We find the numerical agreement between our expectation value and Maxwell's time average, both referred to all N molecules. One should not, however always anticipate agreement between expectation values and time averages. Nonequilibrium problems provide an extreme example; there change with time is crucial, and time averages may not be relevant, or even usefully definable.

6.15. On the reliability of the average distribution. Suppose we focus attention on a region $d^3 v$ around a fixed v in velocity space. Part (c) of problem 6.14 tells us that we should anticipate a statistical spread in the number of particles whose velocities actually are in the region. Similarly, for Zartman's beam, we may ask about the spread in the number of particles with speed in dv around fixed v. A calculation of the probability distribution yields the same binomial distribution, though with $(\mathcal{P}_b(v)) \, dv$ in place of ζ. (For that to be so, we do need to assume that there is only one kind of particle in the beam, not the mixture that Zartman found; for simplicity, let's suppose we're dealing with a one-component beam.) Will the spread be experimentally significant?

Perform a root mean square analysis and compute the ratio, (anticipated spread)/(average value), for arbitrary v and dv. Then use the horizontal separation of the data points for run (2) in figure 6.5.2 to assess the size of dv used by Zartman. The value of N is uncertain, but the amount of bismuth deposited in a run was visible to the naked eye, and so must surely have been greater than 10^{-6} grams. Estimate the numerical value of the ratio for two speeds, one near the peak in Zartman's data, and the other in the region where disagreement continued to plague him. Is a purely statistical explanation of the disagreement plausible?

The reliability of averages is examined further by E. T. Jaynes in section 5 of his Brandeis lectures, cited in the references for Chapter 4.

A MISCELLANY OF GENERAL CONSIDERATIONS

"How is bread made?"

"I know *that*!" Alice answered eagerly. "You take some flour—"

"Where do you pick the flower?" the White Queen asked: "In a garden, or in the hedges?"

"Well, it isn't *picked* at all," Alice explained: "It's *ground*—"

"How many acres of ground?" said the White Queen. "You mustn't leave out so many things."

Lewis Carroll
Through the Looking Glass

This chapter consists of miscellaneous topics, odds and ends of considerable importance that must go somewhere but really fit in nowhere. The first two sections develop computational techniques for calculating certain expectation values with the canonical probability distribution. Though in principle not necessary, the techniques often immeasurably simplify a calculation. The third section illustrates their usefulness and redeems a promise made in Chapter 5. Then comes a jump to a brief discussion of the notions of heat and heat capacity, followed by a further examination of the concept of temperature. The last section has some more to say about the measure of missing information.

7.1. THE PARTITION FUNCTION TECHNIQUE

In this and the following section the primary objective will be the development of a few computational techniques for calculating expectation values when the canonical distribution is applicable. Nothing intrinsically new is to be added to the statistical theory. One should bear that in mind, for the explicit form of some expectation value expressions will change, though their content will not.

For ready reference let us first write down the canonical distribution in the quantum-mechanical form in which we derived it:

$$P_j = \frac{e^{-\beta E_j}}{\sum\limits_{i=1}^{n} e^{-\beta E_i}}. \tag{7.1.1}$$

The symbol β could be replaced by $1/kT$, but we will find the use of β convenient. An energy eigenvalue E_j generally depends on the environment—on the external parameters—and on the number of particles, but not, of course, on the temperature.

A calculation of the expectation value of the total energy takes the form

$$\langle E \rangle = \sum\limits_{j=1}^{n} E_j P_j = \frac{\sum\limits_{j=1}^{n} E_j e^{-\beta E_j}}{\sum\limits_{i=1}^{n} e^{-\beta E_i}}. \tag{7.1.2}$$

We can, as we have twice previously, simplify such an expression by writing the sum in the numerator as the partial derivative with respect to β of a sum like that in the denominator:

$$\sum\limits_{j=1}^{n} E_j e^{-\beta E_j} = \sum\limits_{j=1}^{n} -\frac{\partial}{\partial \beta} e^{-\beta E_j} = -\frac{\partial}{\partial \beta} \left(\sum\limits_{j=1}^{n} e^{-\beta E_j} \right).$$

The use of a partial derivative will avoid any possible ambiguity, for the exponent does depend on the external parameters (through E_j), and we want to extract just E_j from the exponent.

The import of the manipulation is that the calculation of $\langle E \rangle$ can be reduced to operations on a single sum, that of the exponential $\exp(-\beta E_j)$ summed over all energy eigenstates. Since the same sum will arise in other connections, a special symbol and name for it will be convenient. Let us write

$$Z \equiv \sum\limits_{j=1}^{n} e^{-\beta E_j} \tag{7.1.3}$$

and call the function Z the *partition function*. The designation Z for the sum comes from the German name *Zustandsumme* or "state sum" introduced by Planck. The less informative title of "partition function" was devised by Darwin and Fowler. By virtue of its definition, the partition function Z is a function of β (or T), of the external parameters (through the E_j), and of the number of particles (through the number and character of the E_j).

Let us now express $\langle E \rangle$ purely in terms of the partition function and operations on it. The relation

$$\sum_{j=1}^{n} E_j e^{-\beta E_j} = -\frac{\partial}{\partial \beta}\left(\sum_{j=1}^{n} e^{-\beta E_j}\right) = -\frac{\partial Z}{\partial \beta}$$

leads directly to

$$\langle E \rangle = -\frac{1}{Z}\frac{\partial Z}{\partial \beta} = -\frac{\partial \ln Z}{\partial \beta}. \tag{7.1.4}$$

To determine $\langle E \rangle$ we need merely calculate $\ln Z$, in fact, only the part of $\ln Z$ that depends on β. At times this permits a considerable saving in computational effort.

Occasionally one wants to estimate the deviation from $\langle E \rangle$ to be expected for the energy of a system. A measure of this—both useful and convenient—is the root mean square estimate:

$$\Delta E = \langle (E - \langle E \rangle)^2 \rangle^{1/2}. \tag{7.1.5}$$

This is by no means the only possible measure. For example, one could use $\langle |E - \langle E \rangle| \rangle$. The chosen measure has the virtue that it can be written in terms of the partition function, and doing so considerably aids the calculation.

We proceed as we have before in the calculation of a root mean square estimate. First we square ΔE and then multiply out the internal square, using the linearity of the expectation value operation to achieve a more convenient form:

$$(\Delta E)^2 = \langle (E - \langle E \rangle)^2 \rangle = \langle E^2 \rangle - \langle E \rangle^2. \tag{7.1.6}$$

The explicit expression for $\langle E^2 \rangle$ may be juggled around to yield two terms, one of which will subsequently cancel the second term above:

$$\langle E^2 \rangle = \sum_j (E_j)^2 P_j = \sum_j (E_j)^2 \frac{e^{-\beta E_j}}{Z}$$

$$= \sum_j \frac{E_j}{Z}\left(-\frac{\partial}{\partial \beta} e^{-\beta E_j}\right)$$

$$= -\frac{\partial}{\partial\beta}\left(\sum_j \frac{E_j}{Z} e^{-\beta E_j}\right) + \sum_j \left(-\frac{E_j}{Z^2}\frac{\partial Z}{\partial\beta}\right)e^{-\beta E_j}$$

$$= -\frac{\partial\langle E\rangle}{\partial\beta} + \langle E\rangle^2.$$

The second term in the third line comes from differentiating E_j/Z; a glance at equation (7.1.4) and a little mental rearrangement indicate that it is $\langle E\rangle^2$.

When we substitute this two-term expression for $\langle E^2\rangle$ into equation (7.1.6), the terms in $\langle E\rangle^2$ cancel. The end-result can be written in two equivalent ways:

$$(\Delta E)^2 = \begin{cases} -\dfrac{\partial\langle E\rangle}{\partial\beta}, \\[2mm] +\dfrac{\partial^2 \ln Z}{\partial\beta^2}. \end{cases} \qquad (7.1.7)$$

Though the operations may look like sleight of hand, they do show that the calculation of ΔE can be reduced to partial differentiations of the β-dependent part of $\ln Z$.

A bonus comes with this calculation. The lefthand side of equation (7.1.7), as a square, can never be negative. For the righthand side this implies the relation

$$-\frac{\partial\langle E\rangle}{\partial\beta} \geqslant 0.$$

The inequality can be made more informative by expressing the partial derivative in terms of the temperature T rather than the parameter β. We will need the derivative of T with respect to β:

$$\frac{dT}{d\beta} = \frac{d}{d\beta}\left(\frac{1}{k\beta}\right) = -\frac{1}{k\beta^2} = -kT^2.$$

Now the reexpression follows rapidly:

$$-\frac{\partial\langle E\rangle}{\partial\beta} = -\frac{dT}{d\beta}\frac{\partial\langle E\rangle}{\partial T} = -(-kT^2)\frac{\partial\langle E\rangle}{\partial T}.$$

Thus the inequality becomes

$$+kT^2\frac{\partial\langle E\rangle}{\partial T} \geqslant 0. \qquad (7.1.8)$$

If a system is described by the canonical distribution, the expectation value of the total energy is never a decreasing function of the temperature.

A little analysis produces an even stronger statement. First we note that ΔE is greater than zero if two conditions are fulfilled: (1) the system has two (or more) numerically different, finite energy eigenvalues; and (2) the temperature T is not zero. We can show this readily by working with the basic relation

$$(\Delta E)^2 = \sum_{j=1}^{n} (E_j - \langle E \rangle)^2 P_j.$$

The conjunction of the two conditions implies that there are two (or more) different energies E_j for which the corresponding probabilities P_j are nonzero. The difference $(E_j - \langle E \rangle)$ cannot be zero for both, and so the sum must yield a positive result. Virtually any physical system satisfies the first condition; certainly any macroscopic system does. In general, for any system that satisfies the two conditions, the strict inequality holds in relation (7.1.8). One may then assert that the expectation value of the energy is an increasing function of the temperature (when the external parameters are held fixed). This is comfortingly in agreement with commonsense notions of a connection between energy and temperature.

Yet another point can be made. For a system already of macroscopic size, $\langle E \rangle$ grows in proportion to the number N of particles that are present, provided that temperature, density, and any external fields are kept constant. For example, if one doubles the number of particles in the system and concurrently doubles the volume, so that the density remains constant, then $\langle E \rangle$ doubles also; although the surface area does not exactly double, that failure to double has a negligible influence on the growth of $\langle E \rangle$, for it is dominated by the density and volume of the macroscopic system. The derivative $\partial \langle E \rangle / \partial \beta$ must also grow as N, for both $\langle E \rangle$'s that appear in the infinitesimal difference will grow as N. Then equation (7.1.7) implies that ΔE will grow only as $N^{1/2}$. The upshot is the relation

$$\frac{\Delta E}{\langle E \rangle} \sim \frac{N^{1/2}}{N} \sim \frac{1}{N^{1/2}}. \tag{7.1.9}$$

The relation tells us that ΔE will be very small relative to $\langle E \rangle$ when a system is of macroscopic size ($N \simeq 10^{20}$ or so). The import is that a knowledge of the temperature, *together* with the canonical probability distribution, implies an almost certain knowledge of the energy of a macroscopic system. The functional relation between T and $\langle E \rangle$ can be quite complicated—it is seldom a direct proportionality—but the system is exceedingly likely to have an energy close to the estimate $\langle E \rangle$. This, too, is in agreement with common sense and supports the general conclusion about statistical estimates presented at the close of Section 5.2.

7.2. ESTIMATED PRESSURE, MAGNETIC MOMENT, AND THE LIKE

The expectation value of the pressure can also be computed from the partition function. The procedure provides, as we will see, the basis for a powerful technique in the analysis of gases when the intermolecular forces have an appreciable effect. Before we set out to establish that procedure, we should recapitulate the calculation of the pressure directly from the probability distribution, a problem we tackled in Chapter 1. We follow the same line of reasoning, and start, as before, with the connection among energy, pressure, and work done in a small volume change. In order to make maximum use of energy conservation, we specify that the system, in the infinitesimal expansion, exchanges energy with its surroundings only by means of the work done by the pressure. This specification amounts to agreeing that other external parameters, such as a possible external magnetic field, are kept constant. For an infinitesimal expansion dV in volume, the expectation value of the work done on the surroundings is given by the product of the pressure $\langle p \rangle$ and the volume change. Now we apply the principle of energy conservation: the expectation value of the work done on the surroundings plus the expectation value of the energy change of the system must sum to zero. For our infinitesimal volume change we may write

$$\langle p \rangle \, dV + \left\langle \frac{\partial E}{\partial V} \, dV \right\rangle = 0. \tag{7.2.1}$$

Upon dropping the infinitesimal dV, we arrive at the expression for calculating $\langle p \rangle$ for a system in equilibrium:

$$\langle p \rangle = -\left\langle \frac{\partial E}{\partial V} \right\rangle = \sum_{j=1}^{n} \left(-\frac{\partial E_j}{\partial V} \right) P_j. \tag{7.2.2}$$

This is valid for any gas, dilute or not, and holds also for a confined liquid or solid.

When we insert for P_j the probability from the canonical distribution, the relation looks forbidding:

$$\langle p \rangle = \frac{\sum_j \left(-\frac{\partial E_j}{\partial V}\right) e^{-\beta E_j}}{\sum_i e^{-\beta E_i}}. \tag{7.2.3}$$

To simplify, we write the sum in the numerator as a derivative of the partition function:

$$\sum_j \left(-\frac{\partial E_j}{\partial V}\right) e^{-\beta E_j} = \sum_j \frac{1}{\beta} \frac{\partial}{\partial V} e^{-\beta E_j} = \frac{1}{\beta} \frac{\partial Z}{\partial V}.$$

The first step is permissible because β and V are independent variables in the canonical distribution, with V appearing only (implicitly) in the energy eigenvalues E_j. The second step is then a consequence of the definition of Z. The final expression for $\langle p \rangle$ in terms of Z or $\ln Z$ is

$$\langle p \rangle = \frac{1}{\beta Z} \frac{\partial Z}{\partial V} = \frac{1}{\beta} \frac{\partial \ln Z}{\partial V}. \tag{7.2.4}$$

Since Z is a function of T and V, this result enables one to estimate the pressure in terms of those variables.

A few more words in justification of this approach to calculating the pressure are in order. Early in our education we learn to think of gas pressure as being caused by molecular impacts with the container walls. This is an excellent—and true—picture; without it much of the real physics of the laboratory would remain a mystery. It was this image of "little spheres in chaotic motion" that we used successfully in the classical calculation of Section 1.1. The desirability of a different point of view—a work-energy analysis of pressure—arises primarily when one takes into account the intermolecular forces, the effects of which prevent one from saying, blithely, "the molecule travels straight back and forth between opposing walls." The molecule is likely to collide with another molecule on the way and perhaps head back for another impact without having reached the opposite wall. In Section 8.3 we will see that, when the intermolecular forces are not neglected, semiquantitative arguments about the pressure can still be made with the "little sphere" picture, but the arguments are limited in scope. The details of the intermolecular forces can be taken into account more accurately and more consistently if one adopts the present work-energy approach to a pressure calculation.

(The justification for this claim will become visible in Chapter 8.) Admittedly, that is somewhat surprising, for energy arguments tend to hide the details of a process. It is "really" a force that urges a stone to roll down a hill, not the lower potential energy in the valley. But it is precisely the details of the molecular collisions that make the pressure calculation in the "little sphere" picture so difficult. To the extent that these are summarized in an inter-molecular potential energy, the work-energy analysis simplifies the pressure calculation.

We should note that the pressure-energy relation given in equation (7.2.2) is intrinsically quantum mechanical, for it contains the partial derivative of an energy eigenvalue with respect to the volume. By the time the relation has been manipulated to the form in equation (7.2.4), the explicit quantum-mechanical nature has disappeared. One needs only the volume dependence of the logarithm of the partition function in order to calculate $\langle p \rangle$ for a gas, dilute or dense. It is at that point that one may pass to the classical limit, evaluating $\ln Z$ not quantum-mechanically but classically. The procedure is not a swindle, but rather an approximation, one whose validity can be checked. A general prescription for computing the classical approximation to the quantum partition function will be developed in Chapter 9. The partition function stage is often a good stage at which to introduce approx-imations, at least if one cares only about calculating expectation values. The analysis of an imperfect gas in Chapter 8 will provide a typical example.

A generalization lies hidden in the approach we used to establish an ex-pression for $\langle p \rangle$. We examined the response of the system's energy to an infinitesimal change of volume. That examination was sure to be profitable, because we knew that we could relate pressure to the changes in energy and volume. Let us note now that volume is only one of the many external parameters that may be associated with a system. For example, a magnetic field may be present as an external parameter, or there may be an electric field, or both. Will something useful emerge if we examine the response of the system's energy to an infinitesimal change in a different external parameter, for instance, an external magnetic field?

Let us, as usual, take the magnetic field to be directed along the z-axis: $\mathbf{H} = H_z \hat{z}$. The magnetic analog of the partial derivative expression in equation (7.2.2) would be $-\partial E_j / \partial H_z$. There is at least one situation for which we can immediately evaluate the magnetic analog: the system of N particles with permanent magnetic moments over which we labored in Section 5.1. Equation (5.1.3) provides us with an explicit expression for the energy eigenvalue E_j:

$$E_j = -m_j \mu H_z.$$

The factor $m_j\mu$ is the component of the total magnetic moment along the field direction. For conformity with the notation of this section, the symbol H of Section 5.1 has been replaced by H_z. Now we form the magnetic analog:

$$-\frac{\partial E_j}{\partial H_z} = -\frac{\partial}{\partial H_z}(-m_j\mu H_z) = m_j\mu.$$

An identifiable result does emerge: the component of the total magnetic moment along the z-axis in the state ψ_j.

So far we have examined only a single, rather special situation. In Appendix C the question is analyzed with more attention to quantum-mechanical details. Merely stating here the conclusion of that analysis will perhaps be sufficient: the expression $-\partial E_j/\partial H_z$ gives the quantum-mechanical expectation value of the z-component of the total magnetic moment for the energy eigenstate ψ_j.

The conclusion implies that a calculation, with the canonical distribution, of the expectation value of the total magnetic moment of any system can be arranged in full analogy with the calculation of the pressure. Upon denoting the z-component of the total magnetic moment by M_z, we may write an equation for $\langle M_z \rangle$ in terms of the magnetic analog and the probabilities:

$$\langle M_z \rangle = \sum_{j=1}^{n}\left(-\frac{\partial E_j}{\partial H_z}\right)P_j. \tag{7.2.5}$$

Then we may follow the steps that led from equation (7.2.2) to equation (7.2.4) for the pressure. The end result of the process is the relation

$$\langle M_z \rangle = \frac{1}{\beta}\frac{\partial \ln Z}{\partial H_z}. \tag{7.2.6}$$

That the calculation of $\langle M_z \rangle$ can be reduced to operations on the partition function is a great asset in practical calculations. The example in the following section will begin to make that evident.

A similar analysis with a uniform external electric field produces the electric dipole moment of the system. If one investigates the response of an energy eigenvalue to an infinitesimal change in an external parameter, one generally finds something identifiable and useful. This result is physically reasonable. An external parameter provides the experimentalist with an opportunity to "prod" the system in a gentle and controllable fashion. The response is bound to reflect the characteristic properties of the system.

7.3. PARAMAGNETISM REVISITED

This section redeems a promise made at the end of Section 5.1. In addition, it provides an illustration of how the partition function technique can expedite an expectation value calculation. The physical system, the environment, and the simplifying approximations are identical to those of Section 5.1 in all respects but one: the total angular momentum of a single paramagnetic particle is no longer restricted to $\frac{1}{2}\hbar$.

As before, we denote an individual magnetic moment by a vector $\boldsymbol{\mu}$, but now we relate $\boldsymbol{\mu}$ more explicitly to the total angular momentum vector $\hbar\mathbf{J}$ of the particle by the relation

$$\boldsymbol{\mu} = g\mu_B\mathbf{J}. \tag{7.3.1}$$

The symbol μ_B denotes one Bohr magneton,

$$\mu_B \equiv \frac{|e|\,\hbar}{2mc},$$

the charge e and mass m referring to those of an electron. The Bohr magneton provides a useful standard magnetic moment for dealing with atoms. The factor g, the Landé g factor, is dimensionless and of order unity. It expresses the effects of coupling the orbital and spin contributions to the net magnetic moment. These are details. The essential point is that $\boldsymbol{\mu}$ is taken to be proportional to $\mathbf{J}$.

Our ultimate concern lies with $\langle M_z \rangle$ for the system of N paramagnetic particles when in an external magnetic field $\mathbf{H} = H_z\hat{\mathbf{z}}$. That was the quantity compared with experiment in figure 5.1.2. Even with the partition function to help us, we will find it convenient to develop one shortcut by argument. Because we make the assumptions of Section 5.1 and neglect the inevitable mutual interactions among the paramagnetic particles, a laborious direct calculation of $\langle M_z \rangle$ would end up telling us that $\langle M_z \rangle$ is merely N times $\langle \mu_z \rangle$, the latter also computed without regard for the mutual interactions. Calculating $\langle \mu_z \rangle$ will suffice both to give us the correct answer and to illustrate the partition function approach.

We set out, therefore, to calculate the partition function for a single paramagnetic particle. The energy of interaction with the external field is described by

$$\begin{pmatrix}\text{single particle} \\ \text{energy}\end{pmatrix} = -\boldsymbol{\mu} \cdot \mathbf{H} = -g\mu_B H_z J_z. \tag{7.3.2}$$

The admissible values of J_z range in integral steps from $-J$ to J, where J is determined by the total angular momentum of the particle and may be either an integer or half an odd integer. At the moment we need not specify J in any greater detail, though we should note that J is taken, by convention, to be positive. The sum for the partition function will contain $2J + 1$ terms:

$$Z = \sum_{\substack{\text{all energy eigenstates} \\ \text{of the single particle}}} \exp\left[-\beta\binom{\text{energy eigenvalue}}{\text{of the state}}\right]$$

$$= \sum_{m=-J}^{J} \exp[-\beta(-g\mu_B H_z m)]. \tag{7.3.3}$$

The symbol m denotes a summation index that goes from $-J$ to J in integral steps.

To keep the mathematical expressions from getting out of hand, let us write

$$\xi = \beta g \mu_B H_z \tag{7.3.4}$$

as an abbreviation. The partition function takes on a less formidable appearance:

$$Z = \sum_{m=-J}^{J} e^{m\xi}.$$

The sum is that of a finite geometric series and is performed by the usual trick. The maneuver

$$(1 - e^\xi)Z = (1 - e^\xi) \sum_{m=-J}^{J} e^{m\xi} = e^{-J\xi} - e^{(J+1)\xi},$$

in which all terms but two cancel, leads to

$$Z = \frac{e^{-J\xi} - e^{(J+1)\xi}}{1 - e^\xi}.$$

Multiplication, top and bottom, by $-\exp(-\tfrac{1}{2}\xi)$ gives the preferred symmetric form for the single-particle partition function:

$$Z = \frac{e^{(J+\frac{1}{2})\xi} - e^{-(J+\frac{1}{2})\xi}}{e^{\frac{1}{2}\xi} - e^{-\frac{1}{2}\xi}}. \tag{7.3.5}$$

The expectation value $\langle\mu_z\rangle$ is now only a logarithm and a partial differentiation away. Making use of equation (7.2.6), we write

$$\langle\mu_z\rangle = \frac{1}{\beta} \frac{\partial \ln Z}{\partial H_z} = \frac{1}{\beta} \frac{\partial \xi}{\partial H_z} \frac{\partial \ln Z}{\partial \xi}.$$

The indicated differentiations lead to

$$\langle \mu_z \rangle = g\mu_B J B_J(\xi),$$

where all the ugliness is hidden in the *Brillouin function* $B_J(\xi)$:

$$B_J(\xi) \equiv \frac{1}{J}\left[(J + \tfrac{1}{2})\left(\frac{e^{(J+\frac{1}{2})\xi} + e^{-(J+\frac{1}{2})\xi}}{e^{(J+\frac{1}{2})\xi} - e^{-(J+\frac{1}{2})\xi}}\right) - \frac{1}{2}\left(\frac{e^{\frac{1}{2}\xi} + e^{-\frac{1}{2}\xi}}{e^{\frac{1}{2}\xi} - e^{-\frac{1}{2}\xi}}\right)\right].$$

Comfortingly, for $J = \tfrac{1}{2}$ the present expression for $\langle \mu_z \rangle$ does reduce (after much algebra) to the form we found in Section 5.1.

Invoking the argument that the proportionality between $\langle M_z \rangle$ and $\langle \mu_z \rangle$ found in Section 5.1 will hold here as well, we write

$$\langle M_z \rangle = N\langle \mu_z \rangle = Ng\mu_B J B_J(\xi). \tag{7.3.6}$$

As before, the limiting behavior for "low" and "high" temperature is the most useful and readily grasped aspect of the explicit solution. Low temperature means ξ large relative to unity:

$$\xi = \frac{g\mu_B H_z}{kT} \gg 1.$$

The positive exponentials dominate in the Brillouin function, and it approaches the value unity:

$$B_J(\xi) \simeq \frac{1}{J}[(J + \tfrac{1}{2})(1) - \tfrac{1}{2}(1)] = 1, \quad \text{when} \quad \xi \gg 1.$$

The estimate $\langle M_z \rangle$ approaches its maximum possible value:

$$\langle M_z \rangle \simeq Ng\mu_B J.$$

in the low-temperature limit.

The high-temperature limit is more typical of what one sees in the laboratory, even in experiments at liquid helium temperatures. The parameter ξ is then small relative to unity, but one must exercise care in establishing the behavior of the Brillouin function. Each exponential in the denominators must be expanded in a Taylor's series through terms of order ξ^3. For the numerators a similar expansion through order ξ^2 suffices. With the aid of the binomial theorem, one can then combine denominators and numerators. The process, though lengthy, does ensure that one collects all terms that should contribute to the lowest order approximation. The end-result is pleasantly concise:

$$B_J(\xi) \simeq \tfrac{1}{3}(J + 1)\xi, \quad \text{when} \quad \xi \ll 1.$$

The implication for $\langle M_z \rangle$ in the high-temperature limit is

$$\langle M_z \rangle \simeq \frac{N g^2 \mu_B^2 J(J+1)}{3k} \frac{H_z}{T}.$$

We find again the proportionality with H_z and with $1/T$ that emerged for $J = \frac{1}{2}$.

The behavior of $\langle M_z \rangle$ with temperature for $J > \frac{1}{2}$ is qualitatively the same as that for $J = \frac{1}{2}$. Explicit curves for $J = \frac{5}{2}$ and $J = \frac{7}{2}$ were presented in figure 5.1.2.

7.4.   HEAT AND HEAT CAPACITY

Some discussion of heat and heat capacity—both originally thermodynamic notions—belongs in a development of statistical mechanics. The concept of heat forces us to examine our microscopic picture of the interaction between a " system of interest " and the external world. The notion of a heat capacity leads to significant comparisons between laboratory observation and predictions by statistical mechanics.

There are innumerable ways in which one may heat or cool a system. A metal sample might be heated by wrapping it with a lightly insulated wire and then passing a current through the wire. The " heat " developed because of the wire's electrical resistance is (partially) transferred by contact to the sample. In some early studies of paramagnetic systems at low temperature, the samples were heated by irradiation with gamma rays. A can of beer is cooled quite conveniently by putting it in the refrigerator.

The first question we must ask is this: What are the common characteristics of these diverse means of heating and cooling? A list of partial answers might be as follows.

1. There is net transfer of energy (to or from the system).

2. That transfer is not dependent on any change in the system's external parameters.

3. The amount of energy transferred may be controlled and known at the macroscopic level but not at the microscopic level.

For support, let us examine the first example. The temperature of the metal sample rises. Taken together with the negligible change in volume, that implies an increase in the system's energy and hence net energy transfer. Any minor change in volume will be an expansion; hence the increase in the system's energy cannot be attributed to a change in external parameters, for in an expansion against constraining forces the system would lose energy, not gain it. Finally, by monitoring the current and voltage of the electrical heating element, one would know the energy input reasonably accurately. Yet one would certainly be ignorant of the microscopic manner in which the energy of vibrating atoms passed from the wire through the electrical insulation to the metal sample.

A comprehensive definition of heat is difficult to frame. Our purposes will, however, be served by a definition applicable to a restricted, yet common, set of circumstances. We specify that the external parameters associated with the system of interest remain constant. Those parameters might be the volume of a container filled with a dense gas or the magnitude and direction of an external magnetic field imposed on a paramagnetic sample. If there is, nonetheless, an interaction with the external world leading to a transfer of energy, we will refer to the energy thus transferred as heat. There must then be an additional term in the quantum-mechanical energy operator for the system, a term that couples the system to the external world and permits an exchange of energy despite the constancy of the external parameters.

To illustrate the foregoing, let us examine a confined gas. The gas molecules interact with the atoms of the container wall, and the dominant part of this interaction is described by a potential jump at an ideal rigid wall. Thus the shape and volume of the container enter the energy operator as external parameters. The "additional term" describes the residual interaction, the difference between the interaction with the ideal wall and the actual interaction with individual vibrating atoms in a rough wall. The ideal rigid wall, when held fixed, cannot transfer energy to the gas, but the actual vibrating atoms still can. (When a gas molecule hits an ideal rigid wall, its momentum is changed but not its kinetic energy; when it hits an atom, which has comparable mass and is free to change velocity, both its momentum and energy may change.) A net energy transfer due to the interaction of the gas with the vibrating wall atoms is energy transferred as heat.

It is a characteristic of energy transfer as heat that we know neither the small coupling term nor the details of the external world well enough to calculate the effects in microscopic detail. At most we know, at the macroscopic level, the amount of energy transferred or the change in the temperature of the system or both. Because detailed information is lacking, we are unable to follow in a deterministic way the change in a probability distribution for the system.

This raises the question of how one is to describe the system in statistical terms. Of primary concern to us is the situation in which the transfer of energy proceeds very slowly. Experience tells us that the system then remains close to equilibrium. For example, if a gas is slowly heated, no wild, turbulent macroscopic motions arise. In such a situation it is reasonable to continue to use the canonical probability distribution, adjusting the temperature as needed. Indeed, there is little else that one could do. Thus, for the experimentally very relevant case of slow energy transfer as heat, under the stipulation of fixed external parameters, we will continue to use the canonical distribution, allowing the temperature to change as becomes necessary.

Heat Capacity

The transfer of energy as heat leads to correlated changes in the system's energy and temperature. As one grows or declines, so does the other. The relationship of the changes provides a significant test of the statistical theory, as well as of one's understanding of the system's microscopic behavior. In the experimental situation, one most frequently changes the energy of the system by a macroscopically monitored amount and then measures the subsequent temperature change. The reverse point of view is more convenient analytically. One asks, by how much does the system's energy change for a given change in temperature? Let us take the latter point of view and define a *heat capacity* C, at constant external parameters, by the following limiting ratio:

$$C \equiv \lim_{\Delta T \to 0} \frac{\left(\begin{array}{c} \text{the infinitesimal amount of energy} \\ \text{transferred to the system as heat} \\ \text{at constant external parameters} \end{array} \right)}{\Delta T}. \tag{7.4.1}$$

The infinitesimal amount of energy is the amount associated with the change in the temperature when the latter is changed by ΔT. Admittedly, it is perhaps easier to look at the relation in the opposite way: ΔT represents the change in temperature induced by the given infinitesimal amount of energy transferred as heat.

Statistical mechanics enables us to estimate the numerical value and functional dependence of the heat capacity. With the canonical distribution, we may calculate $\langle E \rangle$ at temperatures $T + \Delta T$ and T, form the difference, divide by ΔT, and pass to the limit. Thus the theoretical expression for the heat capacity is simply

$$C = \frac{\partial \langle E \rangle}{\partial T}. \tag{7.4.2}$$

The partial derivative notation is required because we stipulate that the energy transfer is at constant external parameters. This could be made more explicit with parentheses and subscripts like V for constant volume in the gas case, but that would be unnecessarily cumbersome here. The significant point is that we differentiate, with respect to temperature, the expression for $\langle E \rangle$ that emerges from a calculation with the canonical probability distribution. Thus we derive a theoretical expression containing (typically) a dependence on temperature, external parameters, and the size of the system. Incidentally, the inequality presented in relation (7.1.8) tells us that the theoretical heat capacity will never be negative and, indeed, will almost invariably be positive.

> The heat capacity as defined here refers to the entire system. Frequently one finds the expression divided by the volume of the system or by the mass or by the number of constituent particles or, in chemical applications, by the number of moles. Such expressions give a heat capacity per unit volume or per unit mass or per particle or per mole. Often they are called "specific heats." There is nothing intrinsically new in that. Those quantities are merely more useful for the tabulation of physical properties and for comparisons among different systems.

An example of a heat capacity calculation is certainly in order.

The Einstein Model for
the Heat Capacity of a Solid

In a solid, the interaction of a given atom with its neighbors confines the atom to the immediate vicinity of some site in the lattice structure. The atoms can, however, vibrate about their "equilibrium positions" in the lattice, and there is energy, both kinetic and potential, associated with such motion and mutual interaction. Our objective is a quick estimate, based on this picture, of the heat capacity of a solid

A hierarchy of methods, successively increasing in complexity, exists for coping with the mutual interactions of the atoms. The essential ingredient of the simplest realistic method is this: the position-dependent potential energy for any given atom is to be calculated on the assumption that all other atoms are at precisely their equilibrium positions in the lattice structure. This assumption decouples the vibrational motion of the various atoms fom one another. We will make this assumption. If the amplitude of the vibrational motion is not too large, a Taylor's series expansion of the potential energy in powers of the distance from the equilibrium position ought to provide an adequate representation. The first derivative terms will be zero, for they are proportional to the force, evaluated at the equilibrium position, and there the net force is zero. The significant terms in the expansion start with those dependent on the second derivative of the potential; those terms are quadratic in the distance from the equilibrium position. Anticipating vibrations of small amplitude, we stop with the quadratic terms in the expansion of the potential energy. Furthermore, we suppose that we may adequately approximate the environment of any given atom as being spherically symmetric. With these steps we have reduced the expression for the potential energy associated with the restoring forces to that of a three-dimensional harmonic oscillator. The potential energy for each atom is represented by $[-\frac{1}{2}\kappa]$ times [the square of the distance from the equilibrium position in the lattice]. The constant κ is the "spring constant" and is, in principle, determined by the second derivatives of the actual interatomic potential energy, evaluated at the equilibrium position. In this approximation—the *Einstein model*—the atoms vibrate independently of one another and with the same frequency. That frequency is determined by the spring constant and the atomic mass:

$$\nu = \frac{1}{2\pi}\sqrt{\frac{\kappa}{m}}.$$

The real forces between the atoms *are* taken into account, though the treatment is dependent on the approximation detailed at the start of this paragraph.

Since the atoms are taken to vibrate independently of one another, the expectation value $\langle E \rangle$ for the system of N atoms will be N times the estimated energy of a single atom. The latter energy will be three times the expectation value of the energy of a one-dimensional harmonic oscillator. In Section 6.6 we analyzed such an oscillator with quantum mechanics and the canonical probability distribution. Taking over the result from equation (6.6.7), we may write

$$\langle E \rangle = 3N\left(\tfrac{1}{2}h\nu + \frac{h\nu}{e^{h\nu/kT} - 1}\right).$$

To calculate the heat capacity (at constant volume), we need only differentiate $\langle E \rangle$ with respect to the temperature:

$$C = \frac{\partial \langle E \rangle}{\partial T} = 3Nk\left(\frac{h\nu}{kT}\right)^2 \frac{e^{h\nu/kT}}{(e^{h\nu/kT} - 1)^2}.$$

The combination hv/k is the characteristic parameter in the model. Since it has the dimensions of a temperature, it is called the *Einstein temperature* and is abbreviated as Θ_E:

$$\Theta_E \equiv \frac{hv}{k}.$$

After the introduction of the abbreviation, the heat capacity in the Einstein model becomes

$$C = 3Nk\left(\frac{\Theta_E}{T}\right)^2 \frac{e^{-\Theta_E/T}}{(1 - e^{-\Theta_E/T})^2}.$$

The exponentials have been juggled around to facilitate the analysis of the temperature dependence. To that we now turn.

High and low temperature can mean only temperature high and low relative to the Einstein temperature. In particular, high temperature means that the strong inequality $\Theta_E/T \ll 1$ holds. An expansion of the exponentials in this limit leads to

$$C \simeq 3Nk, \quad \text{when} \quad T \gg \Theta_E.$$

This is the classical result; the equipartition theorem of classical statistical mechanics would provide it quite readily.

Of more concern is the low temperature limit: $\Theta_E/T \gg 1$. Then the exponentials are small relative to one, and the heat capacity approaches the form

$$C \simeq 3Nk\left(\frac{\Theta_E}{T}\right)^2 e^{-\Theta_E/T}, \quad \text{when} \quad T \ll \Theta_E.$$

The exponential dominates and, despite the divergent factor in front, the heat capacity becomes much less than the classical value. This is a consequence of the quantization of the energy of a vibrating atom.

The experimental heat capacities do go strongly to zero as the temperature is lowered, providing a general vindication of quantum mechanics and of the applicability of the canonical probability distribution. A fine example is displayed in figure 7.4.1. The observed decrease in the heat capacity is, however, not as strong as exponential decrease. Rather, the heat capacity goes roughly as T^3 at very low temperature. The origin of the difference between observation and the Einstein model lies in the assumption that the potential energy for a given atom can be adequately computed as though all other atoms were at their equilibrium positions. When this assumption is removed, as it is in the Debye model, the T^3 behavior emerges very nicely.

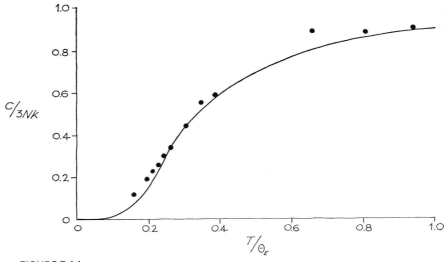

FIGURE 7.4.1

The points represent the experimental heat capacity of diamond. The curve gives the results of a theoretical calculation with the Einstein model; the sole free parameter, the Einstein temperature, was chosen as $\Theta_E = 1325$ K. The now-classic comparison appeared in Einstein's paper, *Annalen der Physik*, **22**, 180 (1907).

The subject of heat capacity will appear again in Chapter 10. There we will compute the contribution from the conduction electrons in a good metallic conductor.

7.5. WHAT IS TEMPERATURE?

We began our discussion of temperature with an intuitive notion of temperature and proceeded to an operational definition with a dilute-gas thermometer. The link between T and $\langle E \rangle$ for the dilute gas (established in Section 4.3) is likely to be quite satisfying. One should not, however, be led astray by that link: a direct proportionality between T and $\langle E \rangle$ is a quite special case. For a dense gas or a liquid, an energy estimate $\langle E \rangle$ will depend significantly on density. After all, the contribution of the mutual attractive and repulsive interactions among molecules does depend on how close together the molecules are likely to be. The connection among $\langle E \rangle$, T, N, and N/V becomes a complicated nonlinear affair. That the potential energy should complicate matters is not unreasonable, but couldn't we still look on the temperature as telling us the typical amount of kinetic energy per particle?

Certainly temperature does that for the dilute gas. Unfortunately, for a dense gas at low temperature, quantum mechanics implies that even the estimated kinetic energy per particle depends on density as well as temperature and that the temperature dependence is not a simple proportionality. (A justification for this assertion will emerge from the calculations in Section 9.6.) The Einstein model for a solid produces a similar negative answer to the question. Quantum mechanics, just like classical mechanics, indicates that the average kinetic energy and the average potential energy of a harmonic oscillator are equal; hence the typical kinetic energy is one-half the total energy. A glance back at Section 6.6 shows that only in the high-temperature limit is the typical kinetic energy per atom in a solid directly proportional to the temperature. At low temperature the connection is quite nonlinear, and dependent on the characteristic frequency v as well.

All this is rather unsettling. The notion from kinetic theory that temperature is a measure of the amount of kinetic energy per particle requires serious modification. The notion is tenable and fruitful for a dilute gas, but not in general. What, then, is temperature?

Appeal to a simple experiment will give us a clue. Suppose we have a block of graphite—just the material of which the "lead" in a lead pencil is made—and suppose we have measured its temperature with a gas thermometer; it is $T_{graphite}$. Suppose further we have a block of common salt and know its temperature, T_{salt}. Finally, suppose we place the two blocks in contact and put them inside a thermos bottle; the last is done to isolate them from the rest of the world. What happens?

Because the materials are rather inert, there will not be any chemical reaction. Whether something does or does not happen depends on the initial temperatures. If the two initial temperatures were numerically the same, nothing happens, but if the initial temperatures were numerically different, then there is change. Specifically, in the latter case energy passes from one block to the other through the surfaces that are in contact. Viewed in mechanical terms, the change is really a net transfer of energy; it is what one normally calls transfer of energy as heat. After some time a mutual equilibrium becomes established. Measurement shows that the two blocks now have equal temperatures.

Is any of this surprising or profound? Probably not surprising, but there is a profound element. The results of the little experiment were *not* built into the operational definition of temperature with the gas thermometer. As far as logic and an open mind are concerned, the results could have been different. (If the two blocks had different temperatures initially, a difference could have prevailed indefinitely.) Hence the specific results really do tell us something about the notion of temperature.

Temperature is intimately associated with a specific type of mutual equilibrium, namely, no net transfer of energy as heat. Even when the graphite and salt blocks have equal temperatures, the kinetic energy per atom is different in the two blocks. If we calculate with the Einstein model, the estimate of the typical kinetic energy will depend on the value of the characteristic frequency v as well as on the temperature. Since v, in turn, depends on the "spring constant" and the atomic mass, v is certain to be different in the two materials. Hence the typical kinetic energy per particle will be different even at the same temperature. A knowledge of temperature, *by itself*, does not enable us to infer much about *amount* of energy. Rather, knowledge of temperature enables us to decide whether or not there will be net *transfer* of energy as heat when two bodies are placed in contact and permitted to exchange energy.

Let's follow up the notion advanced in the last sentence. We consider a system A (which might be a graphite block) and a system C (which might be a small bulb filled with mercury and attached to a narrow glass tube). We put them in contact and wait for mutual equilibrium to become established. Then we observe system C (for instance, observe the height of the mercury in the tube), and afterwards break the contact. Next we extend our considerations to a system B (which might be a salt block) not in contact with system A. We place systems B and C in contact, observe system C for some time, and then break the contact. Let us suppose a special case now: there was no macroscopic change in the condition of system C when it was placed in contact with system B. In other words, after C had been in mutual equilibrium with A, mutual equilibrium with B existed from the outset. Question: can we predict that systems A and B will immediately be in mutual equilibrium —no net transfer of energy as heat—if placed in contact? The empirical response is remarkable: yes. If the internal conditions of systems A and B are such that each will be in equilibrium with system C when placed in contact with C, then systems A and B will be in equilibrium with each other if placed in contact.* This experimental conclusion permits one to use system C as a thermometer, in the sense that observation of system C enables one to predict whether there will be mutual equilibrium between systems A and B. (This empirical conclusion is often promoted to the status of the "zeroth" law of thermodynamics, and with good justification, for it is fundamental to the general notion of temperature.)

---

* If all this seems terribly obvious, note that if electric charges a and c attract one another and if charges b and c also attract one another, then charges a and b repel one another. Relations around a logical triangle are not necessarily the same.

What about the less-special situation in which the condition of system C changes on contact with system B? Experience indicates that we should then expect net energy transfer if systems A and B are placed in contact. Moreover, the specific difference between the two conditions of system C—first when it is in equilibrium with system A and then, separately, with system B— carries information. (The specific difference might be, for instance, that the mercury rises higher when the equilibrium is with A than when it is with B). The sense of the difference can be correlated with the direction that the net transfer of energy between systems A and B would take. We may, if we wish, employ system C as a thermometer and use specific conditions of that system (for instance, specific heights of the mercury) to represent specific temperatures. The essential point is that *differences* in temperature will express an ordering. They will tell us the direction in which there will be energy transfer between systems.

Our view of temperature, *before* we examined the deeper implications of the little experiment, might be summarized in the following fashion.

Purely Operational View. Temperature is whatever a dilute-gas thermometer measures. Experiment shows that macroscopic systems with initially equal temperatures will be in mutual equilibrium—no net transfer of energy as heat—if placed in contact.

Now, however, we might turn the priorities around. That option would generate the following, somewhat different view of temperature.

Empirical Yet Abstract View. There is a measurable property of macroscopic systems that determines, prior to their contact, whether the systems will be in mutual equilibrium—no net transfer of energy as heat—if placed in contact. That property is given the name *temperature*. Prior equality of temperature for two systems implies that mutual equilibrium will hold from the moment of contact. A dilute-gas thermometer provides one means of measuring that property (on a specific scale).

Let us note the merits of the second view of temperature.

1. That view focuses attention on the mutual-equilibrium aspect of temperature.

2. It makes the notion of temperature independent of a specific choice of thermometer, that is, independent of a specific choice of substance, procedure, and scale.

Equality of "temperature" for the graphite and salt blocks as measured with a mercury thermometer or with the resistance of a carbon resistor is— empirically—just as good for predicting mutual equilibrium of the blocks as is equality of "temperature" as measured with a gas thermometer. Different kinds of thermometers measure the same property but possibly on different scales. If one chooses some one specific scale—an arbitrary but practical act—then all thermometers can be calibrated to that scale.

We seem to be moving farther away from a neat mechanical view of what temperature is. Really, we are being forced away. As soon as we recognize that the temperature concept is connected with *net* energy transfer, we can sense that there is an essential statistical element in the background. In a loose sense, temperature is connected with the likelihood of internal rearrangements that will lead to energy transfer (as heat and in macroscopic amounts) between systems. Some kind of microscopic internal rearrangement in each of two systems is likely to occur when they are placed in contact and thus interact with one another. Temperature indicates whether one should expect such rearrangements to generate a net transfer of energy when that possibility is opened up. Partly because of the statistical element, the temperature concept is fruitful only for macroscopic systems, for only with them can one expect the "most likely" behavior of the system to be quite nearly realized by the actual behavior.

That there is such a thing as temperature is indeed remarkable. Its empirical existence indicates that, despite all the microscopic activity and diversity, there is much macroscopic regularity.

Temperature and the Canonical Distribution

Let us confirm that the canonical probability distribution and the relation $\beta = 1/kT$ are, at the very least, consistent with the "empirical yet abstract" view of temperature. We will need to consider two systems, A and B, first separated and then in contact. Let us assume that we know the initial temperature T_A of system A (as measured with a gas thermometer) and that we describe the system with the canonical distribution. We assume the same for system B. The initial temperature T_B need not be equal to T_A. Now we place the two systems in contact, taking care to maintain constant values for the external parameters of each (for instance, volume or external magnetic field). In Section 7.4 we noted that, despite the constancy of the external parameters, interaction between the two systems does exist and permits energy transfer as heat.

How we proceed depends on what we know about that interaction and how we propose to treat it. We could assume that we know the mutual interaction in detail and then try to calculate the evolution of the probability distribution for the combined system. Let us, however, be more realistic and assume that we know only that a mutual interaction exists and that it is relatively weak. This is not much to work with; so let us appeal to experience for one datum. Experience indicates that a mutual equilibrium is generally reached within a reasonable period of time (if it did not exist immediately after contact). This one datum means merely that there ceases to be macroscopic change; it leaves open the question of what happens to temperatures.

When the mutual equilibrium has been attained, we can try to describe the joint system—the combination of systems A and B—with energy eigenstates for the joint system. Since we assume that the mutual interaction is relatively small, energy states of the joint system can be formed by pairing an energy eigenstate of system A (such as $\psi_j{}^A$) with one from system B (such as $\psi_{j'}{}^B$), each to be computed as though the interaction did not exist. Let us denote by $P_{jj'}$, the probability of the inference that the state $\psi_j{}^A\psi_{j'}{}^B$ provides the appropriate description of the joint system, given the background data. Those probabilities are constrained by a normalization condition:

$$\sum_j \sum_{j'} P_{jj'} = 1. \tag{7.5.1}$$

There is, moreover, an energy constraint. Initially we could estimate the energy of system A as $\langle E_A \rangle_{\text{initial}}$ with the aid of the canonical distribution and the initial temperature T_A. A similar situation held for system B. Contact has ensured that energy could be transferred between systems A and B, and so the energy of each may have changed. Nonetheless, the total energy cannot have changed. We may demand, therefore, that the expectation value of the total energy be equal to the sum of the two initial energy estimates:

$$\sum_j \sum_{j'} (E_j{}^A + E_{j'}{}^B) P_{jj'} = \langle E_A \rangle_{\text{initial}} + \langle E_B \rangle_{\text{initial}}. \tag{7.5.2}$$

We are being a little sloppy with the mutual interaction: we invoke it so that energy may be transferred but neglect it when computing actual energies. Provided the mutual interaction is small, that sloppy procedure is reasonable.

Since there are no further constraints, we must turn to Criterion II of Section 3.7 and choose, as the unbiased probability distribution, the probability distribution that maximizes the amount of missing information subject to the two constraints. The mathematical problem is intrinsically the same as the one we solved in Section 4.5. The unbiased probability distribution has the form

$$P_{jj'} = \frac{e^{-\beta(E_j{}^A + E_{j'}{}^B)}}{\sum_i \sum_{i'} e^{-\beta(E_i{}^A + E_{i'}{}^B)}} = \frac{e^{-\beta E_j{}^A}}{\sum_i e^{-\beta E_i{}^A}} \times \frac{e^{-\beta E_{j'}{}^B}}{\sum_{i'} e^{-\beta E_{i'}{}^B}}, \tag{7.5.3}$$

with the parameter β to be determined from equation (7.5.2).

The unbiased probability distribution for the final mutual equilibrium has the form of the product of two canonical distributions, and they have the same value of β. If $T_A = T_B$ initially, so that $\beta_A = \beta_B$ held, then a calculation of β will, of course, yield $\beta = \beta_A = \beta_B$. Initial equality of temperature implies no real change in the probability distributions and hence no change in the estimated energy of either system; there is no net energy transfer. If $T_A \neq T_B$ initially, then we can expect a calculation of β from equation (7.5.2) to yield a value different in general from both β_A and β_B. An expectation value estimate of the energy of system A will differ from the initial estimate, and likewise for system B. We have good reason to believe that there was net transfer of energy. Moreover, the direction of that net transfer is uniquely determined by the sign of the original difference, $T_A - T_B$.

What do we infer from all this? We may say that the canonical probability distribution, which we derived in Chapter 4 with considerable dependence on a gas thermometer, does indeed exhibit properties in agreement with the more general view of temperature. The reciprocal of the parameter β possesses all the qualitative features of "temperature" as presented in the "empirical yet abstract" view.

Adoption of the "empirical yet abstract" view of temperature is desirable, for it gives a *meaning* to temperature in a wide variety of circumstances and avoids the pitfalls of a simple kinetic theory view. One does, of course, then need to extend the *scale* of tempearture beyond the range in which a dilute-gas thermometer provides a practical scale. In Chapter 11 we will address ourselves to a statistical mechanical prescription for extending the scale; in the interim no harm will result if we simply think of a dilute-gas thermometer and reasonable extrapolations therefrom.

Perhaps we may now rest the case for the canonical probability distribution. We derived it carefully when "knowledge of temperature" was taken to mean a measurement with a dilute-gas thermometer, and we have shown that it meshes nicely with the more general view of what temperature is. There are many approaches to the description of a system whose temperature is known (or at least fixed) that lead, via approximation or point of view, to the exponential form characteristic of the canonical probability distribution. Whatever the mode of derivation or generality of argument, the conclusion appears to be that the canonical distribution is uniquely the probability distribution to use in estimating physical quantities when one knows that the system is in equilibrium, and knows the external parameters, the number of particles, the intermolecular forces, and the temperature. Let us proceed on this basis.

7.6. MORE ON MISSING INFORMATION

Chapter 3 introduced the notion of a measure of missing information. The context consisted of a set of probabilities $P_1, \ldots, P_n$ for a set of inferences that are exhaustive and mutually exclusive on a specified hypothesis. We looked for a function $MI(P_1, \ldots, P_n)$ that would measure the "missing information" of the given probability distribution, in the sense that it would express quantitatively the amount of information that must be supplied before we would know the correct inference. Following Shannon, we imposed three conditions on the function MI as required properties. A derivation, based on those conditions, led to the expression

$$MI(P_1, \ldots, P_n) = -K \sum_{j=1}^{n} P_j \ln P_j. \qquad (7.6.1)$$

Except for the open choice of the positive constant K, the function we found is the only function satisfying the conditions; the solution to the problem is unique.

The function MI provided the basis for Criterion II on the assignment of probabilities. In Chapter 4 we used that criterion twice: in deriving the "best" probability distribution for a system in equilibrium when one could specify $\langle E \rangle$, and again in deriving the probability distribution when one knew the temperature of a system. In the preceding section that criterion came into play again. The function MI has played an absolutely essential role, though, significantly, there has been no need to make a specific choice for the positive scale factor K.

There will be times, both in this section and later, when we can learn something by evaluating the MI function for the specific probability distribution describing a specific physical system. There is still no unavoidable need for a definite choice of the constant K. Nonetheless, convention in statistical mechanics prescribes the choice, for K, of Boltzmann's constant k. When we set $K = k$, we will call the measure of missing information the *information-theory entropy* and write it as S_I:

$$S_I \equiv S_I(P_1, \ldots, P_n) \equiv -k \sum_{j=1}^{n} P_j \ln P_j. \qquad (7.6.2)$$

The reasons for this adoption of names, symbols, and units are largely historical and conventional. The discipline of thermodynamics somewhat antedates that of statistical mechanics and, indeed, much of the early work in statistical mechanics was aimed at an understanding of thermodynamic results in terms of a statistical treatment of interacting particles. In pure thermodynamics a function arises that, under some physical circumstances, has many of the properties of the measure of missing information. That function is, however, defined in a vastly different fashion. The definition is framed solely in terms of macroscopic quantities (such as those we would write as T, V, $\langle E \rangle$, or $\langle p \rangle$), and, originally, only the difference in the function for two different equilibrium conditions of a system was defined. By virtue of its definition, the thermodynamic function, which we might write as S_{th}, has the dimensions of (energy)/(temperature) and was given the name *entropy* by Rudolf Clausius in 1865.

If we let 1 and 2 stand for any two equilibrium conditions of a macrocsopic physical system and write S_{th} for the thermodynamic entropy, then the thermodynamic definition is this:

$$(S_{th})_2 - (S_{th})_1 \equiv \int_{\text{``1''}}^{\text{``2''}} \frac{dQ}{T_{th}}.$$

The integration is specified to be taken over any continuous sequence of equilibrium conditions connecting the two end ones. In the integral, dQ stands for an infinitesimal amount of energy absorbed by the system as heat in the process of passing from one equilibrium condition to a neighboring one. The denominator, T_{th}, is the temperature as defined in thermodynamics. The thermodynamic temperature is equivalent to the temperature T that we have defined.

In the applications to come, many of our results connecting macroscopic quantities would be valid as results from thermodynamics if one were to replace S_I, evaluated for the canonical distribution, by S_{th}, and T by the thermodynamic temperature T_{th}. That is the reason for using similar notation. By no means, however, can all our results be so transcribed. Not only confusion but also grief would arise from a wholesale identification of S_{th} with S_I or MI. The two functions have much in common, but they are *not* identical. The function MI or S_I is the more richly endowed of the two. In particular, S_I is applicable to nonequilibrium situations, for which S_{th} is not even defined. Indeed, certain properties of S_I will play an essential role in the discussion of time-dependent situations in Chapter 11. (The relationship of S_I and S_{th} is discussed in Appendix D.) Let us scrupulously preserve the real distinction and always refer to S_I, identical to MI with $K = k$, as the *information-theory entropy*.

Nothing in the definition of S_I restricts the probabilities to those of the canonical distribution. That distribution is, however, the only one with which we have much familiarity. Let us work with the canonical distribution, examine the low- and high-temperature behavior of the probabilities, and assess the implications for S_I.

In Section 4.6 we looked briefly at the ratio P_j/P_i as a function of temperature. Now we return to that subject and examine the behavior of the ratio as the temperature is decreased indefinitely. The ratio has the explicit form

$$\frac{P_j}{P_i} = \frac{e^{-\beta E_j}}{e^{-\beta E_i}} = e^{-(E_j - E_i)/kT} . \tag{7.6.3}$$

Let us suppose for the moment that the physical system has a nondegenerate ground state; that is, if we select the lowest energy eigenvalue, only one state has that energy. Let us call that state ψ_1 with energy E_1. At low temperature and hence small $\langle E \rangle$, we expect that to be the important state. Upon inserting the ground state into the ratio by setting $i = 1$, we have

$$\frac{P_j}{P_1} = e^{-(E_j - E_1)/kT}.$$

Now we examine the behavior of the ratio as the temperature is reduced indefinitely. (Extrapolation beyond the practical range of a dilute-gas thermometer suggests that a temperature of zero kelvin has a meaning. Moreover, since we know now that the parameter T in the canonical distribution has the qualitative properties of a temperature, we are certainly free to ask what happens when we imagine it to go to zero.) As the temperature T goes to zero through positive values, the exponent becomes large and negative, and the ratio goes to zero exponentially:

$$\lim_{T \to +0} \frac{P_j}{P_1} = \lim_{T \to +0} e^{-(E_j - E_1)/kT} = 0, \quad \text{for} \quad j \neq 1.$$

The system settles down into its state of lowest energy. We can codify the behavior of the probabilities by writing

$$\lim_{T \to +0} P_j = \begin{cases} 1 & \text{if} \quad j = 1, \\ 0 & \text{if} \quad j \neq 1, \end{cases}$$

provided the ground state is nondegenerate.

In the nondegenerate case, the information-theory entropy is zero in the limit of zero temperature, for S_I is simply the amount of missing information (with $K = k$), and in the limit there is no further information needed, for one inference has unit probability and the others, zero:

$$\lim_{T \to +0} S_I = 0,$$

provided the ground states is nondegenerate.

The attainability of the limit of zero temperature is a separate question. Let us defer that a moment, and look at the modification in the limiting value of S_I that arises when the ground state is degenerate. For a system of more than two or three particles, it is rather unlikely that there will be more than one state associated with the lowest energy eigenvalue. In dealing with macroscopic systems, one tends not even to entertain the possibility. To be sure, there are times when the *neglect* of some very small forces leads to a degenerate ground state, but that aıises as a matter of analytic convenience rather than as a basic property of the system. There exists no proof, however, that even a macroscopic system necessarily has a nondegenerate ground state, and so the analysis is not purely academic.

We suppose now that there are n' states, ψ_1 to $\psi_{n'}$, all with the same numerical value for the energy, with that being the lowest energy for the system. The canonical distribution ascribes equal probabilities to those states. In the limit of zero temperature, only those states will have a non-zero probability, and hence each will have a probability $1/n'$. Even information that the temperature had been reduced to absolute zero would not be sufficient to establish an ultimate probability distribution, a distribution in which one inference has unit probability and the others zero. For that, additional information would be needed, and so the information-theory entropy will be positive. Indeed, a brief computation yields

$$\lim_{T \to +0} S_I = +k \ln n'$$

when the ground state is n'-fold degenerate. This is not unreasonable: the greater n', the greater our uncertainty about the correct state.

There is general agreement that it is impossible to reduce the temperature of a physical system to absolute zero in a finite number of experimental operations. (This is often codified as one version of the third law of thermodynamics.) Exceedingly low temperatures, of order 10^{-6} K, have been achieved for some special magnetic systems, but the lower the temperature is, the more difficult the problems are.

Still, there does not seem to be a general theorem, derivable from the formal theory of statistical mechanics, that such a reduction is impossible. Rather, the situation appears to be more nearly the following. As lower and lower temperatures are reached, the methods of temperature reduction, based on physical properties of actual macroscopic systems, become less and less efficient. There have been innovations, both in the selection of physical systems and in the methods of temperature reduction, which have produced striking improvement in the temperatures attainable. Yet the actual systems with which an experimenter must work seem to be in league against him. As a new method is pushed to lower temperatures, the method becomes less efficient (and at the same time the experimental problems increase).

In other sections there will be further discussion of low-temperature behavior. Now we turn to the high-temperature limit. As the temperature increases, the exponent in equation (7.6.3) approaches zero for any pair ψ_i, ψ_j, of states, and the ratio of the probabilities approaches unity. The entire canonical probability distribution approaches a " flat " distribution, with no inference being significantly more probable than any other. As the temperature rises, we approach a situation of maximum ignorance concerning the state appropriate for the description of the system. The information-theory entropy is bound to grow correspondingly.

When the distribution approaches a flat distribution, each probability will approach $1/n$, where n denotes the number of distinct energy eigenstates of the system. So we may write

$$\lim_{T \to \infty} S_I = \lim_{\text{each } P_j \to 1/n} -k \sum_{j=1}^{n} P_j \ln P_j = +k \ln n.$$

The value of S_I in the high-temperature limit will always be the maximum attainable, but the value depends on the number of distinct states. Typically, the number n will actually be infinity, and so S_I will grow indefinitely in the high-temperature limit.

Before closing this section, we should consider briefly the question of evaluating S_I at temperatures other than the extremes. Working directly with the probabilities is then rather awkward. Fortunately, the evaluation can be reduced to operations on the logarithm of the partition function. The derivation is pleasantly brief.

First we recall that the normalizing factor in the probability P_j of the canonical distribution is precisely the partition function:

$$P_j = \frac{e^{-\beta E_j}}{\sum_i e^{-\beta E_i}} = \frac{e^{-\beta E_j}}{Z}.$$

Then we turn to the general expression for S_I and insert the last expression for P_j into the logarithmic factor:

$$S_I(P_1, \ldots, P_n) = -k \sum_j P_j \ln P_j$$

$$= -k \sum_j P_j(-\beta E_j - \ln Z) \qquad (7.6.4)$$

$$= +k(\beta \langle E \rangle + \ln Z).$$

This gives us S_I, evaluated for the canonical distribution, in terms of the temperature, the expectation value of the energy, and the logarithm of the partition function. Since we already know how to write $\langle E \rangle$ as a partial derivative of $\ln Z$, the derivation is, in its essentials, complete. Incidentally, the expression for S_I in terms of $\ln Z$ provides the most convenient form for passing from quantum mechanics to a classical limit. In Chapter 8 we will use it for that purpose.

REFERENCES for Chapter 7

An engagingly elementary discussion of temperature is presented by Mark W. Zemansky in a paperback, *Temperatures, Very Low and Very High* (Princeton, N.J.: Van Nostrand, 1964). For Zemansky "very low" means about 10^{-6} K; "very high," about 5×10^7 K. A fair part of his discussion is based on thermo-dynamic notions; we make some contact with them in Section 11.5 and in Appendix D.

R. B. Lindsay explores the notion of temperature in an article, "The Temperature Concept for Systems in Equilibrium," in F. G. Brickwedde, ed., *Temperature, Its Measurement and Control in Science and Industry*, Vol. III, Part 1 (New York: Reinhold, 1962). From the article one may conclude that there are many tenable answers to the question, "What is temperature?" and that no single answer is likely to satisfy everyone. Earlier volumes in the series also merit close examination by anyone trying to decide on his answer to the question.

Kurt Mendelssohn provides a masterful view of the history and contemporary content of low-temperature physics in his paperback, *The Quest for Absolute Zero* (New York: McGraw-Hill, World University Library, 1966). There is not an equation in the book: it is all done with graphs, pictures, and words. As a man who participated in the development of low-temperature physics, Mendelssohn can tell it like it was.

PROBLEMS

7.1. Suppose we shift the zero of the energy scale, so that the energy eigenvalue E'_j for state ψ_j on the new (primed) scale is related to the eigenvalue E_j of the same state ψ_j on the previous scale by

$$E'_j = E_j + E^*,$$

where E^* is the constant amount of shift. For each of the following quantities, state whether it changes and, if so, give the amount: P_j, $\langle E \rangle$, $\langle p \rangle$, S_I, Z, $\ln Z$.

7.2. Once the partition function is known, many statistical estimates follow readily, but this means only that the real problem is calculating the partition function. Suppose someone were generous and gave you the following approximate partition function for a gas:

$$Z = \text{const } (V - Nb)^N (2\pi m/h^2 \beta)^{3N/2} \exp(\beta N^2 a/V).$$

The parameters a and b are positive constants. Calculate the expectation value estimates of energy and pressure and offer some interpretation of the possible physical origin of the terms that appear.

7.3. a. Calculate the quantum-mechanical partition function for a one-dimensional harmonic oscillator. (Most of the hard work has been done for you in Section 6.6). Use the logarithm of Z to compute the expectation value of the energy.

b. How high (in terms of $h\nu$ and k) must the temperature be before the probability that the oscillator is in its ground state has dropped below one-half? Evaluate that temperature for diatomic nitrogen with the data in Section 6.6. What are the implications?

7.4. Suppose the classical energy of a particle constrained to one-dimensional motion $(-\infty \leqslant x \leqslant +\infty)$ is

$$E(x, p_x) = \frac{p_x^2}{2m} + bx^{2n} = KE(p_x) + U(x),$$

with $b > 0$ and n a positive integer. Compute, with the canonical distribution, $\langle KE \rangle$, $\langle U \rangle$, and $\langle E \rangle$. You should be able to do the problem without evaluating a single integral. (Hint: cast the $\langle U \rangle$ calculation into the β derivative of the logarithm of an integral; then change variables in the integral to extract the β dependence.)

7.5. Write down the energy operator $\mathcal{H}$ for a system consisting of a proton and electron interacting both through their mutual Coulomb attraction and with an *external* electric field $\mathbf{E} = E_z\,\hat{z}$. The external field is constant in time and uniform in space. As a first step, show that the electric potential for the external field may be written as

$$\text{(electric potential)} = -z\,E_z + \text{constant}.$$

After you have constructed $\mathcal{H}$, form the quantity $-\partial\mathcal{H}/\partial E_z$ and consider whether that might be the operator for the z-component of the electric dipole moment of the system. Generalize, with the aid of analogy, and indicate how one could compute the expectation value of the electric dipole moment (for a system described by the canonical distribution) from a knowledge of the partition function.

7.6. This problem asks you to examine a certain " classical limit " for the magnetic moment analysis of Section 7.3. When J is very large relative to one, there are many quantum-mechanically allowed " orientations " for the moment relative to the external field (even when we restrict consideration to the energy eigenstates). The classical case, in which a continuous range of orientations (from parallel to antiparallel) is permitted, is being approached. Show that, for $J \gg 1$, the expression for $\langle\mu_z\rangle$ in terms of the Brillouin function approaches the classical result in Problem 6.10. (Write the quantum-mechanical expression in terms of $\mu \equiv g\mu_B J$, which is roughly the magnitude of the magnetic moment, the quantity βH_z, and J. Then take $J \gg 1$ and find the limiting expression.)

7.7. This problem asks you to examine part of the analysis in Section 7.5. If the initial situation has $T_A > T_B$, does the determination of β lead to a final estimated energy for system B greater than the initial estimate?

8

MUTUAL INTERACTIONS
IN A GAS

I have once previously treated the problem of
the dissociation of gases, on the basis of the most general
possible assumptions, which of course I had to specialize at the end.

Ludwig Boltzmann
Lectures on Gas Theory

The interaction of particles among themselves is an inevitable characteristic of physical systems. Nonetheless, we have skirted around that characteristic in our applications of the canonical distribution. The reason is not obscure: explicit treatment of the mutual interactions vastly increases the mathematical complexity of the analysis. Indeed, it is probably safe to say that no realistic three-dimensional physical system with mutual interactions has yet been treated exactly in statistical mechanics. Mathematical complexity in the use of the canonical distribution forces a resort to approximation methods. One is compelled to strike a compromise between fidelity to the physical situation and mathematical tractability.

So much for the prelude. In this chapter our primary objective is to treat a physical system without the disconcerting "neglect of mutual interactions." The physical system is a gas, with its inevitable intermolecular forces, in equilibrium at a known temperature. To achieve the objective we will, how-ever, need to make approximations. One would like to analyze the gas with the quantum-mechanical form of the canonical distribution, but that analysis would be appreciably more difficult than a classical treatment, and so we will use classical statistical mechanics. The analysis in Section 6.1 indicated that a classical treatment is often admissible for a gas, at least for describing the motion of the molecular centers of mass. Since the decision to use classical mechanics precludes an adequate handling of atomic or molecular structure, we will take the molecules to be structureless, except insofar as the structure partially determines the behavior of the intermolecular forces. At the cost of some additional algebra we could deal with molecules having a non-zero spin, but since the rewards would not be worth the effort, we will take the molecules to be spinless. In short, we will work with a classical gas of structure-less molecules. Let us specify that there are N molecules of only one species present in the gas. Lastly, let us take an uncompromisingly classical view and regard the molecules as somehow distinguishable (despite their being all of the same species). This establishes the framework for the chapter.

8.1. THE PERFECT CLASSICAL GAS

This section is intended to generate confidence in the partition function technique. For a preliminary calculation we will neglect the intermolecular forces and work with a perfect classical gas of structureless molecules. Though somewhat artificial, such a gas is a reasonable approximation to a real gas if the latter is sufficiently dilute. We set out to estimate the total energy and pressure by calculating the expectation values of those quantities. By using the partition function, we will be able to do so in just a few steps.

We need the classical expression for the system's total energy as a function of the position and momentum variables. If we denote the molecular mass by m, that energy expression is given by

$$E = \sum_{i=1}^{N} \left(\frac{p_i^2}{2m} + U_{\text{box } i} \right). \tag{8.1.1}$$

As always, the potential energy U_{box} describes the forces that the walls exert to contain the molecules within a volume V. When the center of mass of molecule i is within the box, $U_{\text{box } i}$ is zero; when the center of mass is formally outside, $U_{\text{box } i}$ is infinite. Incidentally, the container shape need not be cubical.

Next we need to transcribe the partition function from quantum-mechanical to classical terms. Here the transcription takes the form

$$Z = \sum_{j=1}^{n} e^{-\beta E_j} \rightarrow$$

$$\int_{\substack{\text{all phase} \\ \text{space}}} \exp\left[-\beta \sum_{i=1}^{N} \left(\frac{p_i^2}{2m} + U_{\text{box } i}\right)\right] \frac{d^2 x_1 \, d^3 p_1 \cdots d^3 x_N \, d^3 p_N}{O(\hbar)^{3N}}. \quad (8.1.2)$$

We replace a summation over all energy eigenstates of the system by an integration over all phase space. The infinitesimal volume element in phase space must be divided by $O(\hbar)^{3N}$ to ensure, in the classical limit, the correct counting of quantum-mechanical states for this set of N molecules, presumed to be distinguishable. (The transcription procedures used in this chapter will be discussed in more detail in Section 9.7.)

The evaluation of the integral goes remarkably rapidly. Since the exponent is a sum of identical terms (except for labels), we may factor the integrand into a product of N identical factors (except for labels). With each such factor, of the form

$$\exp\left[-\beta\left(\frac{p_i^2}{2m} + U_{\text{box } i}\right)\right],$$

we associate the corresponding differentials,

$$\frac{d^3 x_i \, d^3 p_i}{O(\hbar)^3}.$$

The integrations go from minus to plus infinity for each variable. For each factor we have, effectively, an integration over the six-dimensional phase space of a single molecule. The end-result of the factorization is the conclusion that the classical Z for the perfect gas may be written as

$$Z = \left[\int_{\substack{\text{phase space of} \\ \text{one molecule}}} \exp\left[-\beta\left(\frac{p^2}{2m} + U_{\text{box}}\right)\right] \frac{d^3 x \, d^3 p}{O(\hbar)^3}\right]^N.$$

Only the neglect of the intermolecular forces permits this great mathematical simplification, as will later become painfully clear. The integration over the six-dimensional phase space has been performed already. Upon carrying over the results from the beginning of Section 6.4, we arrive at

$$Z = \left[\frac{V}{O(\hbar)^3} \left(\frac{2\pi m}{\beta}\right)^{3/2}\right]^N. \quad (8.1.3)$$

For the ensuing computations we need primarily the dependence of $\ln Z$ on β and on V. After taking the logarithm of both sides of equation (8.1.3) and rearranging for convenience, we find

$$\ln Z = N \ln V - \tfrac{3}{2} N \ln \beta + \ln[(2\pi m)^{\frac{3}{2}N} O(\hbar)^{-3N}].$$

The expectation value of the energy and that of the pressure follow directly by appropriate partial differentiation of $\ln Z$. The derivations are in Sections 7.1 and 7.2, respectively. With astonishing ease we compute those expectation values as

$$\langle E \rangle = -\frac{\partial \ln Z}{\partial \beta} = +\frac{3}{2} N \frac{1}{\beta} = \frac{3}{2} NkT,$$

$$\langle p \rangle = \frac{1}{\beta} \frac{\partial \ln Z}{\partial V} = \frac{1}{\beta} N \frac{1}{V} = \frac{NkT}{V}.$$

We cannot consider the results *per se* much of a triumph, for they were partially built into the theory. Nonetheless, that was for the quantum-mechanical version, so we may look on the results as support for some aspects of the transcription to classical terms.

8.2. THE VALUE OF THE FACTOR $O(\hbar)$

At this point we should perhaps satisfy our curiosity and determine the precise value of the factor $O(\hbar)$ that appears in the classical phase-space limit of quantum mechanics. We can do this by comparing the classical partition function of the preceding section with the quantum-mechanical partition function for a single particle. A calculation of the latter is tractable and, indeed, most of the hard work has already been done in Section 4.6.

For a single spinless particle confined to a cubical box of volume V, an energy eigenvalue E_j has the mathematical form

$$E_j = \frac{\pi^2 \hbar^2}{2mV^{2/3}} (n_x^2 + n_y^2 + n_z^2).$$

To each index j corresponds a specific set of the three positive integers n_x, n_y, n_z. The partition function for a single particle—let us call it $Z(1)$—is given by a summation of $\exp(-\beta E_j)$ over the particle's energy eigenstates. Here such a sum may be written as a sum over all admissible values of the three integers and then simplified:

$$Z(1) = \sum_{n_x=1}^{\infty} \sum_{n_y=1}^{\infty} \sum_{n_z=1}^{\infty} \exp\left[-\beta \frac{\pi^2 \hbar^2}{2mV^{2/3}} (n_x^2 + n_y^2 + n_z^2)\right]$$

$$= \left[\sum_{n_x=1}^{\infty} \exp\left(-\beta \frac{\pi^2 \hbar^2}{2mV^{2/3}} n_x^2\right)\right]^3.$$

The step follows because the form of the exponent and symmetry permit the sum over n_x, n_y, n_z, to be split into a product of three identical sums, each running from one to infinity.

In the classical limit the constant factor in the exponent of the summation will certainly be much less than unity—

$$\beta \frac{\pi^2 \hbar^2}{2mV^{2/3}} \ll 1$$

—because the thermal de Broglie wavelength is then much shorter than a typical linear dimension of the container. Since the lefthand side of the claimed inequality is roughly the square of the thermal de Broglie wavelength divided by the square of an edge length of the cube, the lefthand side will, in the classical limit, be much less than unity. This establishes the inequality, which, in turn, indicates that the summation may satisfactorily be replaced by an integration.

$$\sum_{n_x=1}^{\infty} \exp\left(-\beta \frac{\pi^2 \hbar^2}{2mV^{2/3}} n_x^2\right) \simeq \int_{\eta=0}^{\infty} \exp\left(-\beta \frac{\pi^2 \hbar^2}{2mV^{2/3}} \eta^2\right) d\eta$$

$$\simeq \frac{\sqrt{\pi}}{2} \left(\beta \frac{\pi^2 \hbar^2}{2mV^{2/3}}\right)^{-1/2}.$$

When one inserts this evaluation into the expression for $Z(1)$ and rearranges a bit, the result is

$$Z(1) = \frac{V}{h^3} \left(\frac{2\pi m}{\beta}\right)^{3/2}.$$

To determine the precise value of the factor $O(\hbar)$, we need only compare $Z(1)$ with the classical Z of equation (8.1.3). Either inspection or the setting of N equal to unity in Z implies

$$O(\hbar) = h.$$

For each pair of variables, x and its associated p_x, a factor of $1/h$ is required in computing the number of quantum-mechanical states corresponding to a region $dx\, dp_x$ in the classical phase space.

8.3.   A QUALITATIVE PREVIEW

We turn now to a qualitative examination of the effects of intermolecular forces on the pressure. We look for corrections to the perfect gas-law relation,

$$p_{\text{obs}} = \frac{NkT}{V}.$$

The results will give us an idea of what to expect in the quantitative calculations.

First we should develop a picture of the spatial dependence of the intermolecular forces. That gases do, at sufficiently high density and low temperature, condense to form a liquid, implies the existence of attractive forces between (neutral) molecules. The attraction must, however, become vanishingly small as the separation between molecules increases, for if it did not, the neglect of intermolecular forces in the analysis of a dilute gas would lead to disagreement with observation. We may assume that the attractive intermolecular force effectively vanishes when two molecules have a separation that is large relative to a typical molecular size. There must be a repulsive force as well, for it is difficult to compress a liquid appreciably. The inference is that when two molecules come sufficiently close, the attractive force turns into a repulsive force, as though there were two rather hard spheres in contact. These arguments from a few commonplace physical facts lead to a classification of how the intermolecular force depends on the relative separation of two molecules. That classification is given in table 8.3.1.

Now we may indulge in some admittedly loose arguments about the effects on the pressure. Let us take first the repulsive part of the force, the part that arises from the "finite size" of a molecule. In a classical picture the repulsion at small distances means that interpenetration of two molecules is impossible, and so the volume through which a chosen molecule may roam is less than the volume V of the container. To assess the effect on the pressure, let us note that the molecule, after rebounding from a wall, may head back to the same wall as the result of a repulsive collision with another molecule. It need not make a round trip to the opposite wall and then back. This implies an increase in pressure over that predicted by the perfect-gas law. The notion of an *effective volume* provides a means of incorporating mathematically the influence of the repulsive forces. The effective volume is a volume reduced below the box volume V by a term proportional to the volume actually occupied by the N molecules because of their finite size:

$$V_{\text{effective}} = V - Nb.$$

TABLE 8.3.1

The nature and spatial dependence of the intermolecular force between two neutral molecules.

Relative separation	Nature of force
Zero to order of two "molecular radii"	Repulsive
Order of two "molecular radii" to order of five or ten[a]	Attractive
Larger than order of five or ten "molcular radii"	Effectively zero

[a] The outer limit follows because air at room temperature and pressure behaves very much like a perfect gas. Under those conditions the typical molecular separation is about twenty "molecular radii," that is, about 35 Ångstroms of separation relative to about 3.5 Ångstroms of "molecular diameter."

The symbol b represents a volume of molecular size (and hence b is positive). A replacement of the container volume by the effective volume modifies the pressure expression to

$$\langle p \rangle = \frac{NkT}{V_{\text{effective}}} = \frac{NkT}{V - Nb}.$$

The notation $\langle p \rangle$, rather than p_{obs}, is used because we are now making theoretical estimates of the pressure.

Now for the attractive part of the intermolecular force. When a molecule is quite near a wall, there is a net inward pull on it due to the attractive forces from the nearby molecules behind it. This pull tends to diminish the momentum with which the chosen molecule hits the wall and hence diminishes also the pressure on the wall. The inward pull on a molecule will be proportional to the number of molecules behind it (which pull on it) and that, in turn, will be proportional to the number density N/V. To frame an expression for the reduction in pressure, we note that the pressure depends on the number of molecules near the surface and that this, too, is proportional to N/V. So the (subtractive) correction to the pressure is proportional to $(N/V)^2$. With a positive factor a to represent the effect of the attractive forces, we may combine the repulsive and attractive corrections, and arrive at the estimate

$$\langle p \rangle = \frac{NkT}{V - Nb} - a\left(\frac{N}{V}\right)^2. \tag{8.3.1}$$

Though I would not care to stake my life on these arguments, this result is the justly celebrated *van der Waals equation*. It dates back to van der Waals' dissertation, written in 1873. (There are many equivalent algebraic forms and ours may not be the most common one.) Judiciously employed, the van der Waals equation has had phenomenal success in describing qualitatively the behavior of imperfect gases and even liquids. Although a closer attention to detail in the "derivation" gives more explicit expressions for the coefficients a and b, they are best determined empirically—incidentally, they do depend somewhat on temperature—but none of that need concern us here. For us the relevant points are two disparate ones: (1) that a little thought can lead to physically reasonable corrections to the perfect-gas law, and (2) that the van der Waals equation suggests a kind of series expansion for the corrections.

The second point is not obvious, but one rearrangement shows that both of the van der Waals corrections depend directly on the number density N/V:

$$\langle p \rangle = \frac{NkT}{V}\left(1 - b\frac{N}{V}\right)^{-1} - a\left(\frac{N}{V}\right)^2.$$

For sufficiently small N/V the corrections become negligible, and the expression reduces to the perfect-gas law. If we divide both sides by kT, expand the first term on the right with the binomial theorem, and collect terms of the same power of N/V, we find

$$\frac{\langle p \rangle}{kT} = \frac{N}{V} + \left(b - \frac{a}{kT}\right)\left(\frac{N}{V}\right)^2 + O\left(\frac{N}{V} \times \left[b\frac{N}{V}\right]^2\right). \tag{8.3.2}$$

Let us note in passing that, even for a dilute gas, N/V may be quite a large number. For air at room temperature and pressure, $N/V \simeq 2.4 \times 10^{19}$ molecules per cubic centimeter. The binomial expansion is, of course, necessarily an expansion with a dimensionless quantity, here the ratio bN/V, and for a reasonably dilute gas the ratio will be much less than one. The coefficient b is roughly the volume of a molecule itself, and V/N is the volume per molecule of the container. So bN/V, which may be looked on as $b/(V/N)$, will be quite small in all but high-density gases. For air under room conditions it is of order 10^{-3}.

The expanded form of the van der Waals equation suggests that, for sufficiently small N/V, a series expression for the corrections in ascending powers of N/V should both converge and give a suitable representation of the pressure in an imperfect gas. The coefficients of the powers of N/V will, in general, be temperature-dependent. The general form of the expansion,

$$\frac{\langle p \rangle}{kT} = \frac{N}{V} + B_2(T)\left(\frac{N}{V}\right)^2 + B_3(T)\left(\frac{N}{V}\right)^3 + \ldots, \qquad (8.3.3)$$

is known as a *virial expansion* and the coefficients $B_2(T)$, $B_3(T)$, and so on, as the *virial coefficients*.

> The origin of the designation lies in the work of Clausius. In 1870 he published a paper entitled " On a Mechanical Theorem Applicable to Heat " (as translated from the German). Clausius's theorem related the time-average of the kinetic energy of a group of particles to the forces acting on them, most relevently for us, to the intermolecular forces and the force due to the " box walls." Since the last is directly related to the pressure (by Newton's third law), the theorem—the virial theorem—enabled one to perform a systematic calculation of the effects of intermolecular forces on the pressure. The theorem has been widely applied in astrophysics as well. The word " virial " itself is derived from the Latin *vis* (plural, *vires*) meaning " force." Hence " virial " is an appropriate adjective for the coefficients representing the effects of the intermolecular forces.

Even though N/V may be quite large when measured in units of molecules per cubic centimeter, the virial coefficients themselves are (in the appropriate units) quite small. The series should—one may hope—converge, at least until conditions of density and temperature are such that condensation occurs.

This should suffice for a preview, but before we launch into a calculation of $\langle p \rangle$ from first principles, we should discuss in more detail the behavior of intermolecular forces.

8.4. THE BEHAVIOR OF INTERMOLECULAR FORCES

For the temperatures we envisage, we may restrict consideration to the forces between molecules that, each taken as a unit, are neutral. The probability that a molecule is ionized will be exceedingly small. (Between ions the long-range Coulomb forces, though simple in principle, produce considerable complication in practice. The difficulty of calculations in plasma physics bears witness to this.) With two neutral molecules a force can arise only if the molecules, though over-all neutral, have an uneven distribution of charge leading to an electric dipole moment (or higher moment, such as an electric quadrupole moment). A magnetic interaction, as between a pair of magnetic dipoles, is possible but is generally negligibly small. The electric dipole moment may be a natural, permanent one, as in a water molecule or in the asymmetric diatomic molecules of carbon monoxide and hydrogen chloride. For example, in hydrogen chloride the hydrogen end is slightly positive. Let us leave this case until later and discuss first molecules without such permanent moments.

The noble gases, such as helium and neon, have "monatomic molecules." These are spherically symmetric in their ground states and provide examples of molecules without permanent electric dipole moments. So do symmetric linear molecules, such as diatomic nitrogen and carbon dioxide. The latter have no permanent electric dipole moment, for "it wouldn't know which way to point." The same holds for a molecule like methane (CH_4), in which the hydrogen atoms are symmetrically arranged on the points of a tetrahedron surrounding the carbon atom. Though without permanent dipole moments, these molecules can have a fleeting electric dipole moment. One has only to take a purely classical view of electrons in orbit around the nuclei to see that the inevitable separation of positive and negative charge will result in a rapidly fluctuating dipole moment. Given that such a molecule does momentarily have a dipole moment, the associated electric field will polarize a nearby molecule. As figure 8.4.1 indicates for a few positions of the second molecule relative to the first, the induced dipole will be oriented relative to the inducing dipole in such a way that there will be an attractive force between the two molecules.

In a strictly classical calculation of the interaction between the fleeting dipole and the induced dipole, the force will not, generally, be purely along the line joining the two molecules. When averaged over all orientations of the dipole moment of the first molecule, however, the classical force will be precisely along that line. (A quick argument: after the averaging there is no other "preferred" direction in space.) The proper approach, in any case, is with quantum mechanics. Such a calculation automatically does some "averaging" and leads to the conclusion that the attractive force is purely along the line joining two (initially spherically symmetric) molecules.

For use in the canonical distribution we need the distance dependence of the potential energy, rather than that of the force. Lining up the dipoles as in figure 8.4.2 simplifies the geometric analysis. The electric potential $\varphi(r)$ produced by the first dipole in the diagram falls off with distance as $1/r^2$, giving an electric field $E(r)$ whose magnitude falls off with distance as $1/r^3$. The positive and negative parts, $\pm q$, of the second molecule would contribute with equal magnitude but opposite sign to the consequent potential energy

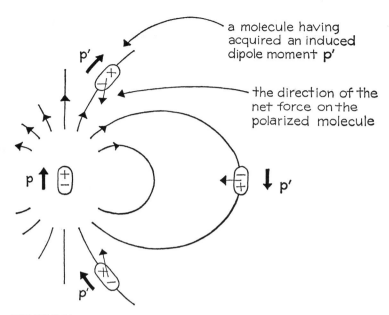

FIGURE 8.4.1

The molecule at left center has a fleeting dipole moment **p**, indicated by
the broad arrow. This dipole moment produces an electric field, some of
the field lines of which are shown. The electric field polarizes a second
molecule in the vicinity; three such "second" molecules are
displayed. The second molecules, now with oppositely charged ends
because they have been polarized, feel a net force because the
electric field from the first molecule has (slightly) different direction and
magnitude at the two ends of the second molecule.

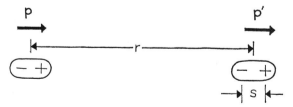

FIGURE 8.4.2

The molecule at the left, with dipole moment **p**,
has polarized the molecule at the right, inducing a
dipole moment **p′**. The dipole moment of the second
molecule is represented by two charges, $\pm q$, separated
by a small distance s.

were it not that the oppositely charged parts are spatially separated, by a distance s, say. It is the variation in electric potential (due to the first dipole) between the $\pm q$ parts of the second dipole that leads to a non-zero potential energy $u(r)$:

$$u(r) = q\varphi(r + \tfrac{1}{2}s) - q\varphi(r - \tfrac{1}{2}s)$$

$$\simeq q\left(\varphi(r) + \frac{s}{2}\frac{d\varphi}{dr}\right) - q\left(\varphi(r) - \frac{s}{2}\frac{d\varphi}{dr}\right)$$

$$\simeq qs\frac{d\varphi}{dr}.$$

The product qs is the magnitude of the induced dipole moment $\mathbf{p}'$. Provided the molecules are not too close, the induced dipole moment will be proportional to the inducing electric field $E(r)$. Thus we may write

$$u(r) \simeq qs\frac{d\varphi}{dr} = p'\frac{d\varphi}{dr}$$

$$\propto E(r)\frac{d\varphi}{dr} \propto \frac{1}{r^3} \times \frac{1}{r^3} = \frac{1}{r^6},$$

that is,

$$u(r) \propto 1/r^6.$$

For two molecules without permanent dipole moments, the leading term in the attractive part of the potential energy goes as $1/r^6$.

Let us return for a moment to figure 8.4.1. A little thought and some sketching will show that the electric field that a second molecule produces (because of its induced dipole moment) in the vicinity of the first molecule tends to increase the dipole moment of the first molecule. We have really a reciprocal affair, each molecule helping to polarize the other. For this reason the interaction is often called an induced-dipole induced-dipole interaction.

Now we turn to the interaction between two molecules with permanent electric dipole moments. The potential energy, for given relative orientation, does not fall off as rapidly as $1/r^6$. Rather, it will go as $1/r^3$, for there is no $1/r^3$ dependence in the magnitude of the second dipole moment itself. The sign of the potential energy depends on the orientation of the two dipole moments relative to one another and to the line joining the two molecular centers of mass. Averaging over all orientations leads to a zero value for the potential energy. If, however, one averages with a weighting proportional to $\exp[-\beta(\text{potential energy})]$, that is, if one computes the expectation value of the potential energy, a non-zero potential energy results. It corresponds to an attractive force. The orientations with attractive force have negative potential energy relative to the repulsive orientations, making the exponent in the weighting factor positive; thus they dominate and determine the sign of the expectation value. The leading term in the expectation value of the potential energy, calculated by expanding the exponential about a value of zero for the exponent, goes as $1/r^6$.

The preceding covers the essential aspects of the attractive part of the potential. For the close-in, repulsive part one can say rather little. For simple interactions, like those between two hydrogen or two helium atoms, some quantum-mechanical calculations are available. They indicate a repulsion roughly expressible as having an exponential dependence. In all but the most refined calculations, it suffices to say that there *is* a repulsive part, for which the associated potential energy becomes very large and positive as soon as the molecules "touch."

A number of semiempirical models portray, in a more or less convincing fashion, the variation in potential energy as the relative separation varies from infinity to zero. The Sutherland potential-energy model is the simplest that incorporates the outer attractive $1/r^6$ part and the known existence of a strong repulsive core. If we denote the relative separation by r and the mutual potential energy by $u(r)$, the model specifies the spatial dependence of $u(r)$ as

$$u(r) = \begin{cases} +\infty, & \text{for } r < r_0; \\ -u_0\left(\dfrac{r_0}{r}\right)^6, & \text{for } r \leqslant r_0. \end{cases} \qquad (8.4.1)$$

This is illustrated in figure 8.4.3. The constant r_0 has the dimensions of a

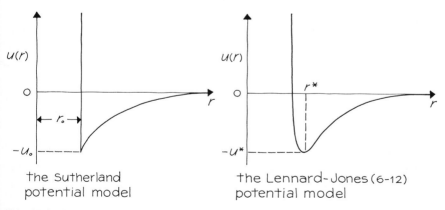

FIGURE 8.4.3

Here we have two more or less convincing models for the intermolecular potential energy of two small molecules without permanent electric dipole moments. On the left is the Sutherland potential, on the right the Lennard-Jones (6-12) potential.

length; one may look on r_0 as being comparable to two "molecular radii." Thus it should be of the order of a few Ångstroms or a few times 10^{-8} cm. The positive constant u_0 is the depth of the attractive potential well. Typically, u_0 ranges between about 0.01 to 0.05 electron volts, and is thus quite small relative to the well depths for intramolecular forces, which tend to be of the order of a few electron volts. To be sure, the ultimate origins of intermolecular and intramolecular forces are the same, but a useful qualitative distinction can be made, just as one distinguishes two interacting (but not bound) hydrogen atoms from the molecule H_2. In the Sutherland model there are two parameters that may be adjusted on the basis of experimental data.

The infinitely sharp rise in the Sutherland potential, though certainly not physically correct, is at times not a bad representation of the short-range repulsive forces. For collisions with low initial kinetic energy, a pair of real molecules will effectively be impenetrable, just like a pair of hard spheres. The Sutherland potential represents this reasonably well and has analytic simplicity. For very high temperature and hence very high initial kinetic energy, however, a pair of molecules will begin to interpenetrate significantly, a behavior that the vertical potential barrier of this model precludes. One can expect the model to fail for the description of a gas at relatively high temperature, and such a failure has been observed for helium and (less clearly) for neon.

More sophisticated are the various versions of the Lennard-Jones potential. In this model the potential energy due to the repulsive forces is represented by another inverse power of r, $\sim 1/r^n$. Various values for the exponent n are possible, as well as for the coefficients of the $1/r^6$ and $1/r^n$ terms. The data seem not to imply a unique choice—though they do indicate that n should be between 8 and 14—because the experiments to date do not strongly depend on the details of the repulsive forces. Often analytic convenience leads to the choice $n = 12$ and the so-called Lennard-Jones (6–12) potential:

$$u(r) = u^* \left[\left(\frac{r^*}{r} \right)^{12} - 2 \left(\frac{r^*}{r} \right)^6 \right]. \qquad (8.4.2)$$

This is again a two-parameter potential. As the diagram in figure 8.4.3 shows, for large r the $1/r^6$ term dominates, and one has the potential for the induced-dipole induced-dipole interaction. For very small r the situation is reversed, with the $1/r^{12}$ term dominating and representing, in some reasonable fashion, the repulsive core. The minimum in the potential energy curve occurs at $r = r^*$. At this point the force changes from repulsion to attraction; thus r^* may be interpreted as a separation of roughly two molecular radii.

There are several other models in use, but these two suffice to display the typical form. A fine compilation is given by Hirschfelder *et al.* in the book listed at the end of this chapter.

8.5. A PRESSURE CALCULATION FOR THE INTERACTING GAS

Now that the preliminaries are over, we can turn our attention to the statistical calculation *per se*. For analytic convenience, and because the method is readily extended and generalized, we will use a partition-function approach. Once the partition function has been calculated to some order of approximation, the pressure estimate will follow directly by a partial differentiation. It should be borne in mind that the partition-function approach is nothing but a mathematical technique for working with the canonical distribution: we are not deserting that probability distribution.

Specific assumptions about the order of approximation and about the details of the intermolecular forces will be made only as we need them. By proceeding in this fashion we will see at each step where we have restricted the generality of the calculation. But one important assumption about the intermolecular forces is needed immediately.

ASSUMPTION I:
We assume that the total intermolecular potential energy depends only on the *positions* of the molecules and not on their relative orientations or velocities.

Rigorously, this assumption restricts the calculation to a gas of spherically symmetric monatomic "molecules," such as helium, neon, and other noble gases, or to a gas of molecules with exceptionally high symmetry, such as methane. The diatomic molecules H_2 and N_2 do, however, act as though nearly spherical. For most small molecules without permanent dipole moments, the assumption is not a bad approximation, and even for small molecules with permanent moments the calculation should give a reasonable estimate. Thus the assumption is not as restrictive as it might seem.

For the classical energy expression we may now write

$$E = \sum_{i=1}^{N} \frac{p_i^2}{2m} + \sum_{i=1}^{N} U_{\text{box } i} + U(x_1, \ldots, x_N), \qquad (8.5.1)$$

The first two terms are familiar. The last is the total intermolecular potential energy under the assumption that it depends on molecular positions only.

For the classical partition function, we use the transcription already employed in Section 8.1 and write

$$Z = \int \exp\left[-\beta\left(\sum_i \frac{p_i^2}{2m} + \sum_i U_{\text{box } i} + U \right) \right] \frac{d^3x_1 \, d^3p_1 \cdots d^3x_N \, d^3p_N}{h^{3N}}.$$

Because the exponent has the form of a sum, the integrand factors into a part that is dependent purely on the momenta and another that is dependent only on the position variables. The momentum integrations can be done just as for a perfect gas. The analogy can be made evident by writing the result as

$$Z = \left(\frac{2\pi m}{h^2 \beta} \right)^{3/2N} V^N \times \left(\frac{Z_U}{V^N} \right). \tag{8.5.2}$$

The V^N has been inserted in order to make the first factor precisely the partition function for a perfect gas. The second factor, an abbreviation, contains the actual spatial integrations:

$$\frac{Z_U}{V^N} \equiv \frac{1}{V^N} \int \exp\left[-\beta\left(\sum_i U_{\text{box } i} + U \right) \right] d^3x_1 \cdots d^3x_N$$

$$= \frac{1}{V^N} \int_V \exp(-\beta U) \, d^3x_1 \cdots d^3x_N. \tag{8.5.3}$$

The step to the second line amounts merely to tidying up the expression: the specification (with a subscript V on the integral sign) that the spatial integration for each particle is to go over the box volume only permits one to drop U_{box} (for it is zero within that volume). If we were to set U to zero, the function Z_U would go to V^N; the ratio Z_U/V^N would go to unity; and thus Z itself would become merely the perfect-gas partition function. As far as the full partition function is concerned, all the effects of the intermolecular forces are contained in the function Z_U.

The problem has been reduced to that of calculating Z_U, but to make further progress we must introduce two additional assumptions about the explicit form and behavior of the potential energy U.

ASSUMPTION 2:

We assume that U is a sum of interaction potentials between pairs of molecules, each such potential being dependent on only the magnitude of the relative separation.

Thus a pair of molecules, labelled j and k, makes a contribution

$$u_{jk} = u(|\mathbf{x}_j - \mathbf{x}_k|)$$

to the total intermolecular potential energy. The latter is a sum of all such contributions:

$$U = \sum_{\text{pairs}} u_{jk}.$$

The summation is to count each distinct pair of molecules once.

The assumption of purely "pair" potentials excludes effects like that of one molecule distorting two others and thus affecting the potential energy between those two. Such a contribution to the energy would require for its description a function of the position variables of all three molecules. For relatively low density, such an effect may be neglected. The form for U adopted by us does, of course, describe the interaction of any chosen molecule with all others simultaneously, though only through "pair" interactions.

ASSUMPTION 3:

The pair intermolecular potential u_{jk} has a dependence on relative separation *qualitatively* like that given by the Lennard-Jones or Sutherland model.

This assumption means merely a high positive inner potential, a negative intermediate region, and a rapid rise to zero beyond that. The last implies that the intermolecular force has a rather short range, on the order of five molecular radii, say.

This much knowledge of the spatial dependence of the potential is an asset in setting up a procedure for evaluating Z_U/V^N. That function now takes the form

$$\frac{Z_U}{V^N} = \frac{1}{V^N} \int_V \exp\left(-\beta \sum_{\text{pairs}} u_{jk}\right) d^3x_1 \cdots d^3x_N$$

$$= \frac{1}{V^N} \int_V e^{-\beta u_{12}} e^{-\beta u_{13}} e^{-\beta u_{23}} e^{-\beta u_{34}} \cdots d^3x_1 \cdots d^3x_N. \tag{8.5.4}$$

The intermolecular potentials, with their dependence on variable combinations like $|x_1 - x_2|$, preclude the trick of factoring the integral into manageable *independent* three-dimensional integrals. To make progress we need to look at the integral in a different way.

We may view the integral, together with the initial factor of $1/V^N$, as giving the average value of the integrand, with the average being taken over all configurations of the molecules. In the course of integration, the position vectors $x_1, \ldots, x_N$ range, independently, over all points within the box, and so the integration as a whole ranges over all geometrically conceivable configurations of the molecules within the box. Each such configuration appears once and only once, and when it does appear, it contributes its value of the integrand to the average. The number of configurations with particle 1 in d^3x_1 around $\mathbf{x}_1$, particle 2 in $\ldots$, and particle N in d^3x_N around $\mathbf{x}_N$ is proportional to the product of the small volumes, $d^3x_1 \cdots d^3x_N$. That product, times the integrand, is the contribution of those configurations to the integral, which is a sum of such contributions. Similarly, the total number of configurations is proportional to V^N, and so dividing the integral by V^N is equivalent to dividing a sum by the total number of terms taken in forming the average. Thus we find

$$\frac{Z_U}{V^N} = \frac{\displaystyle\sum_{\substack{\text{all} \\ \text{configurations}}} \begin{pmatrix} \text{the value of the integrand} \\ \text{for the specific molecular} \\ \text{configuration} \end{pmatrix}}{\begin{pmatrix} \text{the total number} \\ \text{of configurations} \end{pmatrix}},$$

and thereby we justify the claim made in the first sentence of this paragraph. (To be sure, the two proportionality factors that we invoked to relate numbers of configurations to volumes, are infinite, but they are the same, and hence cancel in the division. The cancelation permits us to get away with a sloppy, but intuitive, description of a legitimate average.)

In a moderately dilute gas, most configurations appearing in the integration will have the molecules widely separated. Treating accurately only those configurations and regarding the effect of the others as unimportant provides the lowest approximation to an evaluation of Z_U/V^N. For the widely dispersed configurations, each exponential factor may be set to unity because the inter-molecular potentials will be approximately zero. In lowest approximation, one does this for all configurations. Consequently the average of the integrand will be unity, and we will recover the results for a perfect gas. But we want to do better than that.

Next in importance to the configurations with all molecules widely separated come those containing some molecules close together in pairs. Not only configurations with a single pair but also those with many pairs will show up more often in the integration than configurations with a single close triplet. Figure 8.5.1 and its legend provide justification for this claim. When we seek to treat accurately the configurations with close pairs (as well as the totally dispersed configurations), we may no longer set each exponential factor in equation (8.5.4) to unity. We may, however, assume that when two or more of the exponential factors differ from unity, their values are not related in any way. (We will return to this point in a moment.) The assumed absence

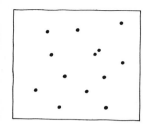

FIGURE 8.5.1
In this sketch of a gas of moderate density, most of the molecules are widely separated, but there is one close pair. Imagine removing some one molecule (not from the pair) and then tossing it back "at random" to generate a new configuration. That molecule is far less likely to produce a close triplet than merely another close pair. Even when a considerable number of pairs are already present, the probability of generating a triplet is small relative to that of getting just a single or another pair.

of correlations in values implies that the average of the product of the exponentials is equal to the product of the separate averages.

Here is a way of seeing the property claimed for the averages. Suppose we are instructed to compute the average of the product $\alpha_1\alpha_2$ of two variables, α_1 and α_2, by summing over a given set of pairs of values (and then dividing by the total number of pairs of values in the set). In the sense used above, absence of correlations means that from a knowledge of the value of one element of a pair (value of α_1, say), we cannot infer anything about the value of the other element (value of α_2). From this we conclude that all values of α_2 which are paired with a given value of α_1 are also paired with every other value of α_1. Holding α_1 fixed and summing over the values of α_2 which are paired with that α_1 then gives α_1 times [average of α_2], after division by the number of terms taken. This holds for each value of α_1. So one may complete the averaging by summing such an intermediate result over all values of α_1, thereby getting [average of α_1] times [average of α_2], again after division by the number of terms taken.

The number of factors in the product in equation (8.5.4) is $N(N - 1)/2$, the number of distinct pairs of molecules. Each factor will have an average equal to that of the factor $\exp(-\beta u_{12})$, for each differs only through the molecular labels. The fruit of the analysis is the intermediate result

$$\frac{Z_U}{V^N} = \left[\frac{1}{V^N} \int_V e^{-\beta u_{12}} \, d^3x_1 \cdots d^3x_N \right]^{\frac{1}{2}N(N-1)}. \qquad (8.5.5)$$

Now we go back to the matter of correlations. Imagine, in the spirit of the sketch used in figure 8.5.1, a typical configuration with two pairs and the other molecules widely dispersed. One pair might consist of particles 1 and 2; the other pair, of molecules 3 and 4. Then, in equation (8.5.4), both $\exp(-\beta u_{12})$ and $\exp(-\beta u_{34})$ would differ from unity (but only they). There is, however, no reason to expect their values to be related. Suppose particles 1 and 2 are very close together, with $u_{12} > 0$ and hence the exponential less than unity. Particles 3 and 4 might also be very close together, or they might be feeling one another's attractive forces, with $u_{34} < 0$ and hence the exponential greater than unity. With just pairs there is no correlation among nonunity values of the exponentials.

The situation is different for configurations with triplets. Let us consider a close triplet, particles 1, 2, and 3, say. If 1 and 2 are very close, with $u_{12} > 0$, and 1 and 3 are very close, with $u_{13} > 0$, then the corresponding two exponentials in equation (8.5.4) are less than unity. The significant point is that then particles 2 and 3 are likely to be very close also, making it probable that $u_{23} > 0$ and hence that the corresponding exponential is less than unity. Triplets do produce correlations. Because of those correlations, the intermediate result in equation (8.5.5) is only an approximation. Configurations with triplets (and groups of higher order) have been included in the averaging but not accurately, for such configurations have been treated as though they did not lead to correlations among those exponentials that differ from unity.

We could go on and treat accurately the configurations with triplets, but for a gas of moderate density doing so should not be necessary, and the mathematics does get more awkward. Let us halt the systematic approximation procedure at this point and codify the decision in a statement.

ASSUMPTION 4:

For a gas of moderate density, we get a good approximation for Z_U/V^N by treating accurately merely those configurations in which all the molecules are widely dispersed and those in which close pairs exist.

The mathematical import of this assumption is that we will use for Z_U/V^N the expression we derived as equation (8.5.5).

The remaining integral can be tidied up a great deal. The integrand is close to unity for most of the range of integration, for only rarely will particles 1 and 2 appear close together. This suggests that we write the integrand as unity plus the difference between the exponential and unity. The expression inside the square brackets in equation (8.5.5) becomes

$$\frac{1}{V^N} \int_V [1 + (e^{-\beta u_{12}} - 1)] \, d^3x_1 \cdots d^3x_N =$$

$$1 + \frac{V^{N-2}}{V^N} \int_V \int_V (e^{-\beta u_{12}} - 1) \, d^3x_1 \, d^3x_2.$$

The integration with integrand 1 gives V^N and hence the isolated 1 in the second line. For the second integrand, each integration with d^3x_i, when $i \neq 1$ or 2, may be done without reference to the integrand. Each such integration gives a factor of V, and together they generate the factor of

V^{N-2}. This leaves the integration over the positions of particles 1 and 2. Further improvement is possible here.

We must remember that u_{12} is a function of the relative separation $|\mathbf{x}_1 - \mathbf{x}_2|$ of the particles. Imagine holding $\mathbf{x}_2$ fixed a few molecular diameters or more inside the box and doing the $\mathbf{x}_1$ integration. The vector relations that we will need are indicated in figure 8.5.2. That integration is effectively an integration

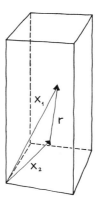

FIGURE 8.5.2
A sketch to illustrate the vector relations.

over the relative separation $r \equiv |\mathbf{x}_1 - \mathbf{x}_2|$ of the two molecules. It may be regarded as an integration with

$$e^{-\beta u(r)} - 1$$

as integrand, with the relative separation ranging from zero to infinity, for the integrand vanishes (to all intents and purposes) before the walls are reached. That integrand is sketched in figure 8.5.3. Polar coordinates provide the most useful means of expressing the integral over relative separation. With their aid we arrive at the expression

$$1 + \frac{V^{N-2}}{V^N} \int_V \left[\int_{r=0}^{\infty} (e^{-\beta u(r)} - 1)4\pi r^2 \, dr \right] d^3 x_2 .$$

Now imagine moving $\mathbf{x}_2$ about within the box and doing the integration with $d^3 x_2$. This means integration over the container volume and results in simply a factor of V. Upon canceling volume factors to the extent possible and inserting the result into equation (8.5.5), we arrive at the approximation,

$$\frac{Z_U}{V^N} = \left[1 + \frac{1}{V} \int_0^{\infty} (e^{-\beta u} - 1)4\pi r^2 \, dr \right]^{\frac{1}{2}N(N-1)} . \tag{8.5.6}$$

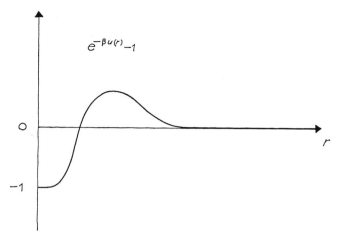

FIGURE 8.5.3
A sketch based on the Lennard-Jones (6-12) potential. As $u(r)$
goes to zero, the integrand goes to zero also, and rapidly.

There may be some uneasiness about moving x_2 close to the wall when the preceding integration assumed a distance of a few molecular diameters or more. The dubious volume involved, of the order of the surface area times a few molecular diameters, is sure in practice to be exceedingly small relative to the total volume V. So the error incurred is negligibly small, and the neglect of a refined calculation for x_2 near the wall is justified. For a common one-liter flask, the ratio of volumes is already of order 10^{-7} and becomes still smaller if V is increased at constant shape.

To calculate $\langle p \rangle$ we will want $\ln Z$. Since Z_U/V^N is a factor in Z, this means we need the logarithm of the righthand side of equation (8.5.6):

$$\ln\left(\frac{Z_U}{V^N}\right) = \tfrac{1}{2}N(N-1)\ln\left[1 + \frac{1}{V}\int_0^\infty (e^{-\beta u} - 1)4\pi r^2 \, dr\right]$$

$$= \tfrac{1}{2}N(N-1)\frac{1}{V}\int_0^\infty (e^{-\beta u} - 1)4\pi r^2 \, dr.$$

The step to the more convenient form in the second line, technically an approximation, is readily justified. The integrand is different from zero only out to a distance of several molecular diameters, and so the significant range of integration is microscopic. The value of the integral, therefore, will be exceedingly small relative to the volume V of the macroscopic container.

The Taylor's series expansion of a logarithm about the value 1 for the argument,

$$\ln(1 + x) = x + O(x^2) \quad \text{when} \quad |x| < 1,$$

may be used with complete confidence and yields the second line above. With the aid of that relation and of equation (8.5.2) we may write out the full expression for $\ln Z$:

$$\ln Z = \ln\left[\left(\frac{2\pi m}{h^2\beta}\right)^{\frac{3}{2}N} V^N \times \left(\frac{Z_U}{V^N}\right) \right]$$

$$= \ln\left(\frac{2\pi m}{h^2\beta}\right)^{\frac{3}{2}N} + N \ln V + \tfrac{1}{2}N(N-1)\frac{1}{V}\int_0^\infty (e^{-\beta u} - 1)4\pi r^2 \, dr.$$

(8.5.7)

To calculate the expectation value of the pressure, we use the method derived in Section 7.2. With the aid of equation (7.2.4) we find

$$\frac{\langle p \rangle}{kT} = \frac{\partial \ln Z}{\partial V}$$

$$= \frac{N}{V} - \frac{1}{2}\frac{N(N-1)}{V^2}\int_0^\infty (e^{-\beta u} - 1)4\pi r^2 \, dr.$$

(8.5.8)

And with this we have arrived: statistical mechanics has enabled us to in- corporate intermolecular forces into an honest, detailed calculation of the pressure in a gas of moderate density. (To be sure, one integral remains, the subject of the physics in the next section.) The range of densities for which the expression for $\langle p \rangle$ is accurate can be described in the terms of kinetic theory. Since we treated configurations with pairs accurately but not those with triplets, the expression should be accurate until so high a density is reached that triple collisions of molecules become important. We can certainly expect the expression to be good for densities much higher than that of air at typical room conditions.

8.6. THE SECOND VIRIAL COEFFICIENT

We may profitably compare our result for $\langle p \rangle$ with the virial expansion, given in equation (8.3.3). Since $N(N-1)$ is tantamount to N^2, the comparison suggests that our approximation procedure has been generating a virial expansion for $\langle p \rangle$, and so it has: an accurate treatment of the configurations with triplets leads to a term in $(N/V)^3$, and so on for groupings of higher order.

The comparison enables us to relate the integral in equation (8.5.8) and the second virial coefficient:

$$B_2(T) = -2\pi \int_0^\infty (e^{-\beta u} - 1)r^2 \, dr. \tag{8.6.1}$$

We may quite conveniently discuss the implications of our expression for $\langle p \rangle$ in terms of the second virial coefficient.

Let us first run some arguments backward. At high temperature and correspondingly large kinetic energies, the weak attractive part of the potential should be relatively unimportant. The strong repulsion between the cores cannot be neglected, and so should dominate, and lead, as in the van der Waals analysis, to an increase in the pressure beyond the perfect-gas law prediction. Thus for high temperature we may expect $B_2(T)$ to be positive. At low temperatures the core should play less of a role. The attractive part will, weakly and momentarily, produce clusters of molecules and will pull back those about to hit the walls. For low temperature we may expect a reduction in pressure and hence a negative value for $B_2(T)$. Further, by interpolation, we should anticipate a zero value for $B_2(T)$ at some intermediate temperature.

Now that we have a qualitative idea of what to expect, let us pick an explicit form for the potential $u(r)$ and calculate $B_2(T)$. For both the Sutherland and the Lennard-Jones potentials, the integral may be evaluated but only after a series expansion of all or part of the exponential. With analytic convenience in mind, let us choose the Sutherland model; it is a good representation for all but the very highest temperatures and kinetic energies.

In the Sutherland case the integration for $B_2(T)$ splits naturally into a part over the inner positive potential region and another over the negative potential region; the latter extends to infinity, though with rapidly vanishing integrand. Upon inserting the Sutherland $u(r)$ into the expression for $B_2(T)$, we have

$$B_2(T) = -2\pi \int_0^{r_0} \left(\exp[-\beta(+\infty)] - 1 \right) r^2 \, dr$$

$$- 2\pi \int_{r_0}^\infty \left(\exp[+\beta u_0(r_0/r)^6] - 1 \right) r^2 \, dr.$$

The first integral yields

$$-2\pi \int_0^{r_0} (0 - 1)r^2 \, dr = \frac{2\pi}{3} r_0^3.$$

To evaluate the second integral, we expand the exponential in a Taylor's series around zero value for the exponent, and then integrate term by term:

$$-2\pi \int_{r_0}^{\infty} \left(\exp[+\beta u_0(r_0/r)^6] - 1 \right) r^2 \, dr$$

$$= -2\pi \int_{r_0}^{\infty} \left(1 + \sum_{n=1}^{\infty} \frac{1}{n!} [\beta u_0(r_0/r)^6]^n - 1 \right) r^2 \, dr$$

$$= -2\pi \sum_{n=1}^{\infty} \frac{[\beta u_0 r_0^6]^n}{n!} \int_{r_0}^{\infty} \frac{r^2}{r^{6n}} \, dr$$

$$= -2\pi \sum_{n=1}^{\infty} \frac{[\beta u_0 r_0^6]^n}{n!} \frac{1}{(6n-3)} \frac{1}{r_0^{6n-3}}$$

$$= -\frac{2\pi}{3} r_0^3 \sum_{n=1}^{\infty} \frac{1}{(2n-1)n!} \left(\frac{u_0}{kT} \right)^n$$

$$= -\frac{2\pi}{3} r_0^3 \left[\frac{u_0}{kT} + \frac{1}{6} \left(\frac{u_0}{kT} \right)^2 + \cdots \right].$$

The particular factorization facilitates the comparison with the first integral. Still keeping the two integrals separate, we arrive at the result that follows from the Sutherland potential:

$$B_2(T) = \left[\frac{2\pi}{3} r_0^3 \right] + \left[-\frac{2\pi}{3} r_0^3 \sum_{n=1}^{\infty} \frac{1}{(2n-1)n!} \left(\frac{u_0}{kT} \right)^n \right]. \qquad (8.6.2)$$

The first contribution is from the repulsive part of $u(r)$, the second from the attractive part.

The Sutherland potential does predict a temperature for which $B_2(T)$ is zero. Inspection of the mathematical limits, $T \to 0$ and $T \to \infty$, indicates that the second term can range in value from minus infinity to zero. Consequently, there must be an intermediate temperature at which the negative second term precisely cancels the positive first term. At that temperature, known as the *Boyle temperature* T_B, the effects on the pressure due to molecular forces vanish (as far as order N^2/V^2). The tendency to increased pressure due to mutual repulsion is balanced by the tendency to decreased pressure due to mutual attraction. The value of the Boyle temperature is determined by looking for the temperature at which the sum in the second term has the value unity. A calculation by successive approximations yields $(u_0/kT_B) = 0.85$. This may be turned around: from an experimental Boyle temperature, the relation may be used to determine u_0, the well depth of the Sutherland potential. For nitrogen the Boyle temperature is 324 K, about 30°C above room temperature. A quick calculation yields a value of 0.024 electron volts for u_0.

At low temperature the attractive part dominates; $B_2(T)$ is negative; and the pressure is reduced below the perfect-gas law value. The transition to the high-temperature behavior is smooth, for one finds that the inequality

$$\frac{dB_2(T)}{dT} > 0$$

holds at all finite temperatures with the Sutherland model. Once past the Boyle temperature to higher temperature, the repulsive part dominates; $B_2(T)$ is positive, and the pressure is increased above the perfect-gas law value. Because of the infinitely hard repulsive core of the Sutherland model, this expression for $B_2(T)$ fails to give a positive maximum for $B_2(T)$ at very high temperature and then a gentle decline, as is actually observed for helium and strongly suggested for neon. On the model the molecules remain impenetrable to arbitrarily high energies.

Comparison with the expansion of the van der Waals equation, given in equation (8.3.2), permits us to interpret in more detail the coefficients b and a introduced in the derivation. For the Sutherland model the identification of the van der Waals coefficient b is clear:

$$b \quad \text{corresponds to} \quad \frac{2\pi}{3} r_0{}^3.$$

The coefficient b was introduced to represent a volume of roughly molecular size; the present correspondence sharpens the estimate of that volume.

The comparison for the van der Waals coefficient a, if we take a to be independent of temperature, requires a sorting out of the $1/kT$ term in the Sutherland result for $B_2(T)$, The procedure yields the correspondence,

$$a \quad \text{corresponds to} \quad \frac{2\pi}{3} r_0{}^3 u_0.$$

Since the attractive part of the potential is vanishingly small for distances greater than a few times r_0, the expression on the right may fruitfully be interpreted as a product: the depth of the attractive well times a measure of the volume over which the attractive force is effective. The coefficient a is, as introduced, an expression for the attractive forces and their range of influence.

This section closes with some values for well depths and effective molecular sizes extracted from observational data on the second virial coefficient. Experimental curves for $B_2(T)$ were compared with calculations based on the Lennard-Jones (6–12) potential, and the two parameters of that model were

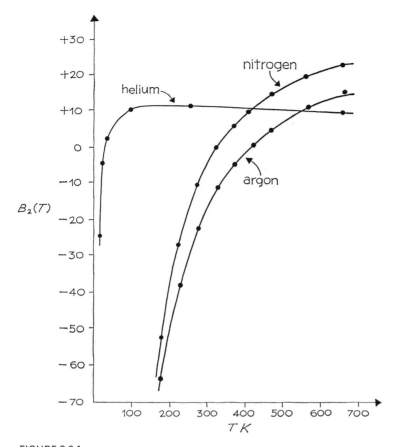

FIGURE 8.6.1

A comparison of some experimental data and theory for the second virial coefficient as a function of temperature. The dots represent experimental points. The curves were calculated with the Lennard-Jones (6-12) potential, the two parameters of the potential model being chosen for best fit. The theoretical curve for helium includes quantum-mechanical corrections to the classical expression, one of which is derived in Section 9.6. The second virial coefficient is given in units of cubic centimeters per mole or, equivalently, 1.660 cubic Ångstroms. The data and curves are from J. O. Hirschfelder, C. F. Curtiss, and R. B. Bird, *Molecular Theory of Gases and Liquids* (New York: Wiley, 1954).

then chosen to give the best fit. We can, from the display in figure 8.4.3, judge that the parameter u^* gives the depth of the attractive well and that r^* is the spatial separation at which repulsion changes to attraction. The latter parameter one *may* interpret further as a reasonable measure of the size of a molecule.

In table 8.6.1, the uniform increase in $r*$ as one proceeds along the periodic table for the noble gases (helium to xenon) is very satisfying. For a comparison with these typical intermolecular energies, note that the energy needed to pull apart the two hydrogen atoms in an H_2 molecule is 4.5 electron volts.

TABLE 8.6.1.

Representative values for the two parameters of the Lennard-Jones (6-12) potential from experimental data on the second virial coefficient.[a]

	Length parameter	Well-depth parameter	
Gas	$r*$(in Å)	$u*$(in ev)	$u*/k$(in K)[b]
He	2.869	0.00088	10.22
Ne	3.12	0.0030	34.9
A	3.822	0.0103	119.8
Kr	4.04	0.0147	171
Xe	4.60	0.0190	221
H_2	3.286	0.0032	37.00
N_2	4.150	0.0082	95.05
O_2	3.88	0.0102	118
CH_4	4.284	0.0128	148.2

[a] The values are from J. O. Hirschfelder, C. F. Curtiss, and R. B. Bird. *Molecular Theory of Gases and Liquids*, (New York: Wiley, 1954).
[b] This column gives $u*$ as an equivalent temperature, defined by $kT_{equivalent} = u*$. The Boyle temperature for each gas is about 3.42 times the equivalent temperature.

The values in table 8.6.1 are physically quite reasonable, but one must bear in mind that they refer to a specific (good)potential model. A fit to the same data with the Sutherland model would lead to significantly different numerical values for the well-depth parameter u_0 of that model. The comparison can be made by noting that there is only one experimental temperature for which the second virial coefficient is zero and that the parameters of each model must be chosen so that, at the observed Boyle temperature, the respective expressions for $B_2(T_B)$ are zero (or closely so). For the Sutherland model this means u_0 must be chosen so that the relation $(u_0/kT_B) = 0.85$ holds. The corresponding condition for the Lennard-Jones (6–12) model works out to be $(u*/kT_B) = 0.292$. Upon forming the ratio, we find

$$\frac{u_0}{u*} = \frac{0.85}{0.292} = 2.9.$$

This means that the well-depth u_0 of the Sutherland model must be about three times as large as the corresponding parameter $u*$ of the Lennard-Jones

(6–12) potential. The interpretation of this somewhat surprising result depends on the infinitely steep rise of the repulsive potential in the Sutherland model. The spatial range (and hence volume) over which the potential is negative is, effectively, smaller in the Sutherland model. To compensate, the well depth of the model must be greater.

The radial parameters are less sensitive to the detailed potential shape. One may reasonably require that the large r tail of the two potentials agree. This means

$$u_0(r_0)^6 \simeq 2u^*(r^*)^6$$

and so

$$r_0 \simeq \left(\frac{2u^*}{u_0}\right)^{1/6} r^* \simeq 0.94r^*.$$

The values of r_0 and r^* are much more closely comparable.

Since the Lennard-Jones (6–12) potential is both theoretically quite reasonable and empirically very successful, one can with considerable confidence take the values given in the table, together with the potential model, as providing a good representation of the actual potential (at least for $r > \frac{1}{2}r^*$ or so).

8.7. THE EFFECT OF INTERMOLECULAR FORCES ON S_I

Thus far we have paused only once, in Section 7.6, to calculate the information-theory entropy of a physical system and the associated probability distribution. Here we will examine S_I for the imperfect classical gas. The internal structure of the molecules will be neglected except insofar as it partially determines the intermolecular forces. Those forces produce additions to the perfect-gas S_I, and the prime reason for performing the calculation is that the additions will give us further insight into the meaning of S_I.

For the formal statement of the information-theory entropy, we use an extended version of equation (7.6.4). Since the canonical distribution is the relevant probability distribution, we have

$$S_I = -k \sum_{j=1}^{n} P_j \ln P_j$$

$$= k\left(-\beta \frac{\partial \ln Z}{\partial \beta} + \ln Z\right). \tag{8.7.1}$$

The relation between $\langle E \rangle$ and $\ln Z$, derived as equation(7.1.4), justifies the extended version, which happens to be the most convenient one for our purposes.

It is the partition function Z and its logarithm that we now need. Let us start with the still rather general expression given in equation (8.5.2):

$$Z = \left(\frac{2\pi m}{h^2 \beta}\right)^{\frac{3}{2}N} V^N \times \left(\frac{Z_U}{V^N}\right) \equiv Z_{\text{perf}} \times \left(\frac{Z_U}{V^N}\right). \tag{8.7.2}$$

The abbreviation Z_{perf} denotes the first factor, the partition function for a perfect gas. In the analysis to follow, we will want to distinguish S_I for a perfect gas from S_I when the molecular forces are taken into account. The factorization will enable us to maintain the distinction easily When we take the logarithm, the product splits into two distinct terms, the perfect-gas logarithm and a correction term produced by the intermolecular forces:

$$\ln Z = \ln Z_{\text{perf}} + \ln\left(\frac{Z_U}{V^N}\right).$$

Now we insert the split expression for $\ln Z$ into equation (8.7.1) and regroup to isolate the two contributions:

$$S_I = k\left[-\beta \frac{\partial}{\partial \beta} \ln Z_{\text{perf}} + \ln Z_{\text{perf}}\right] + k\left[-\beta \frac{\partial}{\partial \beta} \ln\left(\frac{Z_U}{V^N}\right) + \ln\left(\frac{Z_U}{V^N}\right)\right].$$

The first general term is the perfect-gas information-theory entropy; let us abbreviate it as $(S_I)_{\text{perf}}$. The logarithm necessary for working it out is

$$\ln Z_{\text{perf}} = \ln\left[\left(\frac{2\pi m}{h^2 \beta}\right)^{\frac{3}{2}N} V^N\right]$$

$$= -\tfrac{3}{2}N \ln \beta + \text{(a term independent of } \beta).$$

Hence we find

$$(S_I)_{\text{perf}} = k\left\{(-\beta)\left(-\frac{3}{2} N \frac{1}{\beta}\right) + \ln\left[\left(\frac{2\pi m}{h^2 \beta}\right)^{\frac{3}{2}N} V^N\right]\right\}.$$

Though correct, this is not a particularly instructive result. A few manipulations will put it into a form such that we may venture an interpretation. The first term may be combined with the second by writing it, also, as a logarithm:

$$(-\beta)\left(-\frac{3}{2} N \frac{1}{\beta}\right) = \frac{3}{2} N = \frac{3}{2} N \ln e = \ln(e^{\frac{3}{2}N}).$$

Since $\ln e$ is 1, with $e \simeq 2.7$ being the base of the natural logarithms, the maneuver is permissible. In this form the two terms may be combined as a single logarithm. A little judicious rearranging—based on foreknowledge— puts the entire expression into the form

$$(S_I)_{\text{perf}} = k \ln\left\{\left[V \middle/ \left(\frac{h}{\sqrt{2\pi emkT}}\right)^3\right]^N\right\}.$$
(8.7.3)

With $(S_I)_{\text{perf}}$ expressed in this manner, we may *venture* an interpretation. We start with the expression in parentheses and work outward. That ex- pression is a measure of the typical thermal de Broglie wavelength of a molecule; a comparison of the expression with equation (6.1.7) will confirm the identification, modulo factors of order unity. The cube of that expression is then a measure of the volume over which a wave packet, in a quantum- mechanical description of a molecule, would extend. The ratio of the con- tainer volume V to that quantum-mechanical volume—the expression in square brackets—is a measure of the (large) number of such quantum- mechanical volumes in the entire container. For a single molecule the ratio is roughly the number of different regions in which the molecule may be within the container. In different words, the ratio is a measure of the number of ways in which a single molecule may be put into the container when we keep in mind the "spread out" nature of a quantum-mechanical description. For distinguishable molecules, the number of distinct arrangements for the entire gas is the product of the number of possible arrangements for each molecule. Hence we have an interpretation of why the ratio appears raised to the Nth power.* We may interpret the expression in the curly brackets as the number of physically distinct arrangements of N identical, but distinguish- able, noninteracting molecules. If the arrangements are equally probable, the expression for $(S_I)_{\text{perf}}$ is just what we should get. For n equally probable inferences, we found in Section 3.5 that the measure of missing information reduces to $MI = K \ln n$, in agreement with $(S_I)_{\text{perf}}$ when one sets $K = k$.

Thus a reasonable interpretation of the expression for $(S_I)_{\text{perf}}$ is this: the expression gives the amount of information necessary to determine which one of the many possible arrangements actually, at any specific moment, describes the gas.

---

* In Chapter 9 we will see that when one properly treats the actual indistinguishability of identical molecules, a numerical factor (dependent on N alone) is introduced into the logarithm giving $(S_I)_{\text{perf}}$ in the classical limit. But that can wait for the proper place. There are no such effects on $\langle p \rangle$, $\langle E \rangle$, or $B_2(T)$ in the classical limit, nor on the second term in S_I for the imperfect gas the term dependent on the intermolecular forces.

These thoughts are a considerable advance over the bare mathematical expression in equation (8.7.3). There is sure to be much truth and insight in this interpretation, but it does not do full justice to the real possibility of different velocities. The spread in velocities has, of course, been dealt with in computing $(S_I)_{\text{perf}}$, but in a way that is well-buried from sight. Rather than try to unearth it, let us take another approach to interpreting S_I. We'll make progress faster this way.

Let us imagine the $6N$-dimensional phase space to be divided up into small cells, each of "volume" h^{3N}. For N particles, presumed to be distinguishable, h^{3N} is the volume corresponding to a quantum-mechanical state of the entire system; recall that we divided the differentials $d^3x_1 \cdots d^3p_N$ by that volume in arriving at equation (8.5.2). When we talk strict quantum mechanics, we say that S_I gives the amount of additional information needed to determine the appropriate quantum-mechanical state. When we pass to the classical limit, we can say that S_I gives the amount of information needed to determine the cell in which the point classically specifying the system is.

Even if the latter information were provided, we would not know where in the selected cell the point was. We have not gone all the way to an uncompromisingly classical description. This we can see clearly in $(S_I)_{\text{perf}}$, for Planck's constant remains there even in our limit. We have the limit, as physical conditions approach those in which classical reasoning is adequate, of a quantum-mechanical result; we do not have a purely classical result. This is fine, however, for the cells of volume h^{3N}—a small but finite volume—prevent our talking about precisions that the Heisenberg principle asserts are unattainable. The sole "disadvantage" is that h, alone because of its dimensions, is sure to couple momenta and positions in their effects on $(S_I)_{\text{perf}}$. The effect of the spread in velocities is not so much buried as coupled with the effect of the spread in positions, and so long as h remains, there is little point in trying to untangle the effects any further.

Now we turn to the second general term in S_I, the term dependent on the intermolecular forces. We could calculate it explicitly to the same order of approximation that we used in the discussion of the pressure and the second virial coefficient. The expression that would emerge, however, would be anything but transparent. We could make some sense out of the expression only after we had shown that it is negative, intrinsically so, regardless of the detailed behavior of the potential energy between a pair of molecules. With the same expenditure of mathematical effort, we can prove a more general theorem. Let us take the latter option. In a few steps we will prove the following.

THEOREM:

In the classical limit, any forces, represented by a potential energy dependent on the particle positions only, reduce the information-theory entropy below the value it would have at the same temperature and volume in the absence of those forces.

Let us abbreviate the second general term in S_I as

$$\Delta S_I \equiv k\left[-\beta\frac{\partial}{\partial\beta}\ln\left(\frac{Z_U}{V^N}\right) + \ln\left(\frac{Z_U}{V^N}\right)\right] \tag{8.7.4}$$

and remember that Z_U/V^N is given by

$$\frac{Z_U}{V^N} = \frac{1}{V^N}\int_V \exp(-\beta U)\, d^3x_1 \cdots d^3x_N.$$

The theorem concerns the effect of the potential energy U, which is permitted to depend on the particle positions in an arbitrary (physically reasonable) manner. The claim of the theorem is that any such U will yield a negative value for ΔS_I and that therefore the forces so represented will reduce the information-theory entropy.

That is an eminently reasonable conclusion. Suppose the potential U represents a hard-core repulsion between pairs of molecules, and suppose further that we know the position of a single molecule. The existence of the hard-core repulsion enables us to assert that all other molecules lie outside a small but finite region around the center of the molecule whose position we suppose known. Without the repulsive core, the other molecules could be anywhere in the container, including positions arbitrarily close to the center of the specified molecule. The existence of the repulsive core indirectly supplies some information about the spatial distribution of molecules and so should lead to a reduction of the information-theory entropy. But before we consider further physical and information-theoretic interpretations, let us prove the theorem.

Performance of the differentiation indicated in equation (8.7.4) leads to

$$\Delta S_I = k\left[\beta\frac{\int_V e^{-\beta U}U\, d^3x_1 \cdots}{\int_V e^{-\beta U}\, d^3x_1 \cdots} + \ln\frac{\int_V e^{-\beta U}\, d^3x_1 \cdots}{V^N}\right].$$

The differentials for the integration over the positions of all N particles have been abbreviated to $d^3x_1 \cdots$ merely. Since the potential energy U is allowed to be positive or negative in an arbitrary fashion, mere inspection of the expression for ΔS_I is not sufficient for a demonstration that it is negative. Yet, with one trick, the desired proof is rather easy.

The necessary element is the auxiliary function $\Delta S_I(\xi)$, defined by replacing U everywhere in ΔS_I by ξU, with ξ a nonnegative dimensionless parameter:

$$\Delta S_I(\xi) \equiv k\left[\beta\frac{\int_V e^{-\beta\xi U}\xi U\, d^3x_1 \cdots}{\int_V e^{-\beta\xi U}\, d^3x_1 \cdots} + \ln\frac{\int_V e^{-\beta\xi U}\, d^3x_1 \cdots}{V^N}\right].$$

In the limit as ξ goes to 1, the auxiliary function reduces to ΔS_I:

$$\lim_{\xi\to 1} \Delta S_I(\xi) = \Delta S_I.$$

When ξ goes to zero, the auxiliary function goes to zero,

$$\lim_{\xi\to 0} \Delta S_I(\xi) = 0,$$

for the first term vanishes and the argument of the logarithm becomes unity.
The limits permit one to express ΔS_I as

$$\Delta S_I = \int_{\xi=0}^{\xi=1} \frac{d\,\Delta S_I(\xi)}{d\xi}\, d\xi. \tag{8.7.5}$$

Admittedly, this appears at first sight to be no advance. But one need only calculate the indicated derivative to discover the relevance, for the derivative is intrinsically negative (or zero):

$$\frac{d\,\Delta S_I(\xi)}{d\xi} = -k\xi\beta^2\left[\frac{\int_V e^{-\beta\xi U}U^2\, d^3x_1 \cdots}{\int_V e^{-\beta\xi U}\, d^3x_1 \cdots} - \left(\frac{\int_V e^{-\beta\xi U}U\, d^3x_1 \cdots}{\int_V e^{-\beta\xi U}\, d^3x_1 \cdots}\right)^2\right]$$

$$= -k\xi\beta^2\left[\frac{\int_V e^{-\beta\xi U}(U - \langle U\rangle_\xi)^2\, d^3x_1 \cdots}{\int_V e^{-\beta\xi U}\, d^3x_1 \cdots}\right] \leqslant 0.$$

The abbreviation

$$\langle U\rangle_\xi \equiv \frac{\int_V e^{-\beta\xi U}U\, d^3x_1 \cdots}{\int_V e^{-\beta\xi U}\, d^3x_1 \cdots}$$

has been used in the step to the second line. Whenever there is a physical force, the potential $U(x_1, \ldots, x_N)$ will necessarily vary with position. Then the expression in square brackets will be positive, for U can then not everywhere be equal to $\langle U\rangle_\xi$. Except for the endpoint where ξ equals zero, the derivative will be negative, and therefore the value of the integral in equation (8.7.5) will be negative. This completes the mathematical proof: the inequality

$$S_I - (S_I)_{\text{perf}} = \Delta S_I < 0 \tag{8.7.6}$$

holds in the classical limit if there are physical forces represented by a potential energy dependent on particle positions only.

Before we embarked on the proof, we noted that a simple picture of the effect of a hard-core repulsive force between molecules would lead us to expect a decrease in the information-theory entropy, a negative value for ΔS_I. The same qualitative expectation is valid, though perhaps less apparent, for the attractive part of an intermolecular potential. The attractive force encourages an (admittedly fleeting) clustering of molecules into pairs, triplets, and so on. If we know the position of one molecule, we are more likely to find another in some nearby small volume where the intermolecular force is attractive than in some other volume of equal extent where there is no or negligible force. Again, the force indirectly supplies some kind of information about the spatial distribution of molecules.

If we think in terms of the possible spatial arrangements of all N molecules, then we may say that intermolecular forces make some of those spatial arrangements more probable than others. Some are even fully ruled out. A gravitational field does much the same. Since MI attains its maximum value when all inferences are equally probable, such an uneven distribution of probabilities will lead, mathematically, to a diminution of MI.

REFERENCES for Chapter 8

The classic work on intermolecular forces is the massive yet very readable book by Joseph O. Hirschfelder, Charles F. Curtiss, and R. Byron Bird, *Molecular Theory of Gases and Liquids* (New York: Wiley, 1954). Graphs and tables in great profusion complement the theoretical analysis of both the intermolecular forces themselves and their influence on pressure, viscosity, and other experimentally accessible properties. The book is a landmark worth revisiting many times. More recent developments, both experimental and theoretical, are discussed in Joseph O. Hirschfelder, ed., *Intermolecular Forces* (New York: Interscience, 1967). This is a rather advanced, research-level book.

The "different way" of looking at the integral for Z_U in section 8.5 was provided by N. G. Van Kampen, *Physica*, **27**, 783 (1961). The classic on virial expansions is perhaps the book by Joseph Edward Mayer and Maria Goeppert Mayer, *Statistical Mechanics* (New York: Wiley, 1940). The modern theory of imperfect gases was established by H. D. Ursell and J. E. Mayer. The development in the Mayers' book is based on a grand canonical probability distribution, a generalization of the canonical distribution to the case in which one does not specify the precise number of molecules in the system but only the expectation value of that number. (There are also other approaches to the grand canonical distribution.)

PROBLEMS

8.1. a. For a perfect gas of N molecules, calculate the root mean square estimate ΔE of the deviation of the energy from $\langle E \rangle$. Then form the ratio $\Delta E/\langle E \rangle$ and note the dependence on the number of molecules.

b. Do the same with the energy of a single molecule. (One approach consists of using the Maxwell speed probability distribution; there are others.)

c. Comment on the sharpness of the estimates, relative to the expectation values, in the two cases: the first, macroscopic; the second, microscopic.

8.2. For air at room temperature and density, estimate the effect of $B_2(T)$ on $\langle p \rangle$. An order-of-magnitude estimate will suffice.

8.3. Experimental values of the second virial coefficient of gaseous nitrogen (N_2) are given in table P8.3 in units of 10^{-24} cm³ (that is, "cubic Ångströms"). Take $u(r)$ to have the simple form show in figure P8.3, and calculate $B_2(T)$ in

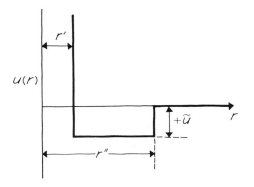

terms of the three parameters and the temperature. Then use the experimental data to estimate good values for the three parameters. Express the (positive) well-depth parameter $\tilde{u}$ in units of both ergs and electron volts.

TABLE P8.3
Values of the Second Virial Coefficient
for N_2

Temperature (K)	$B_2(T)$ (10^{-24}cm³)
100	− 245
200	− 58.0
300	− 7.48
400	+ 15.3
500	+ 27.9

8.4. Use the data provided in problem 8.3 to estimate the van der Waals coefficients a and b of nitrogen on the explicit assumption that the coefficients are independent of temperature. A plot of $B_2(T)$ versus $1/T$ is a help. Incidentally, how good do you judge the assumption to be?

8.5. For $\ln Z$ for a moderately dilute imperfect gas, take the approximate form that we derived in Section 8.5.

a. Set up a general expression for $\langle E \rangle$, given any reasonable potential $u(r)$. (The expression should still have an integral but no derivatives.)

b. Does one find $\langle p \rangle$ proportional to $\langle E \rangle$ to this order of approximation in inverse powers of the volume? Any comments?

c. Evaluate $\langle E \rangle$ for the Sutherland potential. Any comments on the effect of the repulsive part (to this order of approximation)?

8.6. Adopt an intermolecular potential of the shape given in problem 8.3, and use it to compute the first correction dependent on the particle density N/V to the perfect-gas expression for $\langle E \rangle$. Examine the high- and low-temperature behavior. Comment on whether or not the limiting values agree with your physical intuition. (One of the limits does not agree with my intuition.) If you find disagreement, what might be the source?

8.7. If we treat any system of particles with the classical version of the canonical probability distribution and if the interparticle forces depend only on the positions of the particles (effectively, if Assumption 1 of Section 8.5 is valid), then the partition function will have the general form given in equations (8.5.2.3), except for a possible constant factor to compensate for the actual indistinguishability of identical particles.

a. Derive an expression for the heat capacity at constant volume—some integrals will remain—and show that the contribution of the interparticle forces is positive, regardless of the sign of U and of variations in that sign.

b. Since experiment shows that observed heat capacities fall below the value $\frac{3}{2}Nk$ at low temperatures, what conclusions do you draw? (The discrepancy occurs for diamond at a temperature of about 400 K, well above room temperature, and gets worse with lower temperature. For most metals the discrepancy sets in at a lower temperature.)

IDENTICAL PARTICLES AND WAVE-FUNCTION SYMMETRY

Throughout this chapter we will work with the canonical distribution as the probability distribution appropriate for the statistical description of a system of identical particles in equilibrium at a specified temperature. To avoid misunderstanding, this must be stated emphatically: *there is no change in the basic statistical approach, no change in the probability distribution used.* New elements appear because we will look in some detail at the kind of symmetry that a wave function for a system of identical particles must possess. The physical indistinguishability of identical particles makes itself felt in the quantum-mechanical analysis. At the computational level, there will be some innovations: at the appropriate place we will introduce the notion of occupation numbers and will work with their expectation values, which will be computed from the same old canonical probability distribution that we have used all along. This introduction is perhaps an overly stern admonition, but such an admonition may spare one unnecessary grief: do not be misled into thinking that the canonical distribution has been slighted or discarded.

We start the chapter off by examining the implications in quantum mechanics of the physical indistinguishability of identical particles. That examination leads to a discussion of the connection between the symmetry properties of acceptable wave functions and the spins of the particles being described. Then mathematical complexity forces a specialization: an analysis of the tractable (and useful) case of identical particles whose *mutual* inter-actions may be neglected, and the development of a scheme for constructing and specifying energy eigenstates for such a system. At that point the quantum-mechanical notion of occupation numbers makes its appearance. The first two sections constitute the quantum-mechanical background. The primary section is the third: with the canonical probability distribution we compute the expectation values of the occupation numbers for a system with many identical particles. After a digression to develop an integral approximation for the ubiquitous summations, we examine the "classical" and "nearly classical" limits of the occupation number scheme. The formidable apparatus generates a number of significant results, and the reasonableness of the conclusions will give us (hopefully) some confidence in the procedure.

9.1. SPIN AND WAVE-FUNCTION SYMMETRY

In this section we will examine the restrictions that the physical indistin-guishability of identical particles imposes on acceptable wave functions. The restrictions must necessarily concern us, because we seek to describe a physical system, when it is in equilibrium, by means of its energy eigenstates (and the canonical probability distribution).

For a single particle there is no problem. With motion constrained to one dimension (for convenience of illustration), the wave function for a single spinless particle would depend on x and t only, and for time $t = 0$ we might designate it by simply $\psi(x)$. The probability of finding the particle in an infinitesimal spatial interval dx around x would be proportional to

$$\psi^*(x)\psi(x) = |\psi(x)|^2.$$

More exactly, the probability would be equal to

$$|\psi(x)|^2 dx,$$

but for brevity we will drop the infinitesimals and talk about the proportionality with the square of the absolute value of the wave function. This paragraph merely reviews one aspect of quantum mechanics and establishes some notational conventions.

Now suppose we have two different, distinguishable spinless particles, likewise constrained to one-dimensional motion, which interact with one another. (They might be an alpha particle and a π meson, for which there is sure to be strong interaction.) Since the particles are distinguishable, by mass, say, we can meaningfully label the heavier, "particle 1," and the lighter, "particle 2." The wave function for these two interacting particles depends on the variables of both, that is, on x_1 and x_2. At the initial time $t = 0$, the wave function might have the form

$$\psi(x_1, x_2) = C(x_1{}^2 + x_2)e^{-(x_1{}^2 + x_2{}^2)}, \tag{9.1.1}$$

with C being a normalization constant. (This wave function is to be taken as an example, nothing more.) The probability of finding particle 1 in an infinitesimal spatial interval around $x_1 = 1$ and particle 2 in an infinitesimal interval around $x_2 = 3$ would be given by the square of the absolute value of

$$\psi(1, 3) = C(1 + 3)e^{-(1+9)}. \tag{9.1.2a}$$

This probability would differ considerably from the probability of finding particle 1 in an equal infinitesimal spatial interval around $x_1 = 3$ and particle 2 in an equal interval around $x_2 = 1$. The latter probability would be given by the square of the absolute value of

$$\psi(3, 1) = C(9 + 1)e^{-(9+1)}. \tag{9.1.2b}$$

Nonetheless, there is nothing disturbing about this. Although both particles lie somewhere along the same physical line (the physical x-axis), they are distinguishable (by their masses), and the different probabilities refer to physically distinct situations. In one situation the lighter particle is on the positive x side of the heavier; in the other situation, on the negative x side. Figure 9.1.1 illustrates these two physically distinct situations, for which numerically different probabilities are acceptable.

Now we get to the object of the exercise. Suppose we have two *identical, indistinguishable* spinless particles, again constrained to one-dimensional motion and interacting. (One may think of them as two alpha particles.) With two particles the wave function must depend on two variables. After all, the wave function must enable us to answer a physical question like: what is the probability that one particle is near the origin and the other is

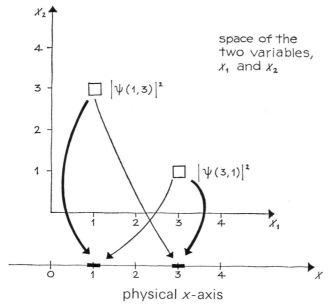

FIGURE 9.1.1
Diagram for two distinguishable particles (spinless and constrained to move in one dimension). Particle 1 is heavier than particle 2; the heavier line refers to the physically heavier particle. The two regions in the x_1, x_2, space refer to physically distinct situations, and so numerically different probabilities are perfectly acceptable.

five centimeters out along the physical positive x-axis? With identical and indistinguishable particles, however, any "labeling" must be a question of mathematics, not physics. Still, since wave functions are mathematical objects, as are the variables on which they depend, we may blithely label one particle, "particle 1," and its twin, "particle 2." For a wave function at $t = 0$ with which to work, we take the same one as in the preceding paragraph:

$$\psi(x_1, x_2) = C(x_1{}^2 + x_2)e^{-(x_1{}^2 + x_2{}^2)} \tag{9.1.3}$$

Notice that we have here:

(1) one probability that is proportional to $|\psi(1, 3)|^2$ for finding particle 1 around $x_1 = 1$ and finding particle 2 around $x_2 = 3$; and

(2) another probability that is proportional to $|\psi(3, 1)|^2$ for finding particle 1 around $x_1 = 3$ and finding particle 2 around $x_2 = 1$,

The numerical difference between these two probabilities *is* disturbing. Although the labels on the two pairs of specifications are different, the

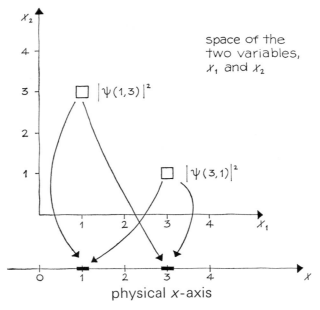

FIGURE 9.1.2
Diagram for two identical, indistinguishable particles
(spinless and constrained to move in one dimension).
The labeling of the particles as 1 and 2 is now purely
mathematical. The two regions in the x_1, x_2, space refer to
the same physical situation: one particle near $x = 1$ on the
physical x-axis and its twin near $x = 3$ on the physical
x-axis. The two formally distinct expressions, $\psi(1, 3)|^2$ and
$|\psi(3, 1)|^2$, for the probability of this single physical situation
should be numerically equal.

physical situation is the same. Each pair specifies the situation: one particle
near $x = 1$ on the physical x-axis and its twin near $x = 3$ on the same physical
axis. The corresponding diagram would be as before, with the essential
distinction that now we *cannot* draw one line heavier than the other, or in
any way distinguish them. The two little squares in the space of the variables
x_1, x_2 represent the *same physical situation*; the two formal expressions for
the probability of that one physical situation should have the *same numerical
value.*

Since nothing in the real world can distinguish between two identical
particles, the wave function should reflect that physical property. In our case
an acceptable wave function should yield the equality

$$|\psi_{\text{acceptable}}(1, 3)|^2 = |\psi_{\text{acceptable}}(3, 1)|^2. \tag{9.1.4}$$

The wave function is evaluated differently—in a purely mathematical sense—but for the same physical situation.

Let us see whether we can patch up the given wave function so that equation (9.1.4) or its equivalent,

$$|\psi_{\text{acceptable}}(1, 3)| = |\psi_{\text{acceptable}}(3, 1)|, \tag{9.1.5}$$

holds. The exponential in equation (9.1.3), being symmetric in x_1 and x_2, is all right. It is the factor of $(x_1{}^2 + x_2)$ that causes the trouble. One way of making the wave function acceptable would be to symmetrize the troublesome factor:

$$\psi_S(x_1, x_2) = C'([x_1{}^2 + x_2{}^2] + [x_2 + x_1])e^{-(x_1{}^2 + x_2{}^2)}.$$

This totally symmetric wave function has the property

$$\psi_S(1, 3) = \psi_S(3, 1)$$

and so certainly satisfies the physical requirement for acceptability:

$$|\psi_S(1, 3)| = |\psi_S(3, 1)|.$$

There is another way of making the wave function acceptable. It consists of antisymmetrizing the initial factor:

$$\psi_A(x_1, x_2) = C'([x_1{}^2 - x_2{}^2] + [x_2 - x_1])e^{-(x_1{}^2 + x_2{}^2)}$$

With this wave function one has the relation

$$\psi_A(1, 3) = -\psi_A(3, 1),$$

but the physical requirement is fulfilled:

$$|\psi_A(1, 3)| = |\psi_A(3, 1)|.$$

The antisymmetric wave function $\psi_A(x_1, x_2)$ vanishes when, numerically, $x_1 = x_2$. Thus ψ_A implies zero probability for finding the two particles in the same infinitesimal spatial interval on the physical x-axis. This property may remind one of the Pauli exclusion principle, which, for electrons in an atom, asserts that there can be no more than one electron in a state with given energy, orbital angular momentum, and orientation for spin and orbital angular momentum.

Without gross additions to the wave function with which we started, there are no other ways of turning it into a wave function acceptable for the description of two identical, indistinguishable particles. Both the symmetric and the antisymmetric wave function satisfy the requirement (9.1.5) that the physical indistinguishability of the two particles imposes. But which kind—symmetric or antisymmetric—should be used for a pair of spinless particles? Or doesn't it matter?

Thanks to the work of Pauli, there is a definite answer. The exclusion principle for electrons was a semiempirical rule enunciated by Pauli in 1925, based on an analysis of atomic spectra.* The culmination of Pauli's work on the problem came in 1939, when, from some very general and reasonable assumptions in relativistic quantum field theory, he established a unique connection between (1) the spins of a set of identical particles and (2) the symmetry properties of wave functions that are acceptable for describing the particles.† Needless to say, we are not about to go into the details of the derivation, but some appreciation of the physical simplicity of the assumptions (as distinguished from the mathematical complexity of the derivations) is possible and eminently worthwhile.

Before we go into the assumptions and conclusions, we should generalize slightly our notation for a two-particle wave function; we need to include all three spatial dimensions and the component of spin (if any) along some specified direction. A set of three position components (as in x, y, z) and the single spin component along a specific direction (if relevant) we will designate with a single letter Q. It may need a subscript as a particle label, just as before we labelled "x" with subscripts: x_1 and x_2. For two identical, indistinguishable particles—*mathematically* labeled as 1 and 2—we write the wave function as $\psi(Q_1, Q_2)$. (It need bear no resemblance to the wave functions with which we have been working.)

As a prelude to the statement of Pauli's results, let us repeat for the generalized wave function $\psi(Q_1, Q_2)$ the arguments that led to equation (9.1.5). The probability of finding particle 1 with values for its variables such that $Q_1 = Q'$ (compare with $x_1 = 1$) and particle 2 with values for its variables such that $Q_2 = Q''$ (compare with $x_2 = 3$) is proportional to the square of $|\psi(Q', Q'')|$. Here Q' and Q'' stand for specific numerical values of the x, y, z,

---

* Remarkably, Pauli arrived at his conclusion *before* the idea of a spinning electron, with intrinsic angular momentum of $\frac{1}{2}\hbar$, had gained acceptance. A. H. Compton (in 1921) was apparently the first to propose the spinning electron idea, but the work which led to acceptance of some such notion was that of Kronig, Uhlenbeck, and Goudsmit. They based their analysis on Pauli's "exclusion principle" paper, in which a fourth quantum number for an electron was introduced without direct physical interpretation. A fascinating historical account is given by B. L. van der Waerden in an article "Exclusion Principle and Spin," a contribution to a memorial volume dedicated to Pauli, *Theoretical Physics in the Twentieth Century*, edited by M. Fierz and V. F. Weisskopf (New York: Interscience, 1960).

† W. Pauli, *Physical Review*, **58**, 716 (1940).

position components and the spin orientation (if the latter is applicable). The probability of finding particle 1 with values for its variables such that $Q_1 = Q''$ and particle 2 with values for its variables such that $Q_2 = Q'$ is proportional to the square of $|\psi(Q'', Q')|$. The two pairs of specifications differ only in that the roles of the twins have been interchanged. Since the two identical particles really are indistinguishable, that interchange is just a matter of mathematics. We are dealing with only one physical situation. The two mathematically different expressions for the probability of that one physical situation must have the same numerical value. Therefore the wavefunction $\psi(Q_1, Q_2)$ must have the property

$$|\psi(Q'', Q')| = |\psi(Q', Q'')| \qquad (9.1.6)$$

if it is to be acceptable for the description of a pair of identical particles. This provides the generalization, to three spatial dimensions and spin, of the relation given in equation (9.1.5).

Following Pauli, we split the particles of physics into two classes.

1. The first class contains the particles whose intrinsic angular momentum is equal to zero or a positive integer times $\hbar$: $0\hbar$, $1\hbar$, $2\hbar$, ... Examples are π mesons ($0\hbar$), photons ($1\hbar$), and, if they exist, gravitons ($2\hbar$). Generically, particles with "integral" spin are called *bosons*, the name arising because S. N. Bose and Einstein first analyzed the statistical consequences of integral spin, that is, the so-called Bose-Einstein statistics.

2. The second class contains the particles whose intrinsic angular momentum is equal to half an odd positive integer times $\hbar$: $\frac{1}{2}\hbar$, $\frac{3}{2}\hbar$, $\frac{5}{2}\hbar$, ... Examples are electrons, positrons, protons, and neutrons, all with $\frac{1}{2}\hbar$ intrinsic angular momentum. Some of the unstable particles of high energy physics have higher values (for instance, the N^* with spin $\frac{3}{2}\hbar$, though whether this is properly an "elementary particle" is open to argument). Generically, particles with "half-odd-integral" spin are known as *fermions*, the designation arising because Fermi and Dirac worked out the statistical consequences of half-odd-integral spin, the so-called Fermi-Dirac statistics.

At times composite particles, such as alpha particles and the stable isotopes of helium, He^4 and He^3, are included in this classification. That inclusion is useful if one can deal with the composite particle as a unit, neglecting the detailed wave function for the individual particles making up the composite. An alpha particle, with two protons and two neutrons, necessarily has integral spin; so it is a boson. The helium isotopes differ. The common isotope He^4, having an alpha particle for the nucleus plus two electrons, is again a boson, but the isotope He^3, like He^4 but missing one neutron, necessarily has half-odd-integral spin, and so is a fermion. The quantum-mechanical consequences of this difference lead to radically different properties at low temperatures, as we will see.

Now we are ready for a statement of Pauli's results on the symmetry properties of wave functions that are acceptable for a set of identical, indistinguishable particles, which may or may not be interacting significantly.

1. Integral spin (bosons): the wave function must be symmetric.

For two identical bosons, this means the equality

$$\psi(Q'', Q') = +\psi(Q', Q''),\tag{9.1.7}$$

whereas all we had been able to require was the relation

$$|\psi(Q'', Q')| = |\psi(Q', Q'')|.$$

For N identical bosons, the wave function will depend on a set of N Q's:

$$\psi(Q_1, Q_2, Q_3, \ldots, Q_N).$$

We may then consider a physical situation in which one particle has position and spin variables equal to Q', another to Q'', a third to Q''', and so on. The statement that the wave function for bosons must be symmetric means the following: the N-boson wave function must be so constructed that these specifications, Q', Q'', Q''', ..., may be inserted into the wave function in any order, as

$$\psi(Q', Q'', Q''', \ldots),$$

or

$$\psi(Q'', Q', Q''', \ldots),$$

or

$$\psi(Q'', Q''', Q', \ldots),$$

and so on, and that the numerical value of the wave function will be the same for all. Any permutation of the order in which Q', Q'', Q''', ..., are inserted leaves the numerical value of the N-boson wave function unaltered.

The assumption in relativistic quantum field theory that Pauli invoked is the following: the observations of physical quantities at two different space-time points should have no influence on one another if the spatial separation of the two space-time points is greater than the speed of light times the temporal separation, that is, if no signal can propagate from one observation and arrive "in time" to influence the other observation. This assumption leads to the (admittedly remarkable) consequence that the wave function for identical bosons must be symmetric.

2. Half-odd-integral spin (fermions): the wave function must be antisymmetric.

For two identical fermions, this means the equality

$$\psi(Q'', Q') = -\psi(Q', Q''),\tag{9.1.8}$$

whereas our previous restriction had gone only as far as the relation

$$|\psi(Q'', Q')| = |\psi(Q', Q'')|.$$

For N fermions we again have a wave function dependent on N Q's:

$$\psi(Q_1, Q_2, Q_3, \ldots, Q_N).$$

Again we consider a physical situation in which one particle has position and spin variables equal to Q', another to Q'', and so on. The statement that the wave function for fermions must be antisymmetric means the following: the N-fermion wave function must be so constructed that its numerical value changes sign when the order in which the specifications Q', Q'', Q''', $\ldots$, are inserted is altered by an *odd* permutation and that its numerical value remains the same when the order is changed by an *even* permutation.

The case with $N = 3$ is sufficiently general to illustrate the full antisymmetry condition. Here is a list of the possible ways of inserting Q', Q'', Q''', into a 3-fermion wave function $\psi(Q_1, Q_2, Q_3)$ and an indication of the sign changes relative to $\psi(Q', Q'', Q''')$.

(a) $\psi(Q', Q'', Q''') = +\psi(Q', Q'', Q''')$.
(b) $\psi(Q'', Q', Q''') = -\psi(Q', Q'', Q''')$.
(c) $\psi(Q'', Q''', Q') = +\psi(Q', Q'', Q''')$.
(d) $\psi(Q''', Q'', Q') = -\psi(Q', Q'', Q''')$.
(e) $\psi(Q''', Q', Q'') = +\psi(Q', Q'', Q''')$.
(f) $\psi(Q', Q''', Q'') = -\psi(Q' \; Q'', Q''')$.

There are precisely 3! distinct ways in which Q', Q'', Q''', may be inserted. In the sequence, (a) to (b) to (c) to $\ldots$, the order of insertion is changed each time by a single transposition of two of the three specifications. The numerical value of the wave function is required to change sign each time. Note that (b), (d), and (f) differ from (a) by an odd number of transpositions (an odd permutation); for them the value of the fermion wave function has a sign opposite to that in (a). To arrive at (c) and (e) from (a), an even number of transpositions (an even permutation) are required; so the numerical value of the fermion wave function is the same as in (a).

Pauli's assumption in field theory is that an acceptable theory should be so formulated that the energy of a set of non-interacting particles is positive. The antisymmetry of the wave functions for identical fermions is one consequence.

Without a rather full development of relativistic quantum field theory, the actual line of deduction from the two physically reasonable assumptions to the respective symmetry properties must remain something of a mystery. But the presentation should remove some of the feeling that the symmetry restrictions are pulled out of thin air (though they do come from the peculiarly rarefied air surrounding axiomatic field theory). The next section may generate a more comfortable feeling about the symmetry properties, for we will see how the antisymmetry for fermion wave functions does imply the Pauli exclusion principle for electrons in an atom.

9.2. PRODUCT WAVE FUNCTIONS

For the description of a system of N identical particles known to be in equilibrium, we will need the energy eigenstates of the entire system and the corresponding energies. We must (in principle) solve the eigenvalue equation

$$\mathcal{H}\psi_j = E_j\psi_j, \tag{9.2.1}$$

where $\mathcal{H}$ denotes the energy operator for the entire system. The energy eigenstate ψ_j for the entire system depends on N Q's:

$$\psi_j = \psi_j(Q_1, Q_2, Q_3, \ldots, Q_N). \tag{9.2.2}$$

It must have the correct symmetry properties, being totally symmetric for integral spin, that is,

$$\psi_j(Q'', Q', Q''', \ldots) = + \psi_j(Q', Q'', Q''', \ldots)$$

for bosons, or being totally antisymmetric for half-odd-integral spin, that is

$$\psi_j(Q'', Q', Q''', \ldots) = - \psi_j(Q', Q'', Q''', \ldots)$$

for fermions. One or the other of these symmetry properties must be imposed on the function in the process of solving the eigenvalue equation; which one is imposed is determined by the spin of the N identical particles.

Finding acceptable, exact solutions to the eigenvalue equation when N is large is, typically, impossible in practice. The cause of the mathematical difficulty lies in the mutual interactions of the particles. But there are times when, to a valid first approximation, one may neglect the interaction of the identical particles among themselves. Then a considerable mathematical simplification is possible, though nothing changes in principle. If one neglects the interactions of the particles among themselves, the energy operator reduces to a sum of N similar operators, one for each of the particles:

$$\mathcal{H} = \sum_{i=1}^{N} \mathcal{H}_i. \tag{9.2.3}$$

For a set of particles with spin zero confined to a box, the energy operator $\mathcal{K}_1$ for particle 1 would be

$$\mathcal{K}_1 = -\frac{\hbar^2}{2m}\left(\frac{\partial^2}{\partial x_1^{\,2}} + \frac{\partial^2}{\partial y_1^{\,2}} + \frac{\partial^2}{\partial z_1^{\,2}}\right) + U_{\text{box}}(x_1, y_1, z_1), \qquad (9.2.4)$$

x_1, y_1, z_1 being a more explicit form for the set of variables compactly designated as Q_1. One could add to this single-particle energy operator a term for interaction with an external electric or magnetic field. One could also insert a gravitational potential energy as mgz_1. Another example of an "external potential" would be the Coulomb potential provided by the nucleus when one deals only with the electrons in an atom.

The significant point is that $\mathcal{K}$ splits into purely a sum of identical (except for the label) single-particle energy operators, the $\mathcal{K}_i$'s. When we neglect the mutual interactions, there is no term in $\mathcal{K}$ containing the variables of two or more particles, whereas there would be such terms if we included a mutual electrostatic potential energy, say. For particles 1 and 2, such a mutual energy would depend on $1/|\mathbf{x}_2 - \mathbf{x}_1|$ and would prevent the decomposition of $\mathcal{K}$ into the special sum.

The simple form of $\mathcal{K}$ when we may neglect the mutual interactions leads one to look (as a first step in constructing ψ_j) for the *single-particle* energy eigenstates φ; they are the solutions of the single-particle equation

$$\mathcal{K}_i\varphi_s(Q_i) = \varepsilon_s\varphi_s(Q_i). \qquad (9.2.5)$$

The subscript s labels the distinct solutions to this eigenvalue equation. We may take the solutions to be normalized to unity and to be mutually orthogonal. The index s will run over the values $s = 1, 2, 3, \ldots$ up to however many distinct single-particle energy eigenstates there are, typically an infinite number. The corresponding single-particle energy is denoted by ε_s.

The present eigenvalue equation is much more tractable than equation (9.2.1). Indeed, we have already worked with its solutions for a single-particle energy operator like that given in equation (9.2.4). The solutions for a spinless particle, first presented in Section 1.2 and now rewritten with a change in notation to conform to the current usage, are the following:

$$\varphi_s(Q_1) = \begin{cases} \left(\dfrac{8}{V}\right)^{1/2} \sin\left(\dfrac{\pi n_x x_1}{L}\right)\sin\left(\dfrac{\pi n_y y_1}{L}\right)\sin\left(\dfrac{\pi n_z z_1}{L}\right), \\ \text{zero outside the box of volume } L^3. \end{cases} \qquad (9.2.6a)$$

Each of the integers n_x, n_y, n_z, should have a subscript s, but for brevity this has been left implicit. As an illustration, $s = 3$ might mean $n_x = 1$, $n_y = 2$, $n_z = 1$. The single-particle energy eigenvalue ε_s is

$$\varepsilon_s = \frac{\pi^2 \hbar^2}{2mV^{2/3}} (n_x{}^2 + n_y{}^2 + n_z{}^2)_s. \tag{9.2.6b}$$

The subscript s on $(n_x{}^2 + n_y{}^2 + n_z{}^2)_s$ indicates that the positive integers are those appropriate for the single-particle φ_s. To finish the illustration, with $s = 3$ one would have

$$\varepsilon_3 = \frac{\pi^2 \hbar^2}{2mV^{2/3}} (1^2 + 2^2 + 1^2).$$

Once the solutions to equation (9.2.5) have been found, an acceptable energy eigenstate ψ_j for the entire system of N identical particles—assumed *not* to be *mutually* interacting—may be constructed in a simple way out of products of the single-particle states φs.

Let us try to construct an energy eigenstate for two identical particles from a pair of different single-particle states, φ_3 and φ_5, say. A combination like $\varphi_3(Q_1)\varphi_5(Q_2)$ would fail to satisfy even out modest physical requirement, for we would find, in general,

$$|\varphi_3(Q'')\varphi_5(Q')| \neq |\varphi_3(Q')\varphi_5(Q'')|,$$

and that is unacceptable for any pair of identical particles.

If the particles have integral spin, a properly symmetrized wave function would be

$$\psi_S(Q_1, Q_2) = \frac{1}{\sqrt{2}} \left[\varphi_3(Q_1)\varphi_5(Q_2) + \varphi_3(Q_2)\varphi_5(Q_1) \right] \tag{9.2.7}$$

The initial numerical factor merely ensures the correct normalization for $\psi_S(Q_1, Q_2)$ if φ_3 and φ_5 are mutually orthogonal and individually normalized. For this wave function, we have

$$\psi_S(Q', Q'') = \frac{1}{\sqrt{2}} \left[\varphi_3(Q')\varphi_5(Q'') + \varphi_3(Q'')\varphi_5(Q') \right]$$

and

$$\psi_S(Q'', Q') = \frac{1}{\sqrt{2}} \left[\varphi_3(Q'')\varphi_5(Q') + \varphi_3(Q')\varphi_5(Q'') \right].$$

So the symmetry requirement for two identical bosons,

$$\psi(Q'', Q') = +\psi(Q', Q''),$$

is satisfied.

If, however, the particles have half-odd-integral spin, then the acceptable wave function would have to be antisymmetric:

$$\psi_A(Q_1, Q_2) = \frac{1}{\sqrt{2}}\left[\varphi_3(Q_1)\varphi_5(Q_2) - \varphi_3(Q_2)\varphi_5(Q_1)\right]. \qquad (9.2.8)$$

For this wave function we have

$$\psi_A(Q', Q'') = \frac{1}{\sqrt{2}}\left[\varphi_3(Q')\varphi_5(Q'') - \varphi_3(Q'')\varphi_5(Q')\right]$$

and

$$\psi_A(Q'', Q') = \frac{1}{\sqrt{2}}\left[\varphi_3(Q'')\varphi_5(Q') - \varphi_3(Q')\varphi_5(Q'')\right].$$

So the antisymmetry requirement for two identical fermions,

$$\psi(Q'', Q') = -\psi(Q', Q''),$$

is satisfied.

Now we can see how the Pauli exclusion principle follows for the fermions. If we try to construct an antisymmetric wave function out of two similar single-particle wave functions—φ_3 and φ_3, not φ_3 and φ_5, say—the wave function vanishes:

$$\frac{1}{\sqrt{2}}\left[\varphi_3(Q_1)\varphi_3(Q_2) - \varphi_3(Q_2)\varphi_3(Q_1)\right] \equiv 0.$$

The exclusive principle holds because one cannot construct an antisymmetric wave function, as is required for the description of a pair of electrons, that would violate the Pauli principle. Any such attempt leads to no wave function at all.

The wave functions given in equations (9.2.7, 8) do have the appropriate symmetry. We should now confirm that they are energy eigenstates of the system of two identical particles and that the total energy is ($\varepsilon_3 + \varepsilon_5$).

The energy operator for the system of two identical particles is

$$\mathcal{H} = \sum_{i=1}^{2} \mathcal{H}_i = \mathcal{H}_1 + \mathcal{H}_2. \qquad (9.2.9)$$

We must bear in mind that $\mathcal{H}_1$ affects (through differentiation, say) only the variable set designated by Q_1; it does not touch the Q_2 variables. The exact opposite is true for $\mathcal{H}_2$.

Since the antisymmetric case, with its minus sign, is likely to be the more worrisome of the two cases, let us work it out; the symmetric case follows the same pattern. We begin by forming the analog of the lefthand side of equation (9.2.1), and then work out the multiplication and operations:

$$\mathcal{H}\psi_A(Q_1, Q_2) = (\mathcal{H}_1 + \mathcal{H}_2)\frac{1}{\sqrt{2}}\left[\varphi_3(Q_1)\varphi_5(Q_2) - \varphi_3(Q_2)\varphi_5(Q_1)\right]$$

$$= \frac{1}{\sqrt{2}}\left[[\mathcal{H}_1\varphi_3(Q_1)]\varphi_5(Q_2) - \varphi_3(Q_2)[\mathcal{H}_1\varphi_5(Q_1)]\right.$$

$$\left. + \varphi_3(Q_1)[\mathcal{H}_2\varphi_5(Q_2)] - [\mathcal{H}_2\varphi_3(Q_2)]\varphi_5(Q_1)\right]$$

$$= \frac{1}{\sqrt{2}}\left[[\varepsilon_2\varphi_3(Q_1)]\varphi_5(Q_2) - \varphi_3(Q_2)[\varepsilon_5\varphi_5(Q_1)]\right.$$

$$\left. + \varphi_3(Q_1)[\varepsilon_5\varphi_5(Q_2)] - [\varepsilon_3\varphi_3(Q_2)]\varphi_5(Q_1)\right]$$

$$= (\varepsilon_3 + \varepsilon_5)\frac{1}{\sqrt{2}}\left[\varphi_3(Q_1)\varphi_5(Q_2) - \varphi_3(Q_2)\varphi_5(Q_1)\right]$$

$$= (\varepsilon_3 + \varepsilon_5)\psi_A(Q_1, Q_2). \tag{9.2.10}$$

So $\psi_A(Q_1, Q_2)$ is indeed a properly antisymmetric energy eigenstate for a system of two identical fermions, with $(\varepsilon_3 + \varepsilon_5)$ being the energy eigenvalue.

A little counting is already worthwhile. For a system of two identical bosons, we may pick any two φ_s functions—the subscripts s may be numerically different or the same—and construct one and only one (symmetric) state ψ of the system with them. Specifying the subscripts on the two functions —3 and 5, say, or 3 and 3—specifies the state ψ.

For a pair of fermions, we may again pick two φ_s functions, but now the two subscripts must be different if we are to be able to construct a non-zero antisymmetric state. A specification of the two subscripts is equivalent to a specification of the (antisymmetric) state ψ. To be sure, there remains an over-all ambiguity of a plus or minus sign in how the state ψ should be constructed from the two different φ_s functions. For example, the order of the two terms in equation (9.2.8) could be reversed, leading to an acceptable wave function differing by an over-all minus sign from the one we have written. That "reversed" wave function does, however, describe the same physical situation, for the over-all sign of a wave function is immaterial. Really only one (antisymmetric) state of the two-fermion system can be constructed from a pair of (different) φ_s functions. For a definite convention, let us agree to write the φ_s with the lower value for the subscript first.

For both bosons and fermions there is a *unique* correspondence between, on the one hand, the number of times each distinct single-particle state appears in the first term of the two-particle wave function and, on the other hand, the entire wave function itself. Thus a two-particle energy eigenstate ψ_j can be completely specified by two statements, as follows.

1. The state ψ_j is symmetric (or antisymmetric).

2. The number of times the single-particle state φ_s appears in the first term of ψ_j is $n_s(j)$, this to be repeated for all s, that is, $s = 1, 2, 3, \ldots$

For fermions each $n_s(j)$ must be either zero or one; the antisymmetry requirement compels this. Here is an illustration of the scheme: for the states given in equations (9.2.7, 8), one has $n_3 = 1$, $n_5 = 1$, and all others zero. The states differ because one state was specified to be symmetric, the other to be antisymmetric. This manner of specifying a state for the entire system will later be exceedingly useful.

Before going on, let us note that the new scheme provides a short way of of indicating the energy E_j of ψ_j. Just as the energy of the states in equations (9.2.7, 8) was simply the sum $\varepsilon_3 + \varepsilon_5$, so the energy of a general eigenstate ψ_j follows by addition when we specify the number of times each φ_s appears in the first term of ψ_j. We may write

$$E_j = \sum_s \varepsilon_s n_s(j), \qquad (9.2.11)$$

with the sum going over the entire set of single-particle states. For the eigenvalue in equations (9.2.7, 8), we get

$$\sum_s \varepsilon_s n_s(j) = 0 + 0 + (\varepsilon_3 \cdot 1) + 0 + (\varepsilon_5 \cdot 1) + 0 + \ldots = \varepsilon_3 + \varepsilon_5.$$

It checks.

Now we venture on to three identical particles. But there is really nothing new. With bosons and a given set of three φ_s functions—φ_3, φ_5, and φ_6, say—there is only one totally symmetric state:

$$\psi_j(Q_1, Q_2, Q_3) = \frac{1}{\sqrt{6}} \left[+\varphi_3(Q_1)\varphi_5(Q_2)\varphi_6(Q_3) + \varphi_3(Q_2)\varphi_5(Q_1)\varphi_6(Q_3) \right.$$

$$+ \varphi_3(Q_2)\varphi_5(Q_3)\varphi_6(Q_1) + \varphi_3(Q_3)\varphi_5(Q_2)\varphi_6(Q_1)$$

$$\left. + \varphi_3(Q_3)\varphi_5(Q_1)\varphi_6(Q_2) + \varphi_3(Q_1)\varphi_5(Q_3)\varphi_6(Q_2) \right].$$

The construction is simple but tedious: one writes down the first term from the specified φ_s functions; then one transposes a pair of Q's and adds the new term; then one transposes a different pair of Q's in the new term and adds that; and so on, for a total of 3! distinct terms. Because one has added all possible permutations, one is guaranteed that

$$\psi_j(Q', Q''', Q'') = + \psi_j(Q', Q'', Q''')$$

and so on. The wave function is totally symmetric.

If two φ_s functions are the same, the labor is reduced by one-half; one needs only three distinct terms. If all three are the same, a single term suffices. Again, a specification of the set of numbers $n_s(j)$ suffices (with the symmetry requirement) to specify the state ψ_j.

For three identical fermions, the difference is only algebraic. We have to ensure that

$$\psi_j(Q', Q''', Q'') = - \psi_j(Q', Q'', Q'''),$$

and so on. The requirement can be handled by a judicious salting of the ingredients with minus signs. As we move from one term to the next by a single transposition, we merely change the sign of the new term relative to the term whence it came. The antisymmetric fermion analog of the preceding symmetric boson state is constructed by changing every other plus sign to a minus sign, starting with the second. (In precise language, one adds all permutations, after a minus sign has been applied to those terms that differ from the first term by an odd permutation.) The presence of the minus signs guarantees both the desired antisymmetry and that the Pauli principle will be satisfied. (If φ_5 is replaced by another φ_3, say, the wave function vanishes identically; one cannot violate the Pauli principle.) Given φ_3, φ_5, and φ_6, only one totally antisymmetric wave function can be constructed: that which has been described.

Furthermore, one may show in the manner used for ψ_A in equation (9.2.10) that the symmetric three-boson function given above, and its antisymmetric three-fermion counterpart, are energy eigenfunctions for the system with total energy to $(\varepsilon_3 + \varepsilon_5 + \varepsilon_6)$.

We may conclude that for a system of three identical particles, the unique correspondence, previously found for two identical particles, between ψ_j and a set of numbers $n_s(j)$ holds (provided, of course, that we specify a symmetric or antisymmetric state).

The generalization is hardly surprising. A unique correspondence holds for a system of N identical particles. For a given set of nonnegative integers $n_s(j)$, such that

$$\sum_{s} n_s(j) = N,$$

there is one and only one totally symmetric state describing N identical bosons. If, in addition, the numbers in the set satisfy the relation

$$n_s(j) \leqslant 1,$$

then there is one and only one totally antisymmetric state describing N identical fermions.

Let us try to summarize the content of this section when shorn of the bulk of its algebra. The real distinction made by integral spin versus half-odd-integral spin lies in the symmetry properties of the wave function for a set of two or more identical particles. The appropriate symmetry must hold whether the identical particles are interacting or not. If the mutual interactions may be neglected, a mathematical simplification appears, though the connection between spin and symmetry remains unaltered. In that limiting case, one may construct an energy eigenstate ψ_j for the entire N-particle system out of simple sums and differences of N-fold products of single-particle energy eigenstates φ_s. A specification of whether the resulting wave function is to be symmetric or antisymmetric plus a specification of the set of numbers $n_s(j)$—the number of times each state φ_s appears in the initial term of ψ_j—is then sufficient to specify ψ_j.

The numbers $n_s(j)$ generally carry the informative name of *occupation numbers*. In a useful sense, each $n_s(j)$ indicates how often the corresponding state φ_s is occupied in the N-particle state ψ_j. Because the wave function ψ_j for the entire system must be properly symmetrized, it does make some sense to say that $n_s(j)$ particles are in state φ_s, but one cannot say *which* of the N identical particles are. Indeed, one should not even pose the question. The N particles are properly described *by ψ_j only*, and that wave function indicates that any one of identical particles may be in state φ_s (provided $n_s(j) \neq 0$). It is difficult to break the habit of thinking of identical particles as, somehow, distinguishable. Yet they are not distinguishable; the labeling is mathematics, not physics. One should not push the picturesque aspect of the occupation number description too far.

Summary in More Algebraic Terms

For bosons the values of $n_s(j)$ may range from zero to the fixed number N of particles in the system:

$$n_s(j) = 0, 1, 2, \ldots, N. \qquad (9.2.12a)$$

For fermions, the values are restricted to zero or one:

$$n_s(j) = 0, 1. \tag{9.2.12b}$$

In both cases the $n_s(j)$ must, of course, sum to the number of particles in the system, that being fixed at N:

$$\sum_s n_s(j) = N. \tag{9.2.13}$$

A specification of a set of numbers $n_s(j)$, *subject to the above conditions*, and a statement of which kind of symmetry is to be imposed, *uniquely* specify a state ψ_j of the entire system.

Since the mutual interactions have been neglected, the energy E_j of the entire system is simply a sum of individual energies ε_s, each weighted according to the number of times φ_s occurs in ψ_j:

$$E_j = \sum_s \varepsilon_s n_s(j). \tag{9.2.14}$$

9.3. NEGLIGIBLE MUTUAL INTERACTIONS

The preceding sections have laid the foundation for the statistical mechanics of identical particles when we may neglect the mutual interactions. In the real world some mutual interactions always exist; so the following treatment is valid only as a limit in which neglect of those is admissible. We can, of course, still include the interaction with "external potentials," such as the walls of a container, an external magnetic field, or, in a sophisticated treatment of conduction electrons, the Coulomb potential of the positive ions in a metallic crystal lattice. Often this level of approximation is astonishingly good. The restrictions on the behavior of a system due to a wave function symmetry can, at sufficiently low temperatures, dominate by a wide margin over the effects of the actual interparticle forces. Qualitatively excellent, and sometimes even quantitatively good, results emerge, as we will see in the applications.

It is essential to bear always in mind that we are merely applying the familiar canonical distribution for the probability that a specific energy eigenstate of the entire system, in equilibrium at some specified temperature, provides the appropriate description of the system. There is really nothing new in the general statistical approach. The new features are (1) the connection between spin and wave function symmetry and (2) the notion of occupation numbers $n_s(j)$ for the appearance of single-particle states φ_s in the state

ψ_j of the N-particle system. Though the occupation numbers will play a prominent role, they arise from the product form of the wave function, not from anything intrinsically new in the statistical approach. One could do without them—though not without the canonical distribution.

For the moment we need not *explicitly* distinguish between bosons and fermions. We just suppose we have a physical system of N identical particles in equilibrium at some known temperature and that we know the energy E_j of each state ψ_j of the system. (Whether ψ_j is symmetric or antisymmetric does depend on whether the particles are bosons or fermions and is vital, but that need not be explicitly indicated at the moment.) As always, the probability that a specific state ψ_j provides the appropriate description of the system is given by the canonical distribution:

$$P_j = \frac{e^{-\beta E_j}}{\sum_i e^{-\beta E_i}}. \qquad (9.3.1)$$

Only after we have written this down should we concern ourselves with product wave functions and the like as a means of simplifying calculations.

Many estimates of quantities of physical interest reduce ultimately to calculations of expectation values. So let us look at a typical calculation, that of $\langle E \rangle$, the estimated total energy. We start, as always, with the relation

$$\langle E \rangle = \sum_j E_j P_j.$$

The sum runs over all energy eigenstates of the system. Now we use the connection among the energy E_j, the single-particle energies ε_s, and the occupation numbers $n_s(j)$. Invoking equation (9.2.14), we insert an equivalent expression for E_j:

$$\langle E \rangle = \sum_j \left[\sum_s \varepsilon_s n_s(j) \right] P_j.$$

Interchanging the order of the j and s summations, we rewrite this as

$$\langle E \rangle = \sum_s \varepsilon_s \left[\sum_j n_s(j) P_j \right].$$

The expression in brackets is itself some kind of expectation value.

The calculation leads one to introduce the notion of the *expectation value of the sth occupation number*. The natural defining relation is this:

$$\langle n_s \rangle \equiv \sum_j n_s(j) P_j$$

$$= \sum_j n_s(j) \frac{e^{-\beta E_j}}{\left(\sum_i e^{-\beta E_i} \right)}. \qquad (9.3.2)$$

In a loose sense, $\langle n_s \rangle$ gives an estimate of the number of particles in the single-particle state φ_s. One must, however, bear in mind that only a properly symmetrized state like ψ_j (formed from products of single-particle states) can correctly describe the entire system of identical particles.

If we can find a convenient scheme for calculating $\langle n_s \rangle$ for bosons and fermions, then a calculation of $\langle E \rangle$ may be greatly simplified as

$$\langle E \rangle = \sum_j E_j P_j = \sum_s \varepsilon_s \langle n_s \rangle. \tag{9.3.3}$$

The calculation is reduced to a summation over the complete set of single-particle states. A knowledge of the single-particle energies and of the expectation values of the occupation numbers enables us to compute $\langle E \rangle$, the expectation value of the energy of the *entire system*. In view of the *unique* correspondence between the states ψ_j and the set of numbers $n_s(j)$, this should not be overly surprising.

Calculations for the expectation value of a pressure or total magnetic moment follow in much the same way. A knowledge of the expectation values of the occupation numbers can be extraordinarily fruitful and, in practice, well-nigh indispensable. It will be to our ultimate advantage to determine those expectation values by performing in a general way the summation defining them in equation (9.3.2). There is, unfortunately, a stumbling block: no one has succeeded in performing the summation exactly when the numbers of particles is large. One can, however, turn large N into an asset and develop an excellent approximation. In fact, the approximation is so good that the distinction between approximation and exact expression is seldom maintained in practical calculations. A knowledge of the $\langle n_s \rangle$'s is so valuable in applying the canonical distribution that we will now devote a little effort to working out the general summation.

First an outline of the procedure. Because the occupation numbers for fermions are restricted to being only zero or one, whereas for bosons there is no such restriction, we will handle fermions and bosons separately. In each case we will first derive an exact relation between

$$\langle n_s \rangle_N \text{ and } \langle n_s \rangle_{N-1}.$$

The subscripts N and $N-1$ mean that the system is taken to have N and $N-1$ identical particles in it, respectively.

Then we will turn to another exact relation. The definition of $\langle n_s \rangle_N$, which is

$$\langle n_s \rangle_N = \sum_j n_s(j) P_j(N), \tag{9.3.4a}$$

where the probability has been written as $P_j(N)$ to emphasize that this is for an N-particle system, and the restriction given in equation (9.2.13), which is

$$\sum_s n_s(j) = N, \tag{9.3.4b}$$

together imply the following:

$$\sum_s \langle n_s \rangle_N = \sum_s \left[\sum_j n_s(j) P_j(N) \right] = \sum_j \left[\sum_s n_s(j) \right] P_j(N)$$

$$= \sum_j N P_j(N) = N \sum_j P_j(N) = N,$$

that is,

$$\sum_s \langle n_s \rangle_N = N. \tag{9.3.5}$$

In loose yet useful terms: if we sum, over all single-particle states, the estimated number of particles in each, we get precisely N, the total number of particles actually present. One would justifiably be disturbed if something like this were not contained in the mathematics.

For a typical laboratory sample, the number N will be extremely large relative to unity, and so there will be negligibly little physical difference between systems with N and with $N - 1$ particles. This will be reflected in the values for $\langle n_s \rangle_N$ and $\langle n_s \rangle_{N-1}$. The sum over s on the lefthand side of equation (9.3.5) is generally over an infinite number of single-particle states. To change the value of the righthand side by one unit when N is very large, the individual $\langle n_s \rangle$'s on the lefthand side need change very little indeed. There is then both physical and mathematical justification for the following approximation:

$$\langle n_s \rangle_{N-1} = \langle n_s \rangle_N \qquad \text{when } N \gg 1. \tag{9.3.6}$$

This *is* an approximation, but for N large relative to unity, it is an excellent approximation. (When N is 10^{10}, which is minute for laboratory conditions, the addition or subtraction of one particle produces less of a relative numerical change in the number of particles than the birth of a single child does on the present world population.)

Once we accept the approximation, we may substitute $\langle n_s \rangle_N$ for $\langle n_s \rangle_{N-1}$ in the exact relation between them, and then solve algebraically for $\langle n_s \rangle_N$, the object of interest. This should suffice for an outline. Let us turn now to the two cases.

Fermions

We set out to calculate the expectation value of the occupation number n_1 for a system of N fermions. Later we will be able to generalize the result to all s, not merely $s = 1$. Because we will find ourselves dealing with the mathematical expressions appropriate to a system of $N - 1$ fermions as well, we will have to be explicit in the notation. So we write, for a start,

$$\langle n_1 \rangle_N = \sum_j n_1(j) \frac{e^{-\beta E_j(N)}}{Z(N)}. \tag{9.3.7}$$

The symbol $E_j(N)$ denotes the energy of the jth state of the N-fermion system, and $Z(N)$ is the partition function for that system.

The clue to progress lies in the recognition that the sum over j, which is a sum over all energy eigenstates of the N-fermion system, may be rewritten as a sum over all *sets* of occupation numbers consistent with

$$n_s(j) = 0, 1 \text{ only} \tag{9.3.8a}$$

and

$$\sum_s n_s(j) = N. \tag{9.3.8b}$$

Thus we may write

$$\langle n_1 \rangle_N = \frac{1}{Z(N)} \sum_{\substack{n_1 \\ n_1 + n_2 + \ldots = N}} \sum_{n_2} \ldots n_1 e^{-\beta(n_1\varepsilon_1 + n_2\varepsilon_2 + \ldots)}. \tag{9.3.9}$$

The energy $E_j(N)$ has been written in terms of the single-particle energies ε_s. The replacement of the j summation by a summation over all *sets* of occupation numbers consistent with equations (9.3.8a, b) is permissible because of the *unique* correspondence between state ψ_j and the set of occupation numbers $n_s(j)$. The dots following the two summation signs indicate more summation signs, associated with n_3, n_4, ..., respectively. The first line of specifications under the summation signs indicates a sum over the values 0 and 1 for each n_s. The second line specifies that the sum of the n_s's in each set must be N, the number of particles in the system.

An example with small numbers may be helpful here. Suppose there are only three single-particle states: $s = 1, 2, 3$ (rather than the typical infinite number). Suppose further that we deal with a system of two identical fermions: $N = 2$. There are three states ψ_i for this two-fermion system, with a unique correspondence between then and three *sets* of occupation numbers (given by the rows in table 8.3.1). Now we may compute $\langle n_1 \rangle_2$ from equation (9.3.9) by

TABLE 9.3.1

	n_1	n_2	n_3	E_j
ψ_1	0	1	1	$\varepsilon_2 + \varepsilon_3$
ψ_2	1	0	1	$\varepsilon_1 + \varepsilon_3$
ψ_3	1	1	0	$\varepsilon_1 + \varepsilon_2$

summing over the (three) admissible sets of occupation numbers. That summation we perform by taking n_1 to be 0 and summing over the admissible values of n_2 and n_3 and then taking n_1 to be 1 and summing over the admissible values of n_2 and n_3.

$$\langle n_1 \rangle_2 = \frac{1}{Z(2)} \sum_{n_1} \sum_{n_2} \sum_{n_3} n_1 \, e^{-\beta(n_1\varepsilon_1 + n_2\varepsilon_2 + n_3\varepsilon_3)}$$
$$\scriptstyle n_1+n_2+n_3=N$$

$$= \frac{1}{Z(2)} [0 \, e^{-\beta(0+\varepsilon_2+\varepsilon_3)} + 1 \, e^{-\beta(\varepsilon_1+0+\varepsilon_3)} + 1 \, e^{-\beta(\varepsilon_1+\varepsilon_2+0)}].$$

Comparison shows that this result is exactly what we would have gotten had we used equation (9.3.7), evaluating that sum with the use of the $n_1(j)$ in the first column of the table and the E_j's in the last column. For large N these direct methods are utterly impractical. Hence the need for some tricks in manipulating the summations.

Now we isolate the summation over the two admissible values of n_1 from the general summation:

$$\langle n_1 \rangle_N = \frac{1}{Z(N)} \sum_{n_1} \left[\sum_{\substack{n_2 \\ n_2+n_3+\ldots=N-n_1}} \ldots n_1 e^{-\beta n_1 \varepsilon_1} e^{-\beta(n_2\varepsilon_2+\ldots)} \right].$$

The summation inside the square brackets goes over all admissible values of occupation numbers *other than* n_1. When n_1 is taken to be zero, the entire square bracket expression vanishes. In effect, we can do the outside sum over n_1 by taking the value of the bracket when n_1 is equal to one:

$$\langle n_1 \rangle_N = \frac{1}{Z(N)} e^{-\beta\varepsilon_1} \left[\sum_{\substack{n_2 \\ n_2+n_3+\ldots=N-1}} \ldots e^{-\beta(n_2\varepsilon_2+\ldots)} \right]. \qquad (9.3.10)$$

The exponential $\exp(-\beta\varepsilon_1)$ has been pulled outside as a separate factor. When n_1 is taken to be 1, the sum of the other occupation numbers must be $N - 1$; hence the particular righthand side for the second line of specifications under the summation sign.

The next move consists of arranging matters so that the summation in the square brackets may go over *all* occupation numbers, n_1 *included*, subject to the proviso that the sum of the occupation numbers themselves is $N - 1$, that is

$$n_1 + n_2 + n_3 + \ldots = N - 1, \tag{9.3.11a}$$

and that each occupation number is restricted to 0 or 1, that is,

$$n_s = 0, 1 \; (s \geqslant 1). \tag{9.3.11b}$$

This objective can be achieved by inserting a factor of $(1 - n_1)$:

$$\langle n_1 \rangle_N = \frac{1}{Z(N)} e^{-\beta \varepsilon_1} \sum_{\substack{n_1 \; n_2 \\ n_1 + n_2 + \ldots = N-1}} \ldots (1 - n_1) e^{-\beta(n_1 \varepsilon_1 + n_2 \varepsilon_2 + \ldots)}.$$

The summation goes once again over *all* occupation numbers, subject now to equations (9.3.11a, b). Let us check the correctness of this move. When n_1 is zero, we get precisely the expression in equation (9.3.10), and when n_1 is one, we get zero; so the maneuver is legitimate.

Now we look at the two parts arising from the $(1 - n_1)$ factor. The summation with the " 1 " term yields $Z(N - 1)$:

$$\sum_{\substack{n_1 \; n_2 \\ n_1 + n_2 + \ldots = N-1}} \ldots e^{-\beta(n_1 \varepsilon_1 + n_2 \varepsilon_2 + \ldots)} = Z(N - 1),$$

$$\tag{9.3.12}$$

since the summation goes, in effect, over all energy eigenstates of a system of $N - 1$ fermions.

The summation with the "n_1" term also produces a recognizable result:

$$\sum_{\substack{n_1 \; n_2 \\ n_1 + n_2 + \ldots = N-1}} \ldots n_1 e^{-\beta(n_1 \varepsilon_1 + n_2 \varepsilon_2 + \ldots)} = \langle n_1 \rangle_{N-1} Z(N - 1).$$

Except for the normalizing factor of $Z(N - 1)$, the summation with the "n_1" term gives the expectation value of the occupation number n_1 in a system of $N - 1$ fermions. (To clarify the identification, one may mentally move the $Z(N - 1)$ factor to downstairs on the lefthand side and then compare with equation (9.3.9). The sole difference is that $N - 1$ here replaces N there.)

Now that we have recognized the two parts of the summation for what they yield, we may tidy up the righthand side of equation (9.3.12) and write

$$\langle n_1 \rangle_N = e^{-\beta \varepsilon_1} \frac{Z(N - 1)}{Z(N)} \left(1 - \langle n_1 \rangle_{N-1} \right). \tag{9.3.13}$$

No approximation has been made; this is an exact result among the indicated quantities. Moreover, nowhere did we use any property of n_1 that would not hold for any n_s in its place. The relation holds in general, with the subscript "1" replaced by "s":

$$\langle n_s \rangle_N = e^{-\beta \varepsilon_s} \frac{Z(N-1)}{Z(N)} \left(1 - \langle n_s \rangle_{N-1} \right). \qquad (9.3.14)$$

We now need a second relation between $\langle n_s \rangle_N$ and $\langle n_s \rangle_{N-1}$, in order to eliminate the latter from the relation just derived. For a typical laboratory sample, the number N will be extremely large, and there will be negligibly little physical difference between a system with N and with $N-1$ particles. Provided that N is significantly larger than unity, we may adopt as an excellent approximation the relation given in equation (9.3.6). That we use to replace $\langle n_s \rangle_{N-1}$ by $\langle n_s \rangle_N$ in equation (9.3.14),

$$\langle n_s \rangle_N = e^{-\beta \varepsilon_s} \frac{Z(N-1)}{Z(N)} \left(1 - \langle n_s \rangle_N \right),$$

and then we solve algebraically for $\langle n_s \rangle_N$:

$$\langle n_s \rangle_N = \frac{1}{\left[\dfrac{Z(N)}{Z(N-1)} \right] e^{\beta \varepsilon_s} + 1}. \qquad (9.3.15)$$

This relation is valid to extremely high accuracy for a large-N fermion system.

One step remains: we must find a means of determining the presently unknown ratio of the two partition functions. The summation relation derived as equation (9.3.5) provides the means. Invoking that relation, we write

$$\sum_s \frac{1}{\left[\dfrac{Z(N)}{Z(N-1)} \right] e^{\beta \varepsilon_s} + 1} = N. \qquad (9.3.16)$$

Given the set of single-particle states and their energies and given β and N, one may, exactly in principle and approximately in most practice, solve for the unknown ratio, thus completing the determination of $\langle n_s \rangle_N$.

This completes the derivation for fermions. Before we go on, however, we should introduce a more useful notation, one that will enable us to distinguish more easily between expressions for fermions and those for bosons. To emphasize that the result in equation (9.3.15) applies to fermions, let us rewrite it with a subscript F in the tidier form

$$\langle n_s \rangle_F = \frac{1}{e^{\alpha + \beta \varepsilon_s} + 1}, \qquad (9.3.17)$$

with the convenient abbreviation

$$e^\alpha \equiv \left[\frac{Z(N)}{Z(N-1)}\right]_F.$$

The parameter α is determined from the rewritten version of equation (9.3.16), namely,

$$\sum_s \frac{1}{e^{\alpha+\beta\varepsilon_s} + 1} = N. \tag{9.3.18}$$

This form tells us that the parameter α is a function of N, β, and the set of single-particle energies:

$$\alpha = \alpha(N, \beta, \varepsilon_s\text{'s}).$$

We may now leave out of the notation an explicit mention of the number of fermions in the system; that is always implicit in the parameter α. To be sure, we must bear in mind that these results are valid only when N is much larger than unity. In practice, that requirement is invariably met.

Bosons

The analysis for bosons is analogous to the preceding one, and the derivation will be kept as nearly parallel as possible. We start by computing the expectation value of the occupation number n_1 for a system of N bosons. The same definitions apply:

$$\langle n_1 \rangle_N = \sum_j n_1(j) \frac{e^{-\beta E_j(N)}}{Z(N)}. \tag{9.3.19}$$

Though the notation indicates nothing different from equation (9.3.7), these expressions refer to a system of N bosons, not N fermions. One could be more explicit, but the equations are already cluttered with subscripts.

The sum over j is a sum over all energy eigenstates of the N-boson system. We now rewrite that sum as a sum over all *sets* of occupation numbers consistent with the restrictions

$$n_s(j) = 0, 1, 2, \ldots, N \tag{9.3.20a}$$

and

$$\sum_s n_s(j) = N. \tag{9.3.20b}$$

In this scheme the major difference from the previous fermion analysis lies in the difference between equation (9.3.20a) and equation (9.3.8a). Changing the form of the summation, we write

$$\langle n_1 \rangle_N = \frac{1}{Z(N)} \sum_{\substack{n_1 \ n_2 \\ n_1 + n_2 + \ldots = N}} \ldots n_1 e^{-\beta(n_1 \varepsilon_1 + n_2 \varepsilon_2 + \ldots)}. \qquad (9.3.21)$$

As before, dots indicate more summation signs, and each n_s runs over the values 0, 1, 2, ..., N. The values are not independent, however, for the second line of specifications indicates that the sum of the n_s's in a set must be N.

Again an example may be helpful. We take two bosons, $N = 2$, and three single-particle states: $s = 1, 2, 3$. There are six states ψ_j for this two-boson system, with a unique correspondence between them and six *sets* of occupation numbers (given by the rows in the table 9.3.2). Now we may compute $\langle n_1 \rangle_2$

TABLE 9.3.2

	n_1	n_2	n_3	E_j
ψ_1	0	0	2	$2\varepsilon_3$
ψ_2	0	1	1	$\varepsilon_2 + \varepsilon_3$
ψ_3	0	2	0	$2\varepsilon_2$
ψ_4	1	0	1	$\varepsilon_1 + \varepsilon_3$
ψ_5	1	1	0	$\varepsilon_1 + \varepsilon_2$
ψ_6	2	0	0	$2\varepsilon_1$

from equation (9.3.21) by summing over the (six) admissible sets of occupation numbers. This is done by taking n_1 to be 0 and performing a summation over the admissible values of the other n_s's, then moving on to $n_1 = 1$ and doing likewise, and finally taking $n_1 = 2$ and summing over the admissible values of the remaining n_s's.

$$\langle n_1 \rangle_2 = \frac{1}{Z(2)} \sum_{\substack{n_1 \ n_2 \ n_3 \\ n_1 + n_2 + n_3 = 2}} n_1 \, e^{-\beta(n_1 \varepsilon_1 + n_2 \varepsilon_2 + n_3 \varepsilon_3)}$$

$$= \frac{1}{Z(2)} [0 + 0 + 0 + 1 \, e^{-\beta(\varepsilon_1 + 0 + \varepsilon_3)}$$

$$+ 1 \, e^{-\beta(\varepsilon_1 + \varepsilon_2 + 0)} + 2 \, e^{-\beta(2\varepsilon_1 + 0 + 0)}].$$

Comparison with the expression in equation (9.3.19), evaluated with the use of the first and fourth columns of the table, indicates that we are correctly performing the summation.

Now the trick. When n_1 is zero in the summation over all consistent values of n_1, there is no contribution, for the factor of n_1 is then zero. Contributions begin only when n_1 is one or greater. So we may make the substitution,

$$n_1 = 1 + n_1',$$

with the new summation variable n_1' having the values

$$n_1' = 0, 1, 2, \ldots, N - 1.$$

The summation takes the form

$$\langle n_1 \rangle_N = \frac{1}{Z(N)} \sum_{\substack{n_1' \ n_2 \\ (1+n_1')+n_2+\ldots=N}} \ldots (1 + n_1') e^{-\beta([1+n_1']\varepsilon_1 + n_2\varepsilon_2 + \ldots)}.$$

Two steps are now in order: we pull out the factor of $\exp(-\beta\varepsilon_1)$, and we take the "1" in the second line of the specifications to the righthand side, which becomes $N - 1$. These steps lead to

$$\langle n_1 \rangle_N = \frac{1}{Z(N)} e^{-\beta\varepsilon_1} \sum_{\substack{n_1' \ n_2 \\ n_1'+n_2+\ldots=N-1}} \ldots (1 + n_1') e^{-\beta(n\varepsilon_1'_1 + n_2\varepsilon_2 + \ldots)}.$$

$$(9.3.22)$$

That there is a prime on n_1' is now really irrelevant. The sum in equation (9.3.22) goes, in effect, over all energy eigenstates of a system of $N - 1$ bosons. When we look at the two parts arising from the $(1 + n_1')$ factor, we can recognize the first term as $Z(N - 1)$, for it is just the energy exponential summed over all states of the $N - 1$ boson system. The second part yields

$$\langle n_1 \rangle_{N-1} Z(N - 1)$$

in a fashion analogous to the fermion case. With these parts recognized, we find that the expression for the expectation value becomes

$$\langle n_1 \rangle_N = e^{-\beta\varepsilon_1} \frac{Z(N-1)}{Z(N)} \left(1 + \langle n_1 \rangle_{N-1}\right). \qquad (9.3.23)$$

Again this is an exact result and did not depend on any properties of n_1 that would not hold for any n_s in its place. So we have, for bosons, the general result

$$\langle n_s \rangle_N = e^{-\beta\varepsilon_s} \frac{Z(N-1)}{Z(N)} \left(1 + \langle n_s \rangle_{N-1}\right). \qquad (9.3.24)$$

Just as we did when working with fermions, we may, for large N, invoke the approximation stated in equation (9.3.6). Then we replace $\langle n_s \rangle_{N-1}$ by $\langle n_s \rangle_N$ in equation (9.3.24) and solve for $\langle n_s \rangle_N$:

$$\langle n_s \rangle_N = \frac{1}{\left[\dfrac{Z(N)}{Z(N-1)}\right]e^{\beta \varepsilon_s} - 1}. \tag{9.3.25}$$

The result is valid to high accuracy for a large-N boson system.

A glance back to the analogous result for fermions, in equation (9.3.15), shows that the "only" difference is a (truly crucial) difference in the sign of the "1" in the denominator.

The summation relation derived as equation (9.3.5) holds for both fermions and bosons. The ratio of the partition functions may, therefore, be determined from the requirement

$$\sum_s \frac{1}{\left[\dfrac{Z(N)}{Z(N-1)}\right]e^{\beta \varepsilon_s} - 1} = N. \tag{9.3.26}$$

The results may be tidied up with an improved notation. To emphasize that the current result refers to bosons, we append a subscript B and rewrite equation (9.3.25) as

$$\langle n_s \rangle_B = \frac{1}{e^{\alpha + \beta \varepsilon_s} - 1} \tag{9.3.27}$$

with the abbreviation

$$e^{\alpha} \equiv \left[\frac{Z(N)}{Z(N-1)}\right]_B.$$

There should be no undue confusion over a repetition of the symbol α, for there is always the crucial $\pm$ sign distinction in the denominator to separate α for fermions from α for bosons.

The parameter α is again determined by the summation relation,

$$\sum_s \frac{1}{e^{\alpha + \beta \varepsilon_s} - 1} = N, \tag{9.3.28}$$

with the definite implication that this α, also, is a function of N, β, and the ε_s's.

Some Remarks

The notion of the expectation value of an occupation number has been introduced because $\langle n_s \rangle_F$ and $\langle n_s \rangle_B$ often greatly simplify calculations. Yet once these quantities have entered the scene, one is tempted to look more closely at them and to ask for interpretations going beyond the ideas that led to their introduction. Here we will note one such interpretation for $\langle n_s \rangle_F$. Significantly, an analogous interpretation for $\langle n_s \rangle_B$ does *not* exist.

The relation defining $\langle n_s \rangle_F$ is

$$\langle n_s \rangle_F = \sum_j n_s(j) P_j. \tag{9.3.29a}$$

This form follows most closely the notation used in equation (9.3.2) and leaves implicit the number N of fermions in the system. The crucial point is that, for fermions, the individual occupation numbers are subject to the restriction

$$n_s(j) = 0, 1 \text{ only.} \tag{9.3.29b}$$

The restriction arises, we recall, because the wave function for fermions must by antisymmetric.

Now let us look more closely at a typical term in the sum over j in equation (9.3.29a). If φ_s does not enter into the state ψ_j, then $n_s(j)$ is zero, and that state contributes nothing to $\langle n_s \rangle_F$. If, however, φ_s does enter into the state ψ_j, then $n_s(j)$ is unity, and the state contributes its probability P_j to the sum. Looked at in this way, $\langle n_s \rangle_F$ is given by the sum of the probabilities (P_j) of those states of the system (ψ_j) in which φ_s appears (once and never more than once):

$$\langle n_s \rangle_F = \sum_{\substack{\text{those states } \psi_j \\ \text{in which } \varphi_s \text{ appears}}} P_j. \tag{9.3.30}$$

So we may loosely regard $\langle n_s \rangle_F$ as giving the probability that the *single-particle* state φ_s is occupied. This is a convenient means of expression, one more suggestive than the (always correct) "expectation value of the occupation number" phrase.

The probabilistic interpretation carries the clear implication that $\langle n_s \rangle_F$ must always have a value between zero and unity, endpoints included:

$$0 \leqslant \langle n_s \rangle_F \leqslant 1. \tag{9.3.31}$$

A careful look at the expression for $\langle n_s \rangle_F$ in equation (9.3.17) confirms this conclusion. The ratio of the partition functions, abbreviated as $\exp(\alpha)$, is necessarily positive, and $\exp(\beta \varepsilon_s)$ cannot be negative. The denominator is, therefore, the sum of unity and a nonnegative quantity. Consequently, $\langle n_s \rangle_F$ can never be greater than unity nor less than zero.

A direct probabilistic interpretation for $\langle n_s \rangle_B$ is *not* possible. The reason lies in the different restriction on the occupation numbers for bosons. The relation defining $\langle n_s \rangle_B$ is

$$\langle n_s \rangle_B = \sum_j n_s(j) P_j, \qquad (9.3.32a)$$

with the number N of bosons in the system left implicit. The restriction on the individual occupation numbers, analogous to equation (9.3.29b), is merely

$$n_s(j) = 0, 1, 2, 3, \ldots, N. \qquad (9.3.32b)$$

Because the wave function for particles with integral spin must be symmetric, values for $n_s(j)$ greater than 0 or 1 are permitted. In a state ψ_j a single-particle state φ_s may appear two or more times. Consequently the sum in equation (9.3.32a) does not yield anything that may be interpreted as a probability.

The last conclusion is reflected in the numerical values that $\langle n_s \rangle_B$ may possess. The sum in the defining relation is a sum of nonnegative quantities. Therefore $\langle n_s \rangle_B$ may not be negative; it has a lower bound of zero, as does $\langle n_s \rangle_F$. The upper bound, however, is not unity. Since $n_s(j)$ may range up to N, so may the numerical value of $\langle n_s \rangle_B$. A look back at equation (9.3.27) shows that the *minus* one in the denominator does make possible a value for $\langle n_s \rangle_B$ that is greater than unity. Indeed, for low temperatures and single-particle states with small energies, $\langle n_s \rangle_B$ does become much larger than unity, as we shall later see. An attempted probabilistic interpretation would be nonsensical. For bosons, one must rest content with the phrase "expectation value of the occupation number."

The very last statement does not quite do justice to the situation prevailing in the literature. The function $\langle n_s \rangle_F$ is often called the *Fermi-Dirac distribution function* and $\langle n_s \rangle_B$, the *Bose-Einstein distribution function*. The names are not totally inappropriate, for, in a sense, the functions do represent the estimated fashion in which particles are distributed over the single-particle states. At times we will use the "distribution function" designation, for it is a quite common term. One should, however, always bear in mind that *only* the properly symmetrized state (formed, perhaps, from single-particle states) correctly describes a system of identical particles.

This section has been something of an algebraic nightmare. An attempt at collecting the ideas, as distinguished from the mathematical results, is in order.

Quantum mechanics tells us that systems of identical particles must be described with wave functions possessing special symmetry properties. When one may neglect the mutual interactions of the particles, energy eigenstates of the system can readily be constructed from single-particle states. That construction introduced the notion of occupation numbers. In this section we recognized that, for such a system, many sums with the probabilities P_j can be reduced to sums with the expectation values of the occupation numbers. (The equivalent expressions for $\langle E \rangle$ showed us this explicitly.) Then we used the canonical distribution to calculate those expectation values for systems of N fermions and of N bosons. Provided N is large, our results are superlative approximations.

9.4. THE OCCUPATION NUMBER SCHEME IN THE CLASSICAL LIMIT

In this and the following few sections, we will examine the occupation number scheme in the classical and near-classical limit. The idea of an occupation number is a distinctly quantum-mechanical notion; so the limit is to be understood in the sense that physical conditions, such as temperature and particle density, approach conditions in which a classical analysis would be adequate or nearly so. The intrinsic quantum nature of the occupation number scheme never disappears.

In the classical and near-classical limit, fermions and bosons can be handled in closely parallel ways. To expedite the analysis and to facilitate comparison, we will work with the two kinds of particles simultaneously. This requires only a careful attention to plus and minus signs. Let us agree that the *upper* sign in a $\pm$ or $\mp$ symbol is to make the expression in which the symbol appears valid for fermions; the *lower* sign is to do the same for bosons. Thus the two distinct expressions for the expectation values of the occupation numbers may be neatly written as

$$\langle n_s \rangle = \frac{1}{e^{\alpha + \beta \varepsilon_s} \pm 1}. \tag{9.4.1}$$

Properly, we should, in the notation, distinguish α for the fermions from α for the bosons, perhaps by writing $\alpha_{(\pm)}$. Such a procedure would badly clutter up the equations. Let us agree to leave α unadorned but to remember that we must maintain a distinction. Then the equation which serves to determine the parameter α becomes simply

$$\sum_s \frac{1}{e^{\alpha + \beta \varepsilon_s} \pm 1} = N. \tag{9.4.2}$$

Let us now investigate the limiting forms of these expressions when the physical situation is such that more or less classical reasoning ought to be admissible. Only for a gas have we worked out in detail a necessary condition for the applicability of a classical treatment; so the present analysis is logically restricted to a gas, though the resulting limits are much more general. A glance at the inequality (6.1.7) tells us that, for fixed environment (fixed volume V), a classical analysis can be valid only if the temperature is sufficiently high and the number of identical particles sufficiently small. We look now for the implications by examining equation (9.4.2) in this classical limit.

The number of single-particle states in the summation will, for a gas, be infinite. That property is implicit in the expression for φ_s in equation (9.2.6a); non-zero spin increases the number still further because of the various possible spin orientations. When the temperature is relatively high, β will be relatively small. So the term $\beta\varepsilon_s$ in the exponential will not significantly affect the size of the denominator until rather large values of ε_s are reached. If the infinite summation is nonetheless to have a relatively small numerical value (for under classical conditions N must be "relatively" small), the many terms with small $\beta\varepsilon_s$ must make individually small contributions. This will be true only if $\exp(\alpha + \beta\varepsilon_s)$ is sufficiently large. Though the argument admittedly lacks rigor, we may tentatively conclude that, for physical conditions to which classical reasoning is applicable, the strong inequality,

$$e^{\alpha + \beta\varepsilon_s} \gg 1 \qquad \text{for all } s, \qquad (9.4.3)$$

will hold.

The inequality will permit us to evaluate the limiting forms for α and $\langle n_s \rangle$. The basic step is the implication:

$$e^{\alpha + \beta\varepsilon_s} \gg 1 \quad \text{implies} \quad e^{\alpha + \beta\varepsilon_s} \pm 1 \simeq e^{\alpha + \beta\varepsilon_s}. \qquad (9.4.4)$$

With this implication, the distinction between fermions and bosons disappears from the mathematics, as it should in a truly *classical* limit. Classical physics knows nothing of wave function symmetry properties.

Indeed, one could begin the analysis of the classical limit by demanding that the proper symmetry of the wave functions ψ_j—symmetric or anti-symmetric—should cease, in that limit, to play a significant role. Since the proper symmetry makes itself felt in the ± 1 distinction in the denominator of $\langle n_s \rangle$, one would argue that in the classical limit the ± 1 must be negligibly small relative to $\exp(\alpha + \beta\varepsilon_s)$, this to be true for all s. One would then arrive at relation (9.4.4), though with the implication now going from right to left. Provided one starts out with the conviction that there is a clssical limit in which wave function symmetry plays only a negligible role, the approach of this paragraph is logically neater and far more convincing.

With the simplification that the classical limit brings, we may readily compute a more explicit expression for the parameter α. The equation for the determination of α becomes

$$N = \sum_s \frac{1}{e^{\alpha + \beta \varepsilon_s} \pm 1} \simeq \sum_s \frac{1}{e^{\alpha + \beta \varepsilon_s}} = e^{-\alpha} \sum_s e^{-\beta \varepsilon_s}.$$

This yields, for the classical limit, the result

$$e^{\alpha} \simeq \frac{\sum\limits_s e^{-\beta \varepsilon_s}}{N}. \qquad (9.4.5)$$

The sum in the numerator goes over infinitely many terms whose magnitude, when β is small, remains close to unity until large values of ε_s are reached. So we may even look on this more explicit expression as confirmation that relatively high temperature and relatively small N imply a large value for $\exp(\alpha + \beta \varepsilon_s)$.

Now we examine the expectation values of the occupation numbers. The strong inequality (9.4.3) directly implies that each $\langle n_s \rangle$ is much less than unity in the classical limit. This is a fact worth bearing in mind, but we can also get a more detailed result. The inequality permits us to expand $\langle n_s \rangle$ as

$$\langle n_s \rangle = \frac{1}{e^{\alpha + \beta \varepsilon_s} \pm 1} \simeq \frac{1}{e^{\alpha + \beta \varepsilon_s}} = \frac{e^{-\beta \varepsilon_s}}{e^{\alpha}}.$$

After inserting the limiting expression for $\exp(\alpha)$, we find

$$\langle n_s \rangle \simeq N \frac{e^{-\beta \varepsilon_s}}{\left(\sum\limits_{s'} e^{-\beta \varepsilon_{s'}} \right)}. \qquad (9.4.6)$$

in the classical limit. This is a pleasantly reasonable result, and one should not stir up a great fuss over it. Yet a few words are in order.

The symmetry or antisymmetry of the system's wave function imposes correlations among the particles, but in the classical limit those correlations should become negligible (though not nonexistent). For fermions one can say more simply that the Pauli principle should cease to have a significant influence on the behavior of the system. A picture of independent, uncorrelated particles becomes tenable. One would then anticipate that $\langle n_s \rangle$ would be directly proportional to the total number of particles present. For a guess at the dependence on energy, one could turn to the reduction of a classical N-particle canonical distribution to a one-particle probability distribution.

We discussed that reduction in Section 6.3 and found that, when interparticle forces are neglected, the classical one-particle probability distribution has the familiar exponential dependence on energy. So one would anticipate that $\langle n_s \rangle$ would be proportional to $\exp(-\beta \varepsilon_s)$ in the classical limit. We do find such a limiting behavior for, loosely, the expectation value of the number of particles in a specified single-particle state. The very real correlations among particles, correlations that are imposed by the symmetry or antisymmetry of the system's wave function, do become negligible under appropriate (and attainable) physical conditions.

One last comment is prompted by the preceding limits. To derive closed-form expressions for $\langle n_s \rangle$, we invoked the approximation

$$\langle n_s \rangle_{N-1} = \langle n_s \rangle_N \qquad \text{when } N \gg 1.$$

Equation (9.4.6), giving $\langle n_s \rangle$ in the classical limit, provides a means of testing the consistency of our procedure. That limiting form implies

$$\frac{\langle n_s \rangle_{N-1}}{\langle n_s \rangle_N} \simeq \frac{(N-1)}{N} = 1 - \frac{1}{N}.$$

For large N the righthand side is exceedingly close to unity, and so we have here an indication that the basic approximation does lead to consistent results.

Sums over single-particle states have appeared several times in this section, though only in a formal way. In the applications we will need to perform them explicitly; so it would be well to interrupt the discussion at this point in order to develop a highly useful integral approximation for such sums. The point is that integrations are generally much easier to perform than summations.

9.5. AN INTEGRAL APPROXIMATION

Our concern is with a sum of the form

$$\sum_s (\text{function of } \varepsilon_s)$$

in which the sum goes over the single-particle energy eigenstates of a particle (possibly with spin) confined to a cubical box. We can always arrange the sum to start with the ground state and to go through the other states in order of increasing energy. If the function varies smoothly with increasing values of ε_s, an approximation of the exact summation by an integration with a continuous variable ε ought to be admissible. The conditions can be made more precise, as follows.

1. The *change* in the function as one goes from one ε_s in the sum to the next must be small *relative* to the function itself at either ε_s (at least when the function has numerically significant values).

2. It must be possible to define the function for continuous ε, and the behavior with continuous ε must be smooth.

To transcribe the indicated sum into an integral with continuous ε, we need to know the number of single-particle states per unit energy interval. For a single *spinless* particle, we write

$$\begin{pmatrix} \text{the number of energy eigenstates} \\ \text{with } \varepsilon_s \text{ such that } \varepsilon < \varepsilon_s \leqslant \varepsilon + d\varepsilon \end{pmatrix} = \omega(\varepsilon)d\varepsilon, \qquad (9.5.1)$$

thereby defining the function $\omega(\varepsilon)$ with the understanding that $d\varepsilon$ is an infinitesimal energy interval. For such a spinless particle the transcription of a sum into an integral takes the form

$$\sum_s f(\varepsilon_s) \rightarrow \int_{\min \varepsilon_s}^{\max \varepsilon_s} f(\varepsilon)\omega(\varepsilon)\, d\varepsilon, \qquad (9.5.2)$$

provided the (otherwise arbitrary) function $f(\varepsilon_s)$ meets the two specifications listed above. The limits of integration are the minimum and maximum values of the single-particle energy. The immediate problem is the determination of $\omega(\varepsilon)$ for a spinless particle; the generalization to include spin will then follow with little effort.

Already in Section 1.2, and again in Section 9.2, we worked with the energy eigenstates and eigenvalues of a single spinless particle confined to a cubical box. The energy eigenvalue is reproduced here:

$$\varepsilon_s = \frac{\pi^2 \hbar^2}{2mV^{2/3}} (n_x^2 + n_y^2 + n_z^2)_s. \qquad (9.5.3)$$

The subscript s on $(n_x^2 + n_y^2 + n_z^2)_s$ indicates that the positive integers are those appropriate for the single-particle state φ_s.

Each of the numbers n_x, n_y, and n_z may range separately over the positive integers: $1, 2, 3, \ldots, \infty$. Specifying a set of three positive integers specifies a single-particle state φ_s and its associated energy ε_s. We may also think of the three integers as specifying a single point in a three-dimensional space with axes labeled n_x, n_y, and n_z. This is illustrated in figure 9.5.1. Because of the restriction to positive integers, only one octant,

$$n_x \geqslant 1, n_y \geqslant 1, n_z \geqslant 1,$$

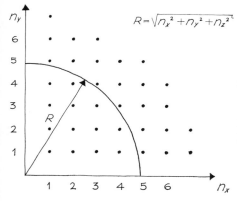

$R = \sqrt{n_x^2 + n_y^2 + n_z^2}$

FIGURE 9.5.1
A slice in the plane $n_z =$ (some positive integer) of the three dimensional mathematical space with axes labeled n_x, n_y, and n_z. Each point indicates a set of values of n_x, n_y, and n_z, and thus each point gives a distinct single-particle state.

of the space is relevant. Except at the edges of the octant, there is one quantum mechanical state per unit volume in this purely mathematical space.

The scheme for determining $\omega(\varepsilon)$ consists of two steps.

1. First we calculate the number of single-particle states with energy less than or equal to ε. Let us abbreviate that number as $\Omega(\varepsilon)$.

2. Then we differentiate $\Omega(\varepsilon)$ with respect to ε to find the number of states per unit energy interval:

$$\frac{d\Omega(\varepsilon)}{d\varepsilon} d\varepsilon = \omega(\varepsilon) \, d\varepsilon.$$

The relation in the second step is valid because $\omega(\varepsilon)$ will give the rate of growth of $\Omega(\varepsilon)$ with respect to ε. Indeed, that relation is a consequence of the proper mathematical procedure for computing the lefthand side of equation (9.5.1). We start by writing the lefthand side in terms of the function Ω, and then we expand:

$$\begin{pmatrix} \text{the number of energy eigenstates} \\ \text{with } \varepsilon_s \text{ such that } \varepsilon < \varepsilon_s \leqslant \varepsilon + d\varepsilon \end{pmatrix} = \Omega(\varepsilon + d\varepsilon) - \Omega(\varepsilon)$$

$$= \frac{d\Omega(\varepsilon)}{d\varepsilon} d\varepsilon + \begin{pmatrix} \text{terms of higher} \\ \text{order in } d\varepsilon \end{pmatrix}.$$

The coefficient of the term linear in $d\varepsilon$ is what we mean by $\omega(\varepsilon)$, and thus the second step is vindicated.

To calculate $\Omega(\varepsilon)$ we note that, since there is one state per unit volume in the n_x, n_y, n_z space, $\Omega(\varepsilon)$ is equal to the volume in the octant out to values of n_x, n_y, n_z, such that

$$\varepsilon = \frac{\pi^2 \hbar^2}{2m V^{2/3}} (n_x^2 + n_y^2 + n_z^2).$$

If we look on $(n_x{}^2 + n_y{}^2 + n_z{}^2)$ as being the square of a radius vector R in the space, then the relevant volume extends out to a value of R such that

$$\varepsilon = \frac{\pi^2 \hbar^2}{2mV^{2/3}} R^2$$

or

$$R = \left(\frac{2m\varepsilon V^{2/3}}{\pi^2 \hbar^2}\right)^{1/2}. \tag{9.5.4}$$

This, too, is illustrated in figure 9.5.1. The volume of the octant is one-eighth the volume of a sphere of radius R. Hence we find that $\Omega(\varepsilon)$ may be expressed as

$$\Omega(\varepsilon) = \frac{1}{8} \times \frac{4\pi R^3}{3} = \frac{\pi}{6}\left(\frac{2m\varepsilon\, V^{2/3}}{\pi^2 \hbar^2}\right)^{3/2}. \tag{9.5.5}$$

In deriving the expression for $\Omega(\varepsilon)$ we have neglected edge effects, both along the plane sides of the relevant volume and over the curved $R = $ (constant) surface. For $n_x \geqslant 10$ or so, and similarly for n_y and n_z, that is a perfectly acceptable neglect. It is the old principle that, for a smooth expansion, a volume grows more rapidly than the surface area, and only on the surface of the relevant octant are we incurring any errors.

To determine $\omega(\varepsilon)$ we differentiate and then tidy up the expression:

$$\omega(\varepsilon) = \frac{d\Omega(\varepsilon)}{d\varepsilon} = \frac{\pi}{6}\left(\frac{2mV^{2/3}}{\pi^2 \hbar^2}\right)^{3/2} \frac{3}{2}\varepsilon^{1/2}$$

$$= \frac{2\pi(2m)^{3/2}}{h^3} V\varepsilon^{1/2}. \tag{9.5.6}$$

This result is actually somewhat more general then the derivation would indicate. It holds for a macroscopic container of any reasonable shape with volume V, not just the cubical one used here.* Ordinarily we may use this result for $\omega(\varepsilon)$ down to $\varepsilon = 0$ with impunity. (A noteworthy exception arises with bosons at low temperature, but more on that at the appropriate time.)

Now we extend the result to cover particles with spin. There will necessarily be a number of distinct states with the same spatial properties and the same

---

* An indication of this is provided by a "classical" counting of states via $(d^3x\, d^3p)/h^3$. Such counting leads to the expression in equation (9.5.6) without any restriction on the shape of the volume.

energy. Those states correspond to different orientations of the spin relative to some fixed spatial direction quite arbitrarily chosen as a quantization axis. If the maximum value of the spin along the chosen axis is $S\hbar$, then there are $2S + 1$ distinct states with the same spatial properties and the same energy. The possible spatial behavior and energy are the same as those for a spinless particle. Therefore, to get the number of states per unit energy interval, we need only multiply $\omega(\varepsilon)$ by $(2S + 1)$:

$$\begin{pmatrix} \text{the number of energy eigenstates} \\ \text{with } \varepsilon_s \text{ such that } \varepsilon < \varepsilon_s \leqslant \varepsilon + d\varepsilon \\ \text{for a particle with spin } S\hbar \end{pmatrix} = (2S + 1)\omega(\varepsilon)$$

$$= (2S + 1)\,\frac{2\pi(2m)^{3/2}}{h^3}\, V\varepsilon^{1/2}.$$

$$(9.5.7)$$

We will refer to this result as the *density of states*. When there is no ambiguity, we will abbreviate it as simply

$$(2S + 1)C\varepsilon^{1/2},$$

with C being a constant that can be read off from the preceding equation. If the particle is a composite one, such as a nucleus, atom, or molecule, we must understand $S\hbar$ to mean the maximum component of the total angular momentum.

A summation over all single-particle states may now be transcribed into an integration by writing

$$\sum_s f(\varepsilon_s) \rightarrow \int_0^\infty f(\varepsilon)(2S + 1)\omega(\varepsilon)\,d\varepsilon. \qquad (9.5.8)$$

This supposes that the energy of the ground state for a single particle is sufficiently close to zero that "min ε_s" is tantamount to zero. For a particle of electron mass in a cube 1 cm on an edge (which implies $V = 1$), the ground-state energy is

$$\text{min } \varepsilon_s = \frac{\pi^2 \hbar^2}{2m}(1^2 + 1^2 + 1^2) = \begin{cases} 1.80 \times 10^{-26} \text{ ergs,} \\ 1.13 \times 10^{-14} \text{ electron volts.} \end{cases}$$

Since the relevant energies of a microscopic physical problem are more nearly of the order of a few electron volts, this is tantamount to zero. For a particle of larger mass the ground state is even closer to zero.

9.6. THE NEARLY CLASSICAL PERFECT GAS

Shortly we will turn to applications at low temperatures and high densities, where the symmetry effects become very pronounced and may, at times, even dominate over interparticle forces. But it is well worthwhile to use the new machinery first in a situation where we have a reasonable idea of what the answers should be like. In this section, therefore, we will calculate the pressure estimate $\langle p \rangle$, through the leading correction due to wave-function symmetry, for a perfect structureless gas of N identical molecules when the physical conditions are such that classical reasoning begins to break down.

We can make a shrewd guess about the physical quantity on which the first correction to the classical $\langle p \rangle$ will depend. In Section 6.1 we arrived at a restriction that must be fulfilled if classical reasoning is to be applicable. Taking the inequality (6.1.7) and rearranging it to give a dimensionless ratio, we may bring the restriction to the form

$$\frac{\hbar/\sqrt{mkT}}{(V/N)^{1/3}} \ll 1.$$

The dimensionless ratio on the left—the ratio of a thermal de Broglie wavelength to the typical separation between molecules—must be small relative to unity if a classical analysis is to be adequate. If the physical conditions change so that the strong inequality becomes more nearly an equality, corrections to the classical value of $\langle p \rangle$ will become significant, and we may reasonably expect that those corrections will depend on the (still small) dimensionless ratio. Such will indeed be the case.

Let us first establish a general expression for $\langle p \rangle$ under the assumptions that the molecules are confined to a cubical box and that the intermolecular forces may be neglected. The *form* for $\langle p \rangle$ will be appropriate for both fermions and bosons, though in *evaluating* it we must maintain a distinction.

Starting with the probabilities P_j for the entire system of N identical molecules, we have for $\langle p \rangle$ the expression

$$\langle p \rangle = \sum_j \left(-\frac{\partial E_j}{\partial V} \right) P_j. \tag{9.6.1}$$

The neglect of intermolecular forces permits us to express the energy E_j in terms of the occupation numbers and the single-particle energies. The expression for $\langle p \rangle$ becomes

$$\langle p \rangle = \sum_j \left(-\frac{\partial}{\partial V} \sum_s \varepsilon_s \, n_s(j) \right) P_j.$$

The relevant partial derivative is

$$\frac{\partial \varepsilon_s}{\partial V} = \frac{\partial}{\partial V}\left[\frac{\pi^2 \hbar^2}{2m}\frac{1}{V^{2/3}}(n_x^2 + n_y^2 + n_z^2)_s\right]$$

$$= -\frac{2}{3}\frac{\varepsilon_s}{V}.$$

(9.6.2)

When we insert this into the preceding equation, the expression for $\langle p\rangle$ becomes

$$\langle p\rangle = \frac{2}{3}\frac{1}{V}\sum_j\left(\sum_s \varepsilon_s n_s(j)\right)P_j.$$

The double summation may be done in either order. Most instructive is the procedure of summing first over j and then over s:

$$\langle p\rangle = \frac{2}{3}\frac{1}{V}\sum_s \varepsilon_s\left(\sum_j n_s(j)P_j\right) = \frac{2}{3}\frac{1}{V}\sum_s \varepsilon_s\langle n_s\rangle$$

$$= \frac{2}{3}\frac{\langle E\rangle}{V}.$$

(9.6.3)

The $\langle n_s\rangle$'s in the first expression will differ between fermions and bosons, but, given our assumptions, $\langle p\rangle$ will always be proportional to the expectation value of the total energy. We will work with the first of the alternative expressions, that in which the distinction between fermions and bosons appears more evidently.

By using a $\pm$ notation, with the *upper* sign always referring to the fermion case, we can deal simultaneously with fermions and bosons. Since $\exp(\alpha + \beta\varepsilon_s) \gg 1$ in the near-classical limit, the inverse of that exponential is much less than one, and we may use it as an expansion parameter. We take $\langle n_s\rangle$ from equation (9.4.1) and expand it, retaining terms through the first that yields a distinction between fermions and bosons:

$$\langle n_s\rangle = \frac{1}{e^{\alpha + \beta\varepsilon_s} \pm 1} = e^{-\alpha - \beta\varepsilon_s}(1 \pm e^{-\alpha - \beta\varepsilon_s})^{-1}$$

$$\simeq e^{-\alpha - \beta\varepsilon_s}(1 \mp e^{-\alpha - \beta\varepsilon_s}).$$

To determine the parameter α, we use equation (9.4.2). After inserting the present expression for $\langle n_s\rangle$ on the lefthand side, we find that α is to be determined from the equation

$$e^{-\alpha}\sum_s e^{-\beta\varepsilon_s} \mp e^{-2\alpha}\sum_s e^{-2\beta\varepsilon_s} = N.$$

It was precisely for coping with sums like these that we developed the integral approximation in the preceding section. If we denote the spin of an individual gas molecule by $S\hbar$, we may use the transcription procedure of equation (9.5.8) and the density of states of equation (9.5.7). The two sums may be approximated by integrals, and the latter evaluated exactly:

$$\sum_s e^{-\beta\varepsilon_s} = \int_0^\infty e^{-\beta\varepsilon}(2S + 1)C\varepsilon^{1/2}\, d\varepsilon = (2S + 1)C\frac{\pi^{1/2}}{2}\beta^{-3/2},$$

$$\sum_s e^{-2\beta\varepsilon_s} = (2S + 1)C\frac{\pi^{1/2}}{2}\beta^{-3/2}\frac{1}{2^{3/2}}.$$

For evaluation, the integrals were reduced, by the variable change $x^2 = \varepsilon$, to those tabulated in Appendix A. After rearrangement, the equation determining α becomes

$$
\begin{aligned}
e^{-\alpha} \mp \frac{1}{2^{3/2}}e^{-2\alpha} &= \frac{2N\beta^{3/2}}{(2S + 1)C\pi^{1/2}} \\
&= \frac{1}{2S + 1}\frac{N}{V}\left(\frac{h}{\sqrt{2\pi mkT}}\right)^3.
\end{aligned}
\tag{9.6.5}
$$

In the expression on the right we find the dimensionless ratio whose appearance we anticipated at the start of this section; indeed, the expression is roughly the cube of that ratio. For the near classical case, the ratio itself will be much less than one and so the cube will be still smaller. Since $h/\sqrt{2\pi mkT}$ is another, roughly equivalent measure of the thermal de Broglie wavelength, let us abbreviate it as λ:

$$\lambda \equiv \frac{h}{\sqrt{2\pi mkT}}.
\tag{9.6.6}$$

Aside from the spin factor of $(2S + 1)$, the righthand side of equation (9.6.5) has the following interpretation: the ratio of the volume of a typical wave-packet (λ^3) to the volume per particle (V/N) in the container. This reinforces the smallness argument: for the near-classical case that we are considering, this ratio of volumes must be small, and so $e^{-\alpha}$ will also be small.

An approximate solution for equation (9.6.5), one consistent with the order of approximation for $\langle n_s \rangle$, is all that we need. The smallness of $e^{-\alpha}$ enables us to expedite the process of solving for it. We may replace $e^{-2\alpha}$ with the square of the lowest-order expression for $e^{-\alpha}$:

$$e^{-\alpha} = \frac{N\lambda^3}{(2S+1)V} \pm \frac{1}{2^{3/2}} e^{-2\alpha}$$

$$\simeq \frac{N\lambda^3}{(2S+1)V} \pm \frac{1}{2^{3/2}} \left(\frac{N\lambda^3}{(2S+1)V}\right)^2 \tag{9.6.7}$$

$$= \frac{N\lambda^3}{(2S+1)V} \left(1 \pm \frac{1}{2^{3/2}} \frac{N\lambda^3}{(2S+1)V}\right).$$

This approximate result is valid through second order in the small quantity $N\lambda^3/V$. (The result follows also by regarding the equation as a quadratic in $e^{-\alpha}$, solving in the usual fashion, and then expanding the square root with the binomial theorem to the necessary order.) Although we could take a logarithm and solve explicitly for α, we will need only $e^{-\alpha}$ in the pressure calculation, to which we now turn.

Invoking equations (9.6.3, 4), we write an expression for $\langle p \rangle$ that maintains through first order, the distinction between fermions and bosons:

$$\langle p \rangle = \frac{2}{3V} \sum_s \varepsilon_s \langle n_s \rangle \simeq \frac{2}{3V} \sum_s \varepsilon_s e^{-\alpha - \beta\varepsilon_s}(1 \mp e^{-\alpha - \beta\varepsilon_s})$$

$$= \frac{2}{3V} e^{-\alpha}\left(\sum_s \varepsilon_s e^{-\beta\varepsilon_s} \mp e^{-\alpha} \sum_s \varepsilon_s e^{-2\beta\varepsilon_s}\right). \tag{9.6.8}$$

The sums are, as before, to be approximated with integrals, which may then be done exactly. The results are

$$\sum_s \varepsilon_s e^{-\beta\varepsilon_s} = (2S+1)C\frac{3}{4}\pi^{1/2}\beta^{-5/2} = (2S+1)\frac{3}{2}\frac{V}{\lambda^3}\beta^{-1},$$

$$\sum_s \varepsilon_s e^{-2\beta\varepsilon_s} = (2S+1)\frac{3}{2}\frac{V}{\lambda^3}\beta^{-1}\frac{1}{2^{5/2}}.$$

Upon inserting these in equation (9.6.8) and factoring, we find

$$\langle p \rangle = (2S+1)\frac{1}{\lambda^3}\beta^{-1}e^{-\alpha}\left(1 \mp \frac{1}{2^{5/2}}e^{-\alpha}\right),$$

and with $e^{-\alpha}$ from equation (9.6.7),

$$\langle p \rangle \simeq \frac{N\beta^{-1}}{V}\left(1 \pm \frac{1}{2^{3/2}}\frac{N\lambda^3}{(2S+1)V}\right)\left(1 \mp \frac{1}{2^{5/2}}\frac{N\lambda^3}{(2S+1)V} + \cdots\right).$$

To tidy this up, we combine the two parentheses by doing the multiplication.

Since we work to first order in the small corrections, the step yields the result

$$\langle p \rangle \simeq \frac{NkT}{V}\left(1 \pm \frac{1}{2^{5/2}}\frac{N\lambda^3}{(2S+1)V}\right). \qquad (9.6.9)$$

The correction does indeed depend on the dimensionless parameter whose appearance we anticipated at the start of this section.

Let us examine the effect of the first-order correction on air at room temperature and pressure. Since each molecule of the dominant constituents, N_2 and O_2, has an even number of spin $\frac{1}{2}\hbar$ particles, the total intrinsic angular momentum of each molecule must be an integral multiple of $\hbar$. Thus molecules of both species, as composite particles, are bosons, and the lower sign in the expression for $\langle p \rangle$ applies. (Actually, our calculation refers to a gas in which all N molecules are of the *same* species—that sameness is essential —but we can afford to be casual here, where we care only about estimating the size of effects.) From the data of Section 6.1, we may estimate the numerical value of the dimensionless parameter as

$$\frac{N\lambda^3}{V} \simeq \left(\frac{8 \times 10^{-10}/\sqrt{2\pi}}{3.5 \times 10^{-7}}\right)^3 \simeq 10^{-9}.$$

Hardly detectable. Under typical room conditions the intermolecular forces vastly swamp this minute symmetry effect. But the aim of this section has not been to derive a startling result; rather, it has been to gain some confidence in the analytical machinery. To lowest order we *do* get

$$\langle p \rangle = \frac{NkT}{V}.$$

Some further remarks are, however, in order. One is prompted by the $\pm$ symbol in the final result for $\langle p \rangle$. In a situation where N, V, and T are experimentally fixed, a fermion gas exerts a greater pressure than a boson gas. At the very beginning of this section, we found a rigorous proportionality between $\langle p \rangle$ and $\langle E \rangle$, when the intermolecular forces are neglected. The implication is that for fixed N, V, and T, a fermion gas has a larger energy than a boson gas. We can find the reason for this in the Pauli principle for fermions. By the restriction

$$\langle n_s \rangle_F \leqslant 1,$$

the Pauli principle "forces" fermions to populate the higher energy single-particle states more heavily than would necessarily happen in a boson gas. At low temperatures the effect becomes extreme, as we will see in the following chapter.

The subject of the other remark is the relation $\beta = 1/kT$, which we found back in Section 4.6. We had used the criterion of maximum missing information to derive the "best" probability distribution for a system in equilibrium when one could specify the expectation value of the energy. The derivation left a parameter β to be determined by the specified numerical value. We applied the probability distribution to the dilute gas of a gas thermometer; in that situation we could justifiably demand that the expectation value of the energy be $\frac{3}{2}NkT$. At that early stage in the development of mathematical apparatus, we restricted the explicit calculation to a single molecule in a container and found the parameter β to be $1/kT$. Now we are in a position to confirm the general validity of that relation.

By combining equations (9.6.3, 9), we may write the calculated value of $\langle E \rangle$ as

$$\langle E \rangle \simeq \frac{3}{2}NkT\left(1 \pm \frac{1}{2^{5/2}}\frac{N\lambda^3}{(2S+1)V}\right).$$

Though mathematical complexity forced some approximations even here, the calculation did deal simultaneously with all N identical particles and did take into account the proper symmetry of the wave functions for the system. For a dilute gas the second term on the right may certainly be neglected relative to the first. Moreover, the neglect of the intermolecular forces is then justifiable. The upshot is that the calculation, in which the relation $\beta = 1/kT$ was used, does produce the correct result for $\langle E \rangle$.

It is true that specifications of large N and of macroscopic container size have gone into this dilute-gas calculation. They justify our expressions for $\langle n_s \rangle$ and our replacement of sums by integrals. This should not, however, be a worrisome point. The temperature concept is a macroscopic notion. The physical systems to which it applies are either macroscopic or—by loose extension—microscopic parts of macroscopic systems. We have merely been using the properties that a dilute-gas thermometer is sure to have. In an important sense, the modest approximations necessary to find $\beta = 1/kT$ are justified by the macroscopic nature of the temperature concept itself.

9.7.   THE CLASSICAL PARTITION FUNCTION FOR IDENTICAL PARTICLES

Before leaving the general domain of the classical limit, we should look at the effect that the indistinguishability of identical particles has on the classical partition function. We will discover that the symmetry properties of wave functions have a permanent effect there.

To find the classical limit for the partition function, we need only two elements. One is the relation defining the parameter α in terms of the ratio of two partition functions. From the definitions following equations (9.3.17, 27), we have

$$e^{\alpha} \equiv \frac{Z(N)}{Z(N-1)}.$$

(9.7.1)

For the sake of brevity, the distinguishing subscripts, F and B, have been omitted. We may usefully look on e^{α} as relating $Z(N)$, the partition function of interest, to $Z(N-1)$:

$$Z(N) = e^{\alpha}Z(N-1).$$

The other element is the approximate expression for e^{α}, valid in the classical limit for both fermions and bosons. Equation (9.4.5) gives that as

$$e^{\alpha} \simeq \frac{\sum\limits_{s} e^{-\beta \varepsilon_s}}{N}$$

(9.7.2)

for a system of N identical particles. The two elements permit us to solve for $Z(N)$ as

$$Z(N) \simeq \frac{\left(\sum\limits_{s} e^{-\beta \varepsilon_s} \right)}{N} Z(N-1).$$

(9.7.3)

At first glance this maneuver seems to provide no advance, but in a similar fashion we may relate $Z(N-1)$ to $Z(N-2)$ and then insert to arrive at the relation

$$Z(N) \simeq \frac{\left(\sum\limits_{s} e^{-\beta \varepsilon_s} \right) \left(\sum\limits_{s} e^{-\beta \varepsilon_s} \right)}{N \qquad (N-1)} Z(N-2).$$

Proceeding in this manner, we may work our way down to $Z(1)$, the partition function for a one-particle system; that would be given exactly by

$$Z(1) = \sum\limits_{s} e^{-\beta \varepsilon_s}.$$

(9.7.4)

So the sequence of substitutions leads to a tidy result:

$$Z(N) \simeq \frac{\left(\sum\limits_{s} e^{-\beta \varepsilon_s} \right)^{N}}{N!}.$$

(9.7.5)

Before we interpret this result, a few words about the approximation are in order. Approximation enters primarily through the expression for e^α in equation (9.7.2), an expression that is valid, in the classical limit, when N is significantly larger than unity. Strictly, we may not continue to use that expression for e^α in the sequence of substitutions all the way down to $Z(1)$. It is true, however, that in using it for all substitutions, we make a relatively minor error. Only for the last hundred substitutions or so is the error worth mentioning. When N itself is of order 10^{20}, that error is negligibly small. Moreover, for virtually all applications of partition functions, one needs only the logarithm of the partition function. Logarithms are amazingly insensitive to changes in the value of their arguments. (Most people regard $\$10^6$ and $\$10^2$ as significantly different; yet the base-ten logarithms, 6 and 2, are not very different.) So an error in a few of the multiplicative substitutions for $Z(N)$ will have little effect on $\ln Z(N)$.

Now we work toward an interpretation of our result for $Z(N)$. In the fully classical limit, we may transcribe a sum over single-particle states into an integral over the classical phase space of a single particle plus a sum over spin orientations:

$$\sum_s e^{-\beta \varepsilon_s} \to \sum_{\text{spin}} \int_{\substack{\text{phase} \\ \text{space}}} e^{-\beta \varepsilon(\mathbf{x}, \mathbf{p}, \text{spin})} \frac{d^3 x \, d^3 p}{h^3}.$$

Here $\varepsilon(\mathbf{x}, \mathbf{p}, \text{spin})$ denotes the classical energy of a single particle, which might depend on the (nonclassical) spin orientation. We have a product of N such factors. So, upon introducing mathematical labels for the N particles, we may transcribe to classical terms as follows:

$$\left(\sum_s e^{-\beta \varepsilon_s} \right)^N \to \prod_{i=1}^{N} \left(\sum_{\text{spin}} \int e^{-\beta \varepsilon} \frac{d^3 x_i \, d^3 p_i}{h^3} \right)$$

$$= \sum_{\substack{\text{spin for} \\ \text{all } N}} \int_{\substack{\text{phase space} \\ \text{for all } N}} e^{-\beta E} \frac{d^2 x_1 \cdots d^3 p_N}{h^{3N}}.$$

The relation

$$E = \sum_{i=1}^{N} \varepsilon(\mathbf{x}_i, \mathbf{p}_i, \text{spin}_i)$$

gives the classical energy of the entire system as a function of the positions, momenta, and spin orientations. The second step, to an integration over the classical phase space for an N-particle system, is merely the reverse of the mathematical operation used in Section 8.1.

Under transcription, the product of summations in equation (9.7.5) acquires a form that appears natural for a classical partition function. What, then, can be the meaning of the $N!$ in the denominator?

To have a picture with which to work, let us take a system of two spinless particles constrained to one-dimensional motion. If we suppress for the moment the coordinates p_1 and p_2, the phase space has a manageable two dimensions, with axes x_1 and x_2. In figure 9.7.1 two small regions, A and B,

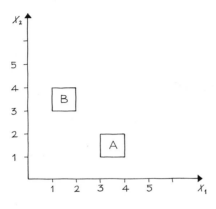

FIGURE 9.7.1
The two small regions of phase space, A and B, are mathematically distinct but, for identical particles, physically the same.

of this phase space are indicated; the regions are given by the specifications

$$A: [3 \leqslant x_1 \leqslant 4, 1 \leqslant x_2 \leqslant 2],$$
$$B: [3 \leqslant x_2 \leqslant 4, 1 \leqslant x_1 \leqslant 2].$$

If the two particles, "1" and "2", are truly different—"1" is a helium atom and "2" is a neon atom, say—then the two regions of phase space do represent very different physical situations. If, however, the two particles are of the *same* kind—each is a helium atom, say—the two regions of phase space do *not* represent different physical situations. Both regions represent a situation in which one helium atom lies between spatial coordinates 3 and 4 on the physical x-axis and another helium atom lies between spatial coordinates 1 and 2 on the same physical axis. Labels like "particle 1" and "particle 2" are a physically meaningless mathematical convenience in a discussion of identical and indistinguishable particles.

Indistinguishability has a strong implication for the counting of quantum-mechanical states in the classical limit. For two identical particles, an integration over our phase space with

$$\frac{dx_1 \, dp_1 \, dx_2 \, dp_2}{h^2}$$

would *overcount* the number of *physically distinct* situations by a factor of 2 (or 2!).

The overcounting notion generalizes readily for a system with N identical particles. For a small region of the $6N$-dimensional phase space, specified by

$$[3 \leqslant x_1 \leqslant 4, \ldots, 7 \leqslant (p_z)_N \leqslant 8],$$

say, there are many physically equivalent regions; they can be found by permuting the labels of the particles in the original specification. (Incidentally, spins can be included in the specification at merely the price of a more elaborate notation.) The number of such permutations is $N!$, and it is by this number that we would be overcounting physically distinct situations in an integration over all phase space.

Now an interpretation of the $N!$ in equation (9.7.5) is available. We start with the rigorously quantum-mechanical description. There the indistinguishability of identical particles compels us to describe the system with wave functions having definite symmetry properties. As we pass to the classical limit, we find that the effects of indistinguishability do not fully disappear. Arrangements in phase space that are mathematically distinct (in that the mathematically labeled particles are differently distributed in the N-particle phase space), but which are physically the same, are not to be counted as distinct in computing a partition function. If, for convenience, we integrate over all phase space, implicitly counting each physically distinct situation $N!$ times, we must compensate by an explicit division with $N!$.

The preceding can be generalized somewhat. We have taken the classical limit for a system of identical particles when the particles were permitted to interact with external potentials but not among themselves. The primary element in the analysis, however, is the indistinguishability of identical particles and the condition of proper symmetry that this imposes on a wave function for the system. Permitting the identical particles to interact among themselves does not alter this primary element, and so should not affect the appearance of $N!$ in the classical limit. From this argument we may conclude that for a system of N identical particles, interacting among themselves or not, the partition function in the classical limit is to be calculated as

$$Z_{\text{classical}} = \sum_{\text{spin}} \int_{\substack{\text{phase} \\ \text{space}}} e^{-\beta E(x_1, \ldots, \, p_N \text{ spins})} \frac{d^3 x_1 \cdots d^3 p_N}{N! \, h^{3N}}.$$

The "spin" summation goes over the spin orientations for all N particles, and the integral goes over the phase space for all N particles. The classical energy in the exponent is permitted to encompass interactions of the particles among themselves.

In Chapter 8 we took an uncompromisingly classical view of a gas and regarded the identical gas molecules as, somehow, distinguishable. Now we can assess how much the results of that chapter must be altered when one drops the uncompromising view and admits that indistinguishability, with its profound consequences in quantum mechanics, may leave residual consequences in the classical limit. The proper classical partition function differs from the partition function we used in Chapter 8 by the factor $N!$ only. The logarithms will differ by a term $\ln N!$. Since the estimated pressure and energy can be computed by *differentiating* the *logarithm* of the partition function with respect to volume and temperature, respectively, the $\ln N!$ term has no influence on those expectation values. The pressure calculation and the virial coefficient analysis are safe. There is, however, one point in the chapter at which the $N!$ does significantly alter results. That point is the calculation of the information-theory entropy of a structureless gas in the classical limit. In the expression for S_I in equation (8.7.1), the logarithm of the partition function appears twice, once with a derivative applied to it and once without. In the latter case the $\ln N!$ term will contribute. The effect is pleasantly easy to work out and to interpret.

We begin by inserting $1/N!$ into equation (8.7.2), so that it now reads

$$Z = \left(\frac{1}{N!} \left(\frac{2\pi m}{h^3 \beta} \right)^{(3/2)N} V^N \right) \times \left(\frac{Z_U}{V^N} \right) = Z_{\text{perf}} \left(\frac{Z_U}{V^N} \right).$$

Incorporating the $N!$ into the partition function for the perfect gas is appropriate, since the $N!$ correction is independent of the presence or absence of intermolecular forces; it must be present in the perfect gas analysis. The steps that led to equation (8.7.3) now lead to

$$(S_I)_{\text{perf}} = k \ln \left\{ \frac{1}{N!} \left[V \middle/ \left(\frac{h}{\sqrt{2\pi emkT}} \right)^3 \right]^N \right\}.$$

In Section 8.7 we noted that the expression in square brackets is (roughly) a measure of the number of ways in which a typical molecule may be put into the container. Then we went on to say that for distinguishable molecules the number of physically distinct arrangements for the entire gas is the product of the number of possible arrangements for each molecule. That explained why the expression in square brackets appeared raised to the Nth power. Now we are no longer treating the identical molecules as, somehow, distinguishable, and so arrangements which previously were physically distinguishable are no longer so. The previous counting of physically distinct arrangements must be corrected by an $N!$ permutation factor—and that is what has been done by the limiting process from the quantum analysis. It is the number of physically distinct arrangements (as opposed to mathematically distinct for identical though mathematically labeled particles) that determines the information-theory entropy. Once we recognize this point, the rest of the interpretation of $(S_1)_{perf}$ proceeds as in Section 8.7.

REFERENCES for Chapter 9

The derivation of $\langle n_s \rangle_F$ and $\langle n_s \rangle_B$ in this chapter is based on a paper by Helmut Schmidt, *Zeitschrift für Physik*, **134**, 430 (1953). References for analyses of the (extremely high) accuracy of our approximation procedure are provided in Chapter 10 among the references on the Bose-Einstein condensation.

   If one is going to spend time on a detailed derivation of the classical partition function (valid for mutually interacting identical particles) as a limit of the quantum-mechanical partition function, one might as well look at a careful treatment. Such is given in adequate detail by Terrell L. Hill in Section 16 of his *Statistical Mechanics* (New York: McGraw-Hill, 1956). The first complete derivation was made by J. G. Kirkwood and published in a relatively brief article: *Physical Review*, **44**, 31 (1933). A reasonable indication of how the limit of classical mechanics is achieved is given by Garrison Sposito, *American Journal of Physics*, **35**, 888 (1967). All these treatments presuppose a familiarity with the expression for Z when written in the more general quantum-mechanical language used in Section 12.3.

PROBLEMS

9.1. Wave function symmetry. We consider a system of two identical particles, each constrained to move along the same physical x-axis between $x = 0$ and $x = L$. The constraints at the endpoints may be imposed with a potential energy $u(x)$:

$$u(x) = \begin{cases} +\infty & \text{for } x \leqslant 0, \\ 0 & \text{for } 0 < x > L, \\ +\infty & \text{for } x \geqslant L. \end{cases}$$

We specify that the two particles do not interact with one another.

a. Show that the single-particle energy eigenstates φ_s have a spatial dependence of the form

$$\varphi_s(x) = \left(\frac{2}{L}\right)^{1/2} \sin\left(\frac{\pi n_s x}{L}\right),$$

with n_s being any positive integer $(1, 2, \ldots)$. Compute the corresponding single-particle energy ε_s for a particle of mass m.

b. Suppose the particles are spinless (and hence are bosons). Construct *explicitly* for this system of two identical bosons the first three states ψ_J in order of increasing energy E_J, that is, the ground state and the two lowest excited states.

c. Suppose now that the particles have spin $\frac{1}{2}\hbar$. Then for each integral value of n there will be two distinct single-particle states, one with spin " up " and another with spin " down." Invent a notation to indicate this and then construct the first three states ψ_J in order of increasing energy for this system of two identical fermions. (Because there is, by assumption, no dependence of the energy on the spin orientation, the three states will not be unique. There will be four states with equal energy directly above the ground state ψ_1.)

9.2. Let us imagine a truly small system of two identical (spinless) bosons described by the canonical probability distribution. Suppose further that the bosons do not interact with one another and that there are only two single-particle energy eigenstates, φ_1 and φ_2, with $\varepsilon_1 < \varepsilon_2$.

a. Construct, symbolically from φ_1 and φ_2, the energy eigenstates ψ_J of the entire two-boson system and their energies E_J. (You should find a total of three.)

b. Calculate the expectation value of the occupation number of single-particle state φ_1, that is, $\langle n_1 \rangle_B$. (Bear in mind that this is not a system with $N \gg 1$.) What are the low-temperature and high-temperature limits?

9.3. An alternate derivation of the expression for $\omega(\varepsilon)\,d\varepsilon$ is the object of this problem. As before, the particle is taken to be spinless and to be confined to a volume V, but now that volume need not be cubical. The energy ε is the kinetic energy (plus a " zero or infinity " contribution from the potential due to the container boundaries). Let us count the number of states by ascribing $d^3x\, d^3p/h^3$ such states to a " volume " $d^3x\, d^3p$ of the single particle's six-dimensional classical phase space.

a. Justify in words the following relations:

$$\omega(\varepsilon)\,d\varepsilon = \int_{\substack{\text{all regions of phase} \\ \text{space for which the} \\ \text{energy lies between} \\ \varepsilon \text{ and } \varepsilon + d\varepsilon}} d^3x\, d^3p/h^3 = \frac{4\pi V}{h^3} \int_{p(\varepsilon)}^{p(\varepsilon + d\varepsilon)} p^2\, dp,$$

where $p(\varepsilon)$, $p(\varepsilon + d\varepsilon)$, designate magnitudes of the momentum for which the energy has the indicated values.

b. For infinitesimal $d\varepsilon$, such that $d\varepsilon/\varepsilon \ll 1$ and terms quadratic (and higher) in $d\varepsilon$ may be neglected, evaluate the last integral and confirm that the result for $\omega(\varepsilon)$ is that derived in Section 9.5.

c. The treatment here of $d\varepsilon$ is mathematically sloppy. How would one tidy it up?

9.4. **A check for consistency.** The derivation of the canonical probability distribution in Chapter 4 was based partially on the demand that $\langle p \rangle = p_{obs}$ for a dilute gas. This is a reasonable demand if $\Delta p/\langle p \rangle$ is very small relative to unity. Let's see whether the theory is consistent.

Let us treat the dilute gas quantum-mechanically as a gas of N identical particles whose mutual interactions we may neglect.

a. Derive (or present some arguments in support of) the result

$$\begin{pmatrix} \text{pressure exerted} \\ \text{if state is } \psi_j \end{pmatrix} \equiv p_j = -\frac{\partial E_j}{\partial V} = +\frac{2}{3}\frac{E_j}{V},$$

which holds, of course, for both fermions and bosons. (The last equality depends on our taking the potential due to the walls to be zero inside the volume V.)

b. Relate $\langle p \rangle$, and the root mean square estimate Δp of deviations, to the corresponding quantities for total energy. Then, using the relation between ΔE and the temperature derivative of $\langle E \rangle$, derived in Section 7.1, derive the exact result

$$(\Delta p)^2 = +\frac{2kT^2}{3V}\frac{\partial\langle p \rangle}{\partial T}.$$

c. Use the result in (b) and the calculation of $\langle p \rangle$ for the nearly classical limit to estimate first Δp and then $\Delta p/\langle p \rangle$, both correct to first order in the difference between fermions and bosons. Is the theory consistent?

9.5. **More Einstein model.** The atoms in a solid, such as diamond (bosons) or solid He^3 (fermions), are certainly indistinguishable. Might not the theoretical expression for the heat capacity depend on whether the atoms are fermions or bosons? At least for the Einstein model, the answer is no. Let us see why.

To start with, let us consider a solid consisting of two identical atoms—mathematically labeled 1 and 2—and two lattice sites—labeled A and B in a physically meaningful manner. Generalization will be easy because there are always equal numbers of atoms and sites in the Einstein model. Section 7.4 described the approximations that are used to calculate the potential energy in the Einstein model; those approximations enable us to construct energy eigenstates for the system from single-particle energy eigenstates. There will be a separate set of three-dimensional harmonic-oscillator wave functions for each of the two sites. Let us replace the old label s on a single-particle state by a pair of labels. One label denotes the lattice site; that label will be either A or B. The other label, written s with one prime or more, denotes both a specific

harmonic-oscillator wave function and a specific spin orientation for the particle. Thus, for example, the single-particle states associated with site A will be
denoted by $\varphi_{As'}$.

Now we construct the energy eigenstates ψ_J for the two-atom system,
bearing in mind the *extra restriction that there is to be one but only one atom
at each site*. We may pick *any* value for s' in $\varphi_{As'}$ and *any* value for s'' in $\varphi_{Bs''}$,
and construct an energy eigenstate of the system with correct symmetry:

$$\psi_J(Q_1, Q_2) = C_{\pm}[\varphi_{As'}(Q_1)\varphi_{Bs''}(Q_2) \pm \varphi_{As'}(Q_2)\varphi_{Bs''}(Q_1)],$$

with $(+)$ if the atoms are bosons and $(-)$ if fermions, The constant $C_{\pm}$ is just
a normalizing factor. There is a vast difference between the wave function for
the bosons and that for the fermions, both because the particles have different
spins and because we need symmetry for the bosons, antisymmetry for the
fermions. The noteworthy point, though, is that the symmetry requirement
has no influence on the energy E_J. We have $E_J = \varepsilon_{s'} + \varepsilon_{s''}$ in each case; the
possible total energies are the same.

a. Pairing all values of s' with all values of s'' will give us all the states ψ_J of the
two atom system. Use this property to show that the following factorization
is permissible:

$$Z(2) \equiv \sum_J e^{-\beta E_J} = \left(\sum_{s'} e^{-\beta\varepsilon_{s'}}\right)\left(\sum_{s''} e^{-\beta\varepsilon_{s''}}\right) = \left(\sum_{s'} e^{-\beta\varepsilon_{s'}}\right)^2,$$

where each sum with s' or s'' goes over the single-particle states associated
with *one lattice site*.

b. Generalize to $Z(N)$, arriving at

$$Z(N) = \left(\sum_{s'} e^{-\beta\varepsilon_{s'}}\right)^N$$

for both fermions and bosons. Note that this is exact, given the Einstein
model, and that no $N!$ appears. The extra restriction mentioned above is at
the root of this. To reiterate, the sum with s' is merely a sum over all single-
particle energy eigenstates associated with one lattice site. It is *not* a sum over
all the single-particle states of the system (having N lattice sites).

c. Since a three-dimensional harmonic oscillator is equivalent to three one-
dimensional oscillators, we have

$$\sum_{s'} e^{-\beta\varepsilon_{s'}} = (2S+1) \sum_{n_1=0}^{\infty} \sum_{n_2=0}^{\infty} \sum_{n_3=0}^{\infty} \exp\left(-\beta h\nu[(n_1 + \tfrac{1}{2}) + (n_2 + \tfrac{1}{2}) + (n_3 + \tfrac{1}{2})]\right).$$

The factor $(2S + 1)$ arises because $2S + 1$ different spin orientations can be
associated with each harmonic-oscillator wave function.

Evaluate the above summation (which may be factored and done exactly),
form $Z(N)$, and use that to compute the heat capacity on the Einstein model.
Do you find that the heat capacity depends on whether the N atoms are bosons
or fermions?

9.6. More paramagnetism. The system consists of N "spatially fixed," identical (and hence indistinguishable) paramagnetic particles whose mutual interactions may be neglected. The particles do, however, interact with an external magnetic field $H_z \hat{z}$. Show that the partition function for the system is the N-fold product of the single-paramagnet partition function derived in Section 7.3. To show that this is so for both bosons and fermions, you will probably want to make an analysis similar to that in the preceding problem. Then take $J = \frac{1}{2}$, and calculate the expectation value of the total magnetic moment and the information-theory entropy. The latter will be needed in Chapter 11.

APPLICATIONS OF
THE QUANTUM
DISTRIBUTION FUNCTIONS

This chapter has three natural divisions. The first deals with the free-electron model as an application of the Fermi-Dirac distribution function, $\langle n_s \rangle_F$. The next part looks at the low-temperature behavior of a system of identical spin-zero particles, with liquid helium in mind; here the Bose-Einstein distribution function, $\langle n_s \rangle_B$, for a fixed number of bosons, comes into play. Finally, photons in a heated cavity—black-body radiation—are considered as a situation in which the number of identical bosons is not fixed. The treatment requires a brief return to the derivation of the Bose-Einstein distribution function.

10.1. THE FREE-ELECTRON MODEL FOR METALS

In a good metallic conductor, the electrons may, at least roughly, be grouped into two classes: (1) those electrons that remain tightly bound to the positive nuclei and thus form, with the nuclei, a background lattice of positive ions;

and (2) those electrons that are relatively free to move throughout the entire metal and thus are responsible for the electrical conductivity.

In this and the next few sections we will be concerned with the electrons in the latter group, loosely called the conduction electrons. As a first approximation in studying them, we will treat them as independent of the positive ions and as not interacting among themselves. Since electrons have spin $\frac{1}{2}\hbar$, the preceding means that we treat the conduction electrons as a perfect fermion gas of N identical electrons. They are confined to a volume V (the size of the metallic sample) by the positive ions, but that is the only cognizance we take of the ions.

In view of the Coulomb interaction between the conduction electrons and the positive ions, and among the conduction electrons themselves, it is by no means evident that such a treatment is useful as even a first approximation. Remarkably, many of the results of this "free electron" model are in excellent qualitative agreement with observation. For example, the model gives the correct temperature dependence for the contribution that the conduction electrons make to the metallic heat capacity. The numerical coefficients provided by the model may, however, be off by a factor of two or more. At the end of the applications, we will look briefly into the question of why, for the quantities we will have calculated, such good qualitative agreement is possible. For the moment let us accept the free-electron model as the simplest approximate treatment in which we use antisymmetric wave functions, as demanded for a system of identical fermions.

In Section 6.1, on the admissible domain of a classical analysis, we estimated some parameters for a system of conduction electrons. We concluded that the inequality necessary for the applicability of a classical treatment, far from being fulfilled, was actually reversed. Taking over those estimates, we can form the now-familiar dimensionless parameter and find

$$\frac{N\lambda^3}{V} \equiv \frac{N}{V}\left(\frac{h}{\sqrt{2\pi mkT}}\right)^3 \simeq 3{,}000$$

for room temperature and a contribution of one conduction electron from each atom in a typical crystal lattice. For a classical analysis to be adequate, the parameter must have a value much less than unity. Quite obviously we must take quantum mechanics and its antisymmetric fermion wave functions into account.

Previously we have argued that a classical treatment is generally acceptable for a gas if the particle density is not too high and the temperature not too low. A contribution of one conduction electron from each atom implies a conduction-electron density comparable to that of the molecules in a liquid like water. So density is not likely to be the crucial factor here. The very small mass of the electron makes the thermal de Broglie wavelength very large already at room temperature, as large as the thermal wavelength that molecules acquire only at exceedingly low temperatures. The implication is that, for a gas of conduction electrons, room temperature is already a very low temperature. With this as a hint, let us go to the extreme and look first at the behavior of a fermion system at absolute zero.

In Section 7.6 we noted that in the limit as the temperature goes to absolute zero, a physical system is sure to be found in its ground state. Let us see what this implies about $\langle n_s \rangle_F$ and the parameter α, the latter being determined by the equation

$$\sum_s \langle n_s \rangle_F = \sum_s \frac{1}{e^{\alpha + \beta \varepsilon_s} + 1} = N. \tag{10.1.1}$$

Simply letting β go to infinity (for T going to zero) appears to lead to nonsense, for then each $\langle n_s \rangle_F$ appears to go to zero. The sum apparently cannot yield N. (Even the single-particle ground state is not exempt, for ε_1 is greater than zero, though the approximation $\varepsilon_1 \simeq 0$ is ordinarily satisfactory. Besides, one single-particle state cannot contribute more than unity to the sum.) Yet when the N-particle system is in its ground state ψ_1, the Pauli principle implies that N distinct single-particle states must be occupied, in fact, that set of N of them with the lowest total energy. So $\langle n_s \rangle_F$ ought to be unity out to some value of ε_s, then zero beyond that.

The apparent difficulties of the preceding paragraph imply that we have not been sufficiently careful: the low-temperature behavior of the parameter α might be singular, and enough so to resolve our difficulties. The derivation of $\langle n_s \rangle_F$ implies that e^{α} is positive, for it is the ratio of two partition functions, but for fermions the parameter α itself may be either positive or negative. In the classical limit we did find α to be positive, but now only a singular negative value of α will enable us to satisfy equation (10.1.1). To have, nonetheless, a positive quantity with which to work and one that is nonsingular in the limit of zero temperature, let us follow convention and define a parameter μ by

$$-\beta\mu \equiv \alpha. \tag{10.1.2}$$

At the moment the parameter μ is little more than a convenient substitute for α, one that splits off part of the temperature dependence.

Upon inserting $-\beta\mu$ in place of α, we have

$$\langle n_s \rangle_F = \frac{1}{e^{\beta(\varepsilon_s - \mu)} + 1}. \tag{10.1.3}$$

The Pauli principle tells us that as the temperature goes to zero, $\langle n_s \rangle_F$ should become equal to unity from ε_1 out to some much higher single-particle energy and then zero beyond, thus giving, for the expectation values of the occupation numbers, the values of the occupation numbers of the ground state ψ_1. The expression for $\langle n_s \rangle_F$ will exhibit this behavior if

$$\lim_{T \to 0} \mu \equiv \mu_0 > 0, \tag{10.1.4}$$

that is, if in the limit as the temperature goes to absolute zero, the parameter μ has a *finite positive* limit. Then we will have the following pattern for the limit of $\langle n_s \rangle_F$:

$$\lim_{T \to 0} \frac{1}{e^{\beta(\varepsilon_s - \mu)} + 1} = \begin{cases} 1 & \text{if} \quad \varepsilon_s < \mu_0, \\ 0 & \text{if} \quad \varepsilon_s > \mu_0. \end{cases}$$

In the indicated limit, β goes to infinity, and the value of $\langle n_s \rangle_F$ depends critically on whether $(\varepsilon_s - \mu_0)$ is negative or positive. Thus we can arrange matters so that $\langle n_s \rangle_F$ has the zero-temperature behavior demanded by the Pauli principle. Indeed, all that remains is to compute the numerical value of μ_0, which is determined by imposing equation (10.1.1).

Once again, integration is easier than summation. The energy step between one ε_s and the next is sufficiently small (for a reasonable volume V) that we may replace a summation over states φ_s by an integration with a continuous ε and a density of states factor. Let us, for convenience, abbreviate $\langle n_s \rangle_F$ in the continuous ε limit as $F(\varepsilon)$:

$$F(\varepsilon) \equiv \frac{1}{e^{\alpha + \beta \varepsilon} + 1} = \frac{1}{e^{\beta(\varepsilon - \mu)} + 1}. \tag{10.1.5}$$

The function $F(\varepsilon)$ is often called the *Fermi function*. Curve (a) in figure 10.1.1 gives the run of $F(\varepsilon)$ against ε for $T = 0$, $\mu = \mu_0$. Let us note, incidentally, that the Fermi function, when evaluated for $\varepsilon = \mu$, has the value $\frac{1}{2}$, regardless of the temperature:

$$F(\varepsilon) \Big|_{\varepsilon = \mu} = \frac{1}{e^{\beta(\text{zero})} + 1} = \frac{1}{2}.$$

A calculation of μ_0 now goes readily. From the condition (10.1.1) on α, which now becomes a condition on μ or μ_0, we have

$$N = \int_0^\infty F(\varepsilon)(2 \cdot \tfrac{1}{2} + 1)\omega(\varepsilon) \, d\varepsilon.$$

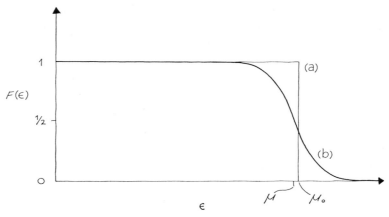

FIGURE 10.1.1
(a) The behavior of the Fermi function at $T = 0$. Single-particle states with single-particle energies ε_s up to $\varepsilon_s = \mu_0$ are occupied; those with higher energy are unoccupied. This pattern gives the ground state ψ_1 for the entire system of N identical fermions. (b) The Fermi function for a temperature T such that the sequence of inequalities $0 < T \ll T_F(0)$ is fulfilled. For clarity the variation in $F(\varepsilon)$ has been exaggerated.

The value $S = \frac{1}{2}$, appropriate for an electron with its spin of $\frac{1}{2}\hbar$, has been inserted in the density of states factor. The energy of the single-particle ground state is so close to zero for a macroscopic volume that we may safely set the lower limit of integration equal to zero. Since the Fermi function is zero for $\varepsilon > \mu_0$ when $T = 0$ and is then equal to unity for $\varepsilon < \mu_0$, the integral becomes

$$N = \int_0^{\mu_0} 1 \times 2 \times \frac{2\pi(2m)^{3/2}}{h^3} V\varepsilon^{1/2} \, d\varepsilon.$$

The integration is straightforward; the solution for μ_0, which enters through the upper limit of integration, is

$$\mu_0 = \frac{h^2}{2m} \left(\frac{3}{8\pi} \frac{N}{V} \right)^{2/3}. \tag{10.1.6}$$

Before we estimate a numerical magnitude, let us calculate the expectation value $\langle E \rangle$ of the total energy. The computation goes rapidly:

$$\langle E \rangle = \sum_j E_j P_j = \sum_s \varepsilon_s \langle n_s \rangle_F = \int_0^\infty F(\varepsilon)\varepsilon 2\omega(\varepsilon) \, d\varepsilon$$

$$= \int_0^{\mu_0} 1 \times \varepsilon \times 2 \times \frac{2\pi(2m)^{3/2}}{h^3} V\varepsilon^{1/2} \, d\varepsilon,$$

the last expression being appropriate at $T = 0$. The result of the integration, when expressed in terms of μ_0, is

$$\langle E \rangle = \tfrac{3}{5} N \mu_0 \qquad \text{at } T = 0. \tag{10.1.7}$$

In the Fermi function at zero temperature, the parameter μ_0 represents the cutoff energy above which $F(\varepsilon)$ is zero. The result for $\langle E \rangle$ implies another interpretation: μ_0 indicates (roughly) the typical energy per particle, $\langle E \rangle / N$, at zero temperature. Because μ_0 and, more generally, μ are characteristic single-particle energies, each is often called the *Fermi energy*.

Now a numerical estimate. For a contribution of one conduction electron from each atom and a typical metallic lattice spacing of 3×10^{-8} cm (3Å), we have

$$\frac{N}{V} = 3.7 \times 10^{22} \text{ conduction electrons per cubic centimeter,}$$

and therefore

$$\mu_0 = \begin{cases} 6.5 \times 10^{-12} \text{ ergs,} \\ 4.1 \qquad\qquad\quad \text{electron volts.} \end{cases}$$

In comparison with a typical thermal energy, for example, $kT = 0.025$ electron volts at room temperature, the Fermi energy is huge, greater by two orders of magnitude. This surprising size is a consequence of the Pauli principle: single-particle energy states φ_s up to $\varepsilon_s \simeq 4$ electron volts must be occupied even at absolute zero.

The preceding tells us that even the ground state ψ_1 of the fermion system has much kinetic energy. Let us remember that in the limit $T \to 0$, all probabilities vanish except the probability P_1 for the ground state, which has the value one: $P_1 = 1$. In that limit, $\langle E \rangle$ reduces to merely $E_1 \times 1$. The result for $\langle E \rangle$ in equation (10.1.7), together with the numerical estimate for μ_0, indicates a very large kinetic energy for the ground state of the conduction electrons. This is an unavoidable consequence of antisymmetric wave functions for the fermion system or, more simply, of the Pauli principle.

In microscopic statistical physics, a characteristic energy may always be converted into a characteristic temperature by division with Boltzmann's constant k. For the *Fermi temperature* of an electron gas at a physical temperature of absolute zero, we find

$$T_F(0) \equiv \frac{\mu_0}{k} = \frac{1}{k} \frac{h^2}{2m} \left(\frac{3}{8\pi} \frac{N}{V} \right)^{3/2}$$
$$\simeq 47{,}000 \text{ K,} \tag{10.1.8}$$

the last following from the estimate for μ_0. The characteristic temperature is so much larger than the physical $T = 300$ K of room temperature that the latter is virtually absolute zero for an electron gas. This conclusion confirms the usefulness of our starting the discussion with the absolute-zero limit.

Two remarks are in order here. The first is that the Fermi temperature at absolute zero does depend on the density N/V of conduction electrons and that therefore $T_F(0)$ will vary somewhat from one good conducting metal to another. The second remark concerns the physical temperature that stands on the borderline between the adequacy and inadequacy of a classical treatment, in particular, the temperature for which the dimensionless parameter $N\lambda^3/V$ equals one. The expression for that temperature is

$$\frac{h^2}{2\pi mk} \left(\frac{N}{V}\right)^{2/3}.$$

Comparison with $T_F(0)$ shows that the two temperatures are nearly the same. The implication is that until the physical temperature is much larger than the Fermi temperature at absolute zero, a classical analysis will be inadequate. There is not much chance of achieving classical conditions in a good metallic conductor.

Now we move away from absolute zero and consider more common temperatures: room temperature at 300 K or the temperature of a hot tungsten filament at about 1,000 K. These temperatures are still well below the characteristic temperature at absolute zero; we have now a sequence of inequalities:

$$0 < T \ll T_F(0). \tag{10.1.9}$$

The system need no longer be in its ground state. There will be a non-zero probability that a different state with higher energy will provide the appropriate description of the system. Nonetheless, the physical temperature is still *relatively* low; so only states with an energy close to the ground-state energy E_1 will acquire a probability worthy of serious attention. Such a state ψ_j will have some admixture of single-particle states φ_s from beyond the previous $\varepsilon_s = \mu_0$ cutoff, and so $\langle n_s \rangle_F$ will no longer be zero when ε_s is greater than μ_0, though it will remain small.

Toward the end of Section 9.3, we noted that we may give $\langle n_s \rangle_F$ an interpretation going beyond the phrase "the expectation value of the sth occupation number." We may view it as the probability that the single-particle state φ_s is occupied (as part of some antisymmetric state ψ_j for the N-fermion system). This view permits us to restate the conclusion of the preceding paragraph: when the temperature satisfies the inequalities (10.1.9), some of the single-particle states φ_s beyond μ_0 will have a small but non-zero probability of being occupied. The Fermi function will be smeared out a bit at what was, for $T = 0$, the energy cutoff. Since the smearing is a thermal effect, we should expect the range to be of order kT. Individual electrons can acquire extra energy of roughly this amount and move up to states φ_s beyond the previous $\varepsilon_s = \mu_0$ cutoff. In so doing, they reduce the probability that the single-particle states immediately below the zero-temperature cutoff are occupied. This is indicated by curve (b) in figure 10.1.1.

The value of the Fermi energy μ will shift a little from its value μ_0 at absolute zero. We can determine the direction of the shift by remembering that in the classical limit the parameter α is positive. Since μ is defined by $-\beta\mu \equiv \alpha$, the parameter μ will head toward a negative value. This implies the inequality

$$\mu < \mu_0 \text{ when } T > 0; \qquad (10.1.10)$$

that is, μ becomes smaller as the temperature rises from absolute zero. The point at which $F(\varepsilon) = \frac{1}{2}$ shifts toward lower energy. Later we will calculate the shift for the temperature range being considered now. It is much less than the $O(kT)$ spread of the region in which $F(\varepsilon)$ falls from almost unity to virtually zero.

On the basis of this qualitative analysis we can estimate the increase in $\langle E \rangle$, relative to its value at absolute zero, when the temperature is finite but still much less than $T_{\mathrm{F}}(0)$. The single-particle states φ_s whose probability of occupancy is increased at all significantly (from zero) are those with ε_s such that

$$\mu_0 < \varepsilon_s \leqslant \mu_0 + kT.$$

The number of those states is of order $2\omega(\mu_0)kT$, the product of the density of states at $\varepsilon = \mu_0$ and the energy range $O(kT)$. There is a compensatory decrease in the probability of occupancy on the lower energy side of μ_0, for a range of order kT. We may reasonably say that the number of electrons affected by the temperature increase is of order $2\omega(\mu_0)kT$ and that each acquires an extra energy of order kT. This implies that $\langle E \rangle$ may be roughly estimated as

$$\langle E \rangle \simeq \langle E \rangle \Big|_{T=0} + 2\omega(\mu_0)(kT)^2. \qquad (10.1.11)$$

The contribution to the metallic heat capacity from the conduction electrons will be roughly

$$\frac{\partial \langle E \rangle}{\partial T} \simeq 4\omega(\mu_0)k^2 T. \qquad (10.1.12)$$

The factor $\omega(\mu_0)$ may be written, with a little algebraic manipulation, as

$$\omega(\mu_0) = \frac{2\pi(2m)^{3/2}V}{h^3}(\mu_0)^{1/2}$$

$$= \frac{3}{4}\frac{N}{\mu_0} = \frac{3N}{4kT_F(0)}.$$

The first equality follows directly from the general expression for $\omega(\varepsilon)$ in equation (9.5.6); the second uses the explicit result for μ_0 in equation (10.1.6). When the last expression for $\omega(\mu_0)$ is inserted above, the estimate for the contribution to the heat capacity becomes

$$\frac{\partial\langle E\rangle}{\partial T} \simeq 3Nk\frac{T}{T_F(0)}. \tag{10.1.13}$$

This rough value is smaller, by the impressive factor

$$\frac{2T}{T_F(0)} \simeq \frac{2T}{50,000},$$

than the contribution of $\frac{3}{2}Nk$ that the classical equipartition theorem would imply. There is, moreover, a dependence on the temperature. That linear dependence is well confirmed in experiments at low temperatures, in the range 0.1 to 1.0 kelvin, where the small electronic contribution to the heat capacity is not masked by the contribution from (the lattice vibrations of) the positive ions.

10.2. QUANTITATIVE CALCULATIONS WITH $F(\varepsilon)$

This section provides quantitative support for the conclusions about μ and $\langle E\rangle$ that were arrived at in the preceding section by qualitative arguments. The range of the physical temperature is restricted here, as it was there; the temperature T must satisfy the sequence of inequalities

$$0 \leqslant T \ll T_F(0). \tag{10.2.1}$$

Providing the support requires much effort, and we merely sharpen results already established, but the mathematical techniques are instructive and worth noting in their own right.

For a quantitative calculation of $\langle E\rangle$, we need the integral in the relation

$$\langle E\rangle = \int_0^\infty F(\varepsilon)\varepsilon 2\omega(\varepsilon)\,d\varepsilon$$

for $T \neq 0$. But the dependence of the Fermi function on μ means that we must also determine μ from the relation

$$N = \int_0^\infty F(\varepsilon) 2\omega(\varepsilon) \, d\varepsilon.$$

Because of the uncooperative form of $F(\varepsilon)$ for $T \neq 0$, these integrals cannot be done exactly. When the temperature is much less than $T_F(0)$, there is, however, a trick that enables one to approximate the integrals in a systematic fashion. The trick depends on a certain property of the Fermi function: under the specified temperature conditions, the derivative of $F(\varepsilon)$ with respect to ε is almost zero except in a narrow region around $\varepsilon = \mu$; that is,

$$\frac{dF(\varepsilon)}{d\varepsilon} \simeq 0 \qquad \text{except when } \varepsilon \simeq \mu.$$

A glance back at figure 10.1.1 will confirm this contention. The trick consists of manipulating the integrals so that $dF/d\varepsilon$, rather than $F(\varepsilon)$, appears. Then the integrand is close to zero except for $\varepsilon \simeq \mu$, and one can expand a still recalcitrant integrand about $\varepsilon = \mu$ for a highly satisfactory approximation.

Since there are two different integrals that we must ultimately evaluate, let us establish a general approximation procedure. The integrals of interest are of the form

$$\int_0^\infty F(\varepsilon) g(\varepsilon) \, d\varepsilon,$$

with $g(\varepsilon)$ a smooth and easily integrated function. To make it easier to follow the integration by parts that leads to the $dF/d\varepsilon$ form, let us first define an auxiliary function $G(\varepsilon)$ by

$$G(\varepsilon) \equiv \int_0^\varepsilon g(\varepsilon') \, d\varepsilon'$$

and note the property

$$\frac{dG(\varepsilon)}{d\varepsilon} = g(\varepsilon).$$

Then, with an integration by parts in the second step, we manipulate the integral to the desired form:

$$\int_0^\infty F(\varepsilon) g(\varepsilon) \, d\varepsilon = \int_0^\infty F(\varepsilon) \frac{dG}{d\varepsilon} \, d\varepsilon$$

$$= F(\varepsilon) G(\varepsilon) \Big|_0^\infty - \int_0^\infty \frac{dF(\varepsilon)}{d\varepsilon} G(\varepsilon) \, d\varepsilon.$$

In the expression preceding the remaining integral the contribution from the lower limit is zero because $G(0) = 0$ and $F(0)$ is finite (in fact, very nearly unity). For the functions that we will consider—$g(\varepsilon)$ growing no faster than some power of ε—the upper limit contributes nothing because $F(\varepsilon)$ goes to zero exponentially as ε goes to infinity. Thus there is no contribution from either endpoint, and we have achieved a reduction to an integral in which only $dF/d\varepsilon$ appears:

$$\int_0^\infty F(\varepsilon)g(\varepsilon)\,d\varepsilon = -\int_0^\infty \frac{dF(\varepsilon)}{d\varepsilon}\,G(\varepsilon)\,d\varepsilon. \tag{10.2.2}$$

The minus sign is all right, for $dF/d\varepsilon$ is negative when not zero.

Since $dF/d\varepsilon$ is significantly different from zero in only a small region, we can get a satisfactory approximation to the exact result by expanding $G(\varepsilon)$ about the point where $|dF/d\varepsilon|$ has its maximum value. The maximum is found by looking for the value of ε for which the equality

$$\frac{d}{d\varepsilon}\left(\frac{dF(\varepsilon)}{d\varepsilon}\right) = 0$$

holds. Not too surprisingly, the point turns out to be given by $\varepsilon = \mu$. So we expand $G(\varepsilon)$ about that point in a Taylor's series:

$$G(\varepsilon) = G(\mu) + \left[\frac{dG}{d\varepsilon}\right]_\mu (\varepsilon - \mu) + \ldots + R_{n'}(\varepsilon)$$

$$= \sum_{n=0}^{n'} \frac{1}{n!}\left[\frac{d^n G}{d\varepsilon^n}\right]_\mu (\varepsilon - \mu)^n + R_{n'}(\varepsilon).$$

The subscript on the square brackets indicates that the derivative is to be evaluated at $\varepsilon = \mu$. Since our purposes will be well-served by the first few terms in the expansion, $(n' + 1)$ terms, say, the series has been terminated at $n = n'$ and the remainder written as $R_{n'}(\varepsilon)$. We are now to insert this expansion into equation (10.2.2). When we do that, we may drop the remainder term, for it is small when $\varepsilon \simeq \mu$ and its value elsewhere is unimportant, for then $dF/d\varepsilon$ is so small. After the insertion and an interchange of the order of integration and summation, we arrive at

$$\int_0^\infty F(\varepsilon)g(\varepsilon)\,d\varepsilon \simeq -\sum_{n=0}^{n'} \frac{1}{n!}\left[\frac{d^n G}{d\varepsilon^n}\right]_\mu \int_0^\infty \frac{dF}{d\varepsilon}(\varepsilon - \mu)^n\,d\varepsilon. \tag{10.2.3}$$

The integrals now needed are of the form

$$\int_0^\infty \frac{dF}{d\varepsilon}(\varepsilon - \mu)^n\,d\varepsilon = \int_0^\infty \frac{-\beta e^{\beta(\varepsilon - \mu)}}{(e^{\beta(\varepsilon - \mu)} + 1)^2}(\varepsilon - \mu)^n\,d\varepsilon.$$

After the variable change $x = \beta(\varepsilon - \mu)$, which implies $dx = \beta\, d\varepsilon$, the integral takes on a cleaner appearance:

$$\int_{\varepsilon=0}^{\infty} \frac{dF}{d\varepsilon}\, (\varepsilon - \mu)^n\, d\varepsilon = -\beta^{-n} \int_{x=-\beta\mu}^{\infty} \frac{e^x}{(e^x + 1)^2}\, x^n\, dx.$$

Still, except for a few values of n, the integral cannot be done exactly in closed form, and so further approximation is in order. For the low temperatures of interest, the lower limit of integration is very large and negative:

$$x = -\beta\mu \simeq -\frac{\mu_0}{kT} = -\frac{T_{\!F}(0)}{T} \ll -1.$$

We may safely extend the lower limit to $-\infty$, for the integrand vanishes exponentially for large negative x. The numerical error thereby incurred will be negligible, and the integral may then be evaluated analytically for each value of n. Thus we approximate by writing

$$\int_{\varepsilon=0}^{\infty} \frac{dF}{d\varepsilon}\, (\varepsilon - \mu)^n\, d\varepsilon \simeq -\beta^{-n} \int_{x=-\infty}^{\infty} \frac{e^x}{(e^x + 1)^2}\, x^n\, dx \equiv -\beta^{-n} I_n. \quad (10.2.4)$$

The second step defines I_n as a convenient abbreviation.

For $n = 0$, there is no trouble:

$$I_0 = \int_{-\infty}^{\infty} \frac{e^x}{(e^x + 1)^2}\, dx = -\frac{1}{e^x + 1}\Bigg|_{-\infty}^{\infty} = +1.$$

All the integrals with n odd are zero, for the function

$$\frac{e^x}{(e^x + 1)^2} = \frac{1}{(e^x + 1)} \cdot \frac{e^x}{(e^x + 1)} = \frac{1}{(e^x + 1)(e^{-x} + 1)}$$

is even in x, whereas x^n, for n odd, is odd in x. The integrand as a whole is thus odd, and the integral from minus to plus infinity vanishes. For $n = 2$, the calculation is most easily handled by resort to a table of definite integrals:

$$I_2 = \int_{-\infty}^{\infty} \frac{e^x}{(e^x + 1)^2}\, x^2\, dx = \frac{\pi^2}{3}.$$

The next non-zero I_n is I_4, but we will cut off the series after I_3, that is, take $n' = 3$. Remembering the factor $-\beta^{-n}$ from equation (10.2.4), we insert these results into equation (10.2.3):

$$\int_0^{\infty} F(\varepsilon)g(\varepsilon)\, d\varepsilon \simeq +G(\mu) + 0 + \frac{1}{2}\beta^{-2}\frac{\pi^2}{3}\left[\frac{d^2 G}{d\varepsilon^2}\right]_{\mu} + 0$$

$$\simeq \int_0^{\mu} g(\varepsilon)\, d\varepsilon + \frac{\pi^2}{6}(kT)^2\left[\frac{dg}{d\varepsilon}\right]_{\mu}. \quad (10.2.5)$$

The first term is almost the $T = 0$ result, but it differs slightly from that because the upper limit is μ, not μ_0. Had we carried the series further, the next new term would have been a term of order T^4. It is worth noting that for most functions $g(\varepsilon)$, though not for all, the series is only an asymptotic series, just as is Stirling's approximation for n-factorial.

Calculation of the Shift in μ from μ_0

Now we can compute the shift in μ away from μ_0 under the temperature conditions specified by the inequalities (10.2.1). For this computation the relevant integral is

$$N = \int_0^\infty F(\varepsilon) 2\omega(\varepsilon)\, d\varepsilon,$$

and from that we identify $g(\varepsilon)$ as here being

$$g(\varepsilon) = 2\omega(\varepsilon).$$

An abbreviation,

$$\omega'(\varepsilon) \equiv \frac{d\omega(\varepsilon)}{d\varepsilon},$$

and an appeal to the approximation scheme given by equation (10.2.5) yield

$$N \simeq \int_0^\mu 2\omega(\varepsilon)\, d\varepsilon + \frac{\pi^2}{6}(kT)^2 2\omega'(\mu).$$

The remaining integral gives something close to N. We may extract N from it by introducing an auxiliary limit of integration

$$N \simeq \int_0^{\mu_0} 2\omega(\varepsilon)\, d\varepsilon + \int_{\mu_0}^\mu 2\omega(\varepsilon)\, d\varepsilon + \frac{\pi^2}{6}(kT)^2 2\omega'(\mu).$$

Since the first integral yields precisely N, it cancels the lefthand side:

$$0 \simeq \int_{\mu_0}^\mu 2\omega(\varepsilon)\, d\varepsilon + \frac{\pi^2}{6}(kT)^2 2\omega'(\mu).$$

If μ is close to μ_0, the remaining range of integration is small, and we may evaluate the slowly varying integrand at $\varepsilon = \mu_0$, a useful procedure since we do know μ_0. Similarly, we may replace $\omega'(\mu)$ by $\omega'(\mu_0)$. These steps lead to

$$0 \simeq 2\omega(\mu_0)(\mu - \mu_0) + \frac{\pi^2}{6}(kT)^2 2\omega'(\mu_0).$$

Solved for $(\mu - \mu_0)$, this approximate equation gives

$$\mu - \mu_0 \simeq -\frac{\pi^2}{6}\frac{\omega'(\mu_0)}{\omega(\mu_0)}(kT)^2. \tag{10.2.6}$$

The density of states factors may be simplified. Since $\omega(\varepsilon)$ has the form

$$\omega(\varepsilon) = C\varepsilon^{1/2},$$

we find

$$\omega'(\mu_0) = \frac{1}{2}C\mu_0^{-1/2} = \frac{1}{2}\frac{\omega(\mu_0)}{\mu_0}.$$

The final result for μ is

$$\mu \simeq \mu_0 - \frac{\pi^2}{12}\frac{(kT)^2}{\mu_0}. \tag{10.2.7}$$

As predicted, the shift is toward a negative value for μ, but for room temperature the shift is really quite small:

$$\frac{\mu - \mu_0}{\mu_0} \simeq -\frac{\pi^2}{12}\left(\frac{kT}{\mu_0}\right)^2 = -\frac{\pi^2}{12}\left(\frac{T}{T_F(0)}\right)^2$$

$$\simeq -O(2 \times 10^{-4}).$$

The final estimate is with as low a realistic value for $T_F(0)$ as 20,000 K. On dimensional grounds a dependence of the shift on the ratio $T/T_F(0)$ might have been expected. With the correction starting off quadratically, it is sure to be quite small.

Calculation of $\langle E \rangle$

Now we work out the first temperature-dependent term in $\langle E \rangle$; we already have reason to believe that it will go as T^2. The relevant integral is

$$\langle E \rangle = \int_0^\infty F(\varepsilon)\varepsilon 2\omega(\varepsilon)\,d\varepsilon,$$

and the new identification of $g(\varepsilon)$ is

$$g(\varepsilon) = 2\varepsilon\omega(\varepsilon).$$

Appeal to the general approximation scheme yields

$$\langle E \rangle \simeq \int_0^\mu 2\varepsilon\omega(\varepsilon)\,d\varepsilon + \frac{\pi^2}{6}(kT)^2\left[\frac{d}{d\varepsilon}\left(2\varepsilon\omega(\varepsilon)\right)\right]_\mu. \tag{10.2.8}$$

The integral we treat as before:

$$\int_0^\mu 2\varepsilon\omega(\varepsilon)\, d\varepsilon = \int_0^{\mu_0} 2\varepsilon\omega(\varepsilon)\, d\varepsilon + \int_{\mu_0}^\mu 2\varepsilon\omega(\varepsilon)\, d\varepsilon$$

$$\simeq \langle E\rangle\Big|_{T=0} + 2\mu_0\omega(\mu_0)(\mu - \mu_0).$$

For the derivative we have

$$\left[\frac{d}{d\varepsilon}\left(2\varepsilon\omega(\varepsilon)\right)\right]_\mu = [2\omega(\varepsilon) + 2\varepsilon\omega'(\varepsilon)]_\mu$$

$$\simeq 2\omega(\mu_0) + 2\mu_0\,\omega'(\mu_0).$$

The evaluation at μ_0, rather than at μ, is permissible because the derivative term already has a T^2 coefficient. Inserted into equation (10.2.8), these evaluations give

$$\langle E\rangle \simeq \langle E\rangle\Big|_{T=0} + 2\mu_0\,\omega(\mu_0)(\mu - \mu_0) + \frac{\pi^2}{6}(kT)^2[2\omega(\mu_0) + 2\mu_0\,\omega'(\mu_0)].$$

When we write for $(\mu - \mu_0)$ its explicit expression from equation (10.2.6), we find that it is canceled by the second term in the square brackets, and so we arrive at the result

$$\langle E\rangle \simeq \langle E\rangle\Big|_{T=0} + \frac{\pi^2}{3}\omega(\mu_0)(kT)^2. \tag{10.2.9}$$

The dependence of $\langle E\rangle$ on T^2 and on $\omega(\mu_0)$ confirms the essential correctness of the qualitative estimate in the preceding section. For the contribution of the conduction electrons to the metallic heat capacity (at constant volume), we have

$$\frac{\partial\langle E\rangle}{\partial T} \simeq \frac{2\pi^2}{3}\omega(\mu_0)k^2 T = \begin{cases} \dfrac{\pi^2}{2}Nk\,\dfrac{T}{T_F(0)}, \\[2ex] \left[\left(\dfrac{8\pi^4}{3}\right)^{2/3}\dfrac{mk^2 V}{h^2}\left(\dfrac{N}{V}\right)^{1/3}\right]T. \end{cases} \tag{10.2.10}$$

The equivalent expressions on the far right are given to make two points.

1. The contribution is smaller by a factor of order $T/T_F(0)$ than the contribution $\frac{3}{2}Nk$ that a classical calculation with the equipartition theorem would yield.

2. Once N/V is estimated, the free-electron model gives a definite numerical value for the coefficient in the linear dependence on temperature.

The prediction of a linear dependence on temperature is well-confirmed. The situation with the numerical coefficients is dubious at best: for silver the agreement is quite good (within 5 per cent); for copper, sodium, and mercury, fair (factor of 2); and for cobalt or platinum, poor (factor of 10 or 20). The good agreement for silver is likely to be fortuitous. Though the free-electron model correctly predicts the temperature dependence, its neglect of the interaction with the positive ions leads to considerable numerical discrepancy. Nonetheless, the model does generally yield a result of the appropriate order of magnitude, and that—with the correct temperature dependence—is no small triumph, since a classical calculation fails dismally.

10.3. THERMIONIC EMISSION

Why heat the cathode of a vacuum tube? Why cool a photomultiplier? The ready answers are these: to increase the emission of electrons in the former and to decrease the undesired spontaneous emission in the latter. In this section we will go into some of the details, in particular, the major dependence on temperature of the spontaneous emission of electrons from a metal.

For the electrons inside the metal, we continue to use the free-electron model, though now we must look more carefully at the " walls of the box." The attractive Coulomb interaction with the positive ions of the metal does generate a force tending to contain the conduction electrons, but the jump in potential at the edge of the metal cannot really be taken as infinitely high. A reasonable model for the single-particle potential is indicated in figure 10.3.1. We may continue to choose a zero point for the potential so that

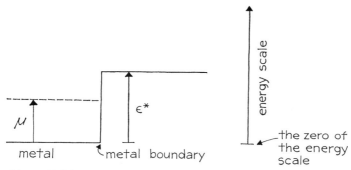

FIGURE 10.3.1

A model for the potential that a single conduction electron "sees" in the vicinity of the boundary of a metal. The dashed line represents a typical electron energy, that of the Fermi energy μ at the temperature of the material. Only a quite atypical electron can escape.

inside the metal the potential is zero. At the boundary the potential jumps to the finite value ε^* and remains constant outside. Thus ε^* is the minimum energy that an electron must possess in order to escape the metal. This is the simplest potential model that can begin to do justice to the complicated real situation.

We seek first an expression for the estimated number of electrons that emerge, per second, from a patch of unit area on the plane metallic surface. Multiplication by the electronic charge will then give the current density.

Let us look a moment at the problem in classical terms. Suppose an electron has a momentum in the infinitesimal volume d^3p around $\mathbf{p}$. The region of the metal in which the electron may be and still succeed in hitting the chosen surface patch within one second is that shown in figure 10.3.2. A necessary

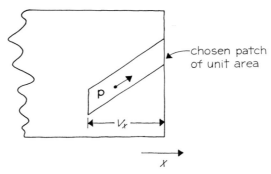

FIGURE 10.3.2
A view of the region within the metal in which an electron with $\mathbf{p}$ in d^3p may be and still hit the chosen patch of unit area within one second.

proviso is that v_x, equal to p_x/m, be positive. The values of p_y and p_z need not be constrained.

The classical picture suggests a procedure for performing a quantum-mechanical calculation. Let us first count the number of different single-particle states in which the successful electron of the preceding paragraph might be. A counting with $(2S + 1)d^3x\, d^3p/h^3$ single-particle states in a volume $d^3x\, d^3p$ of the classical phase-space of a single electron is the easiest method (and perfectly adequate when the volume of the metal is macroscopic). The "ordinary" volume, corresponding to integration with d^3x over the spatial region in the diagram, is given by $(v_x) \cdot (1) \cdot (1) = v_x$, because the surface patch has unit area. Since $S = 1/2$, there are $2v_x\, d^3p/h^3$ quantum states in which the electron might be and still succeed in hitting the chosen patch within one second. All these states have the same energy, for $\mathbf{p}$

is in a single infinitesimal volume (in momentum space). Now we ask for the estimated number of such successful electrons. That number we can get by multiplying with $F(\varepsilon)$, the expectation value of the number of electrons in a single-particle energy eigenstate with energy ε. So we may estimate the number of electrons (with $\mathbf{p}$ in d^3p) that will hit the chosen surface patch within one second as

$$F(\varepsilon)\frac{2v_x\,d^3p}{h^3}.$$

To estimate the emission current density $\mathfrak{J}$, we multiply by the electronic charge and integrate with d^3p:

$$\mathfrak{J} = \int eF(\varepsilon)\frac{2v_x}{h^3}\,d^3p. \tag{10.3.1}$$

This amounts to summing over all possible ways in which an electron might be heading for the patch and hit it within one second.

The limits of integration must now be specified. For p_y and p_z, they are $-\infty$ to $+\infty$. The situation for p_x is more complicated. Certainly negative p_x must be excluded. Furthermore, the electron must have sufficient momentum in the x-direction so that the retarding force (represented by the jump ε^* in potential) does not send it back into the metal. Since the p_y, p_z, components are unaffected by the wall force as the electron hits, we may put the restriction in energy terms as

$$\frac{p_x{}^2}{2m} \geqslant \varepsilon^*,$$

together with $p_x > 0$. Now we may write the integral for $\mathfrak{J}$ with explicit limits of integration:

$$\mathfrak{J} = \int_{\sqrt{2m\varepsilon*}}^{\infty}\int_{-\infty}^{\infty}\int_{-\infty}^{\infty}\frac{2e}{h^3}F(\varepsilon)v_x\,dp_x\,dp_y\,dp_z. \tag{10.3.2}$$

Because of the generally intractable nature of $F(\varepsilon)$, the p_x integration is most easily done with a change of variables to an integration over ε. With p_y, p_z, held fixed, we have

$$d\varepsilon = \frac{1}{2m}d(p_x{}^2 + p_y{}^2 + p_z{}^2) = \frac{1}{2m}2p_x\,dp_x = v_x\,dp_x.$$

So we may replace $v_x\,dp_x$ by $d\varepsilon$, provided we suitably adjust the limits of integration on the first integral sign. The lower limit becomes

$$\varepsilon = \frac{1}{2m}\left((\sqrt{2m\varepsilon*})^2 + p_y{}^2 + p_z{}^2\right) = \varepsilon^* + \frac{(p_y{}^2 + p_z{}^2)}{2m}.$$

The upper limit remains $+\infty$. Then $\mathfrak{J}$ becomes

$$\mathfrak{J} = \frac{2e}{h^3} \int_{\varepsilon*+\frac{(p_y{}^2+p_z{}^2)}{2m}}^{\infty} \int_{-\infty}^{\infty} \int_{-\infty}^{\infty} F(\varepsilon)\, d\varepsilon\, dp_y\, dp_z.$$

Now the first integration may be done exactly:

$$\int \frac{1}{e^{\beta(\varepsilon-\mu)}+1}\, d\varepsilon = -\frac{1}{\beta} \ln\{e^{-\beta(\varepsilon-\mu)}+1\}\bigg|_{\varepsilon*+\frac{(p_y{}^2+p_z{}^2)}{2m}}^{\infty}$$

The contribution of the upper limit is zero, for $\ln\{0 + 1\} = 0$. After inserting into $\mathfrak{J}$ the contribution from the lower limit, we find

$$\mathfrak{J} = \frac{2e}{h^3\beta} \int_{-\infty}^{\infty} \int_{-\infty}^{\infty} \ln\left\{1 + \exp\left[-\beta\left(\varepsilon* - \mu + \frac{(p_y{}^2 + p_z{}^2)}{2m}\right)\right]\right\} dp_y\, dp_z.$$

What shows up as a characteristic parameter is $\varepsilon* - \mu$, the difference between the potential jump and the Fermi energy. A typical conduction electron has an energy of order μ. Consequently $\varepsilon* - \mu$ is a measure of the extra energy that an atypical electron must possess in order to escape. (The parameter $\varepsilon* - \mu$ must be positive; if $\varepsilon*$ were not greater than μ, the electrons would gush forth.) The numerical value of $\varepsilon* - \mu$ depends significantly on the metal and on whether or not the surface has been coated with a thin layer of another material to reduce, in effect, the potential jump. A value of 4 to 5 electron volts would be typical for a clean, uncoated surface. A coating of barium or an oxide layer can reduce the difference to between 1 and 2 electron volts. That the numbers fall in the low electron-volt range is physically reasonable. We have found the Fermi energy μ to be of that order, and the jump $\varepsilon*$ must be something like an ionization potential, perhaps 5 or 10 electron volts. There is no reason to expect near-cancelation in the difference, because the former depends primarily on the electron density N/V, whereas the latter is strongly dependent on the positive ions and surface details. Thus a difference also of the order of a few electron volts is to be expected.

At this point an approximation of the integral is in order. The combination $\beta(\varepsilon* - \mu)$ is sizeable. For room temperature and a modest difference of 2.5 electron volts, we have

$$\frac{(\varepsilon* - \mu)}{kT} = \frac{2.5}{0.025} = 100.$$

Even for a hot filament, the ratio is about 10. Since the ratio appears in an exponential with a minus sign, we can be sure that the exponential in the integral will be very much less than unity over the entire p_y, p_z, integration range. To an excellent approximation, we may expand the logarithm about unity for its argument and retain merely the first term:

$$\mathfrak{J} \simeq \frac{2e}{h^3\beta} e^{-\beta(\varepsilon*-\mu)} \int_{-\infty}^{\infty} \int_{-\infty}^{\infty} e^{-\beta(p_y^2+p_z^2)/2m} \, dp_y \, dp_z .$$

The remaining integrals are like those we met when analyzing a classical gas. The final expression for $\mathfrak{J}$ is then

$$\mathfrak{J} \simeq \left(\frac{4\pi emk^2}{h^3}\right) T^2 e^{-(\varepsilon*-\mu)/kT}. \tag{10.3.3}$$

This estimate for the emission current density is usually called the *Richardson-Dushman equation*, with the characteristic parameter $\varepsilon* - \mu$ being called the *work function*. Let us note that because both Planck's constant and the Fermi energy appear, the explicit form is definitely a quantum-mechanical result.

Richardson's work antedated the Pauli principle and antisymmetric wave functions for systems of electrons; so his was a purely classical treatment. That led to $T^{1/2}$ in place of T^2 and to $\varepsilon*$ rather than $\varepsilon* - \mu$ in the exponent, together with a different over-all numerical coefficient. Since the exponential dominates, and since the value of the parameter appearing in it is generally determined by fits to data, Richardson's classical expression gives about as good an agreement on temperature dependence as the correct quantum-mechanical expression. Thermionic emission does not provide a sensitive test for wave-function symmetry effects.

The simple quantum-mechanical theory predicts a universal over-all numerical coefficient, whose value (in convenient units) is 120 ampere/cm^2-K^2. This aspect is not borne out by the observations. The experimental coefficients are generally smaller, by a factor of two or even 100. Reasons for this are not hard to find. We have not worried about purely quantum-mechanical reflections from the potential rise for particles that, classically, would be sure to escape. The non-zero probability of reflection from a finite potential barrier would reduce the coefficient. Moreover, the potential barrier itself is a little too simple to guarantee a quantitatively correct result. At the very least it should be smoothed off at the corners, which would affect a detailed computation of the reflection probability (though not our calculation). Surface effects appear to produce a bump near the boundary on what we assumed was a flat potential outside the metal. Electrons with $p_x^2/2m$ greater than ε^* can still escape, but may have to "tunnel" through the potential bump. This would significantly reduce the over-all coefficient.

> It should be confessed here that the entire treatment has been a little loose. If electrons are continually leaving the metal, then the situation is not quite an equilibrium one. Furthermore, we have used a mixture of classical and quantum reasoning to get the expression for $\mathfrak{J}$ in equation (10.3.1). Such looseness is typical of the methods one is compelled to use in order to solve real problems.

But the temperature dependence is what we really set out to calculate. Here the observations and theory agree very well. Indeed, since one can expect $\varepsilon^* - \mu$ to change very little with temperature, the numerical values for the work function are generally taken from fits to the data with the Richardson-Dushman equation. It is a matter then of a two-parameter fit: the over-all numerical coefficient and the difference $\varepsilon^* - \mu$.

Finally, to confirm the good sense of heating a cathode or cooling a photomultiplier, let us look at the value of $T^2 \exp[-(\varepsilon^* - \mu)/kT]$ for some typical temperatures. To avoid unfairly overemphasizing the temperature dependence in the exponential, let us take a low value for the work function: $\varepsilon^* - \mu = 1.5$ electron volts, corresponding to a good coated surface. Liquid nitrogen temperature (77 K) may be taken for the cooled photomultiplier. For the hot filament a typical value of 1,000 K may be used, though this is still on the low side. (A tube filament if hot enough to glow.) For further comparison, the room temperature case has been inserted between these in table 10.3.1. The moral of the story is this: exponential dependence is an extraordinarily strong dependence.

TABLE 10.3.1

Typical values for components of the Richardson–Dushman thermionic emission equation with $\varepsilon^* - \mu$ $= 1.5$ electron volts. The temperatures are in kelvin.

T	T^2	$(\varepsilon^* - \mu)/kT$	$T^2 e^{-(\varepsilon^* - \mu)/kT}$
77	5.9×10^3	226	3.9×10^{-95}
300	9.0×10^4	58	5.7×10^{-21}
1,000	1.0×10^6	17	2.8×10^{-2}

10.4. THE SUCCESS OF
THE FREE-ELECTRON MODEL

Before we leave the topic of the free-electron model for conduction electrons, we should examine the reasons for its success. One might think that the very large value of the Fermi energy may account for the qualitative success of the model. After all, if the typical kinetic energy of an electron—here roughly the Fermi energy—is very large relative to a typical Coulomb potential energy, then the latter should have little effect. There are interactions both between the electrons and the positive ions and also among the electrons themselves. For both interactions the order of magnitude of the potential energy is given by the square of the electronic charge divided by a characteristic distance, the lattice spacing, say. This gives as an estimate

$$\begin{pmatrix}\text{the absolute value of} \\ \text{the Coulomb potential} \\ \text{energy}\end{pmatrix} \simeq \frac{e^2}{\begin{pmatrix}\text{lattice} \\ \text{spacing}\end{pmatrix}} \simeq \frac{e^2}{(V/N)^{1/3}} \simeq 5 \text{ electron volts.}$$

This potential energy is comparable to the Fermi energy; so the explanation does not lie here. Indeed, the interaction between the conduction electrons and the positive ions is responsible for the large differences in electrical conductivity among various metals and between "good conductors" and "good insulators."

The reasons for the qualitative success are more subtle. They lie in the restricted nature of the questions we asked: the temperature dependence of the contribution to the heat capacity by the conduction electrons; the temperature dependence of the thermionic emission current. A glance back at equation (10.1.12) or (10.2.10) shows that the question about heat capacity can be answered even if we have only a qualitative knowledge of the behavior of the density of states. What shows up is $\omega(\mu_0)$, the density of states at the $T=0$ Fermi energy. When the interaction between the conduction electrons and the positive ions is treated in detail, the single-particle states are different, and the density of states is different from that for the free-electron model. But so long as the new density of states has a behavior near the Fermi energy roughly like that for the free-electron model, the analysis will go through as before. All that is necessary is that the new density of states be a smooth, not too rapidly changing function of energy near the Fermi energy. The temperature dependence is governed primarily by the Pauli principle and the existence of a Fermi energy; only the numerical value of the coefficient depends sensitively on the approximation used to calculate the density of single-particle states.

For thermionic emission, too, the detailed behavior of $\omega(\varepsilon)$ is unimportant. Because of the potential barrier at the edge of the metal, only electrons with relatively large kinetic energies are even candidates for a contribution to the emission current. For energies significantly above the Fermi energy, the exponential in the Fermi function dominates over the density of states factor in an integral for the current. Any reasonable density of states will yield a thermionic current dominated by an exponential of the form $\exp[-(\varepsilon^* - \mu)/kT]$. Only the temperature dependence of the coefficient multiplying that exponential can be appreciably influenced by the density of states factor, and that coefficient is relatively unimportant.

Thus far the mutual interactions of the electrons have been omitted from the explanations. To the extent that wave functions for electrons extend over the entire metal sample, the negative charge is distributed more or less uniformly. Rather than thinking of one electron as interacting with $N - 1$ point electrons, we may think of it as interacting with a rather smooth smear of negative charge. As the one electron wanders through the metal sample, it sees roughly the same negative charge distribution and so has a roughly constant interaction with it. The previous analysis with single-particle states can cope with this interaction by merely shifting the energy of each single-particle state by the same fixed amount. The general shift will have no observable effect. Thus the property that the negative charge may be regarded as smeared out (as far as the electron-electron interactions are concerned) ensures the approximate validity of the calculations in the preceding sections.

10.5. THE BOSE-EINSTEIN CONDENSATION

We turn now to bosons, particles that have integral spin and totally symmetric wave functions. Because we will later want to compare our results with some experimental properties of the spinless helium isotope He^4, let us specify at the outset that the bosons we consider have zero spin. Then the algebraic questions of spin orientations drops out of the problem without our losing any significant generality in the results.

If, for fixed number N of bosons, volume V, and particle mass m, the temperature T is large, so that the strong inequality

$$\frac{N\lambda^3}{V} = \frac{N}{V}\left(\frac{h}{\sqrt{2\pi mkT}}\right)^3 \ll 1$$

holds, then we are in the classical or nearly classical domain. That temperature region we have already analyzed. So, just as we did with fermions, let us go to the opposite extreme and look at the system of N spinless bosons—not mutually interacting—in the limit of absolute zero. The probabilities of the canonical distribution reduce to

$$P_j = \begin{cases} 1 & \text{if } j = 1 \\ 0 & \text{if } j \neq 1 \end{cases} \qquad \text{at } T = 0. \qquad (10.5.1)$$

At absolute zero we can be certain that the system is in its ground state ψ_1.

With φ_1 being the single-particle state with the lowest single-particle energy, we may construct the (totally symmetric) ground-state wave function as

$$\psi_1(Q_1, Q_2, \ldots, Q_N) = \varphi_1(Q_1)\varphi_1(Q_2) \cdots \varphi_1(Q_N). \qquad (10.5.2)$$

The permutations typically needed in constructing a totally symmetric state for the boson system are unnecessary here, because all N of the single-particle states are φ_1 states. For a cube of volume $V = L^3$, the explicit form for $\varphi_1(Q_1)$ is

$$\varphi_1(Q_1) = \left(\frac{8}{V}\right)^{1/2} \sin\left(\frac{\pi x_1}{L}\right) \sin\left(\frac{\pi y_1}{L}\right) \sin\left(\frac{\pi z_1}{L}\right). \qquad (10.5.3a)$$

The single-particle energy eigenvalue is

$$\varepsilon_1 = \frac{\pi^2 \hbar^2}{2m V^{2/3}} (1^2 + 1^2 + 1^2) = \frac{3\pi^2 \hbar^2}{2m V^{2/3}}, \qquad (10.5.3b)$$

provided we take the potential energy U_{box} due to the wall forces to be zero inside the box (and, of course, plus infinity outside). With this zero for the energy scale (our conventional one), the energy E_1 of the ground state of the system is

$$E_1 = N\varepsilon_1.$$

The expectation values of the boson occupation numbers at absolute zero now follow readily:

$$\langle n_s \rangle_{\text{B}} = \sum_j n_s(j) P_j$$

$$= n_s(1) = \begin{cases} N & \text{if } s = 1 \\ 0 & \text{if } s > 1 \end{cases} \qquad \text{at } T = 0. \qquad (10.5.4)$$

These simple expressions emerge because the system is certain to be in its ground state, ψ_1, and because that state is formed from an N-fold product of φ_1.

The preceding results indicate that $\langle n_1 \rangle_B$ becomes exceedingly large and important at low temperature. In the limit of absolute zero, the single-particle state φ_1 is the only such state that matters. At high temperatures, however, the state φ_1 will not be significantly more important than any other single-particle state φ_s with $s > 1$. In particular, we already know (from Section 9.4) that in the classical domain $\langle n_s \rangle_B$ will be much less than one for each s. This suggests a question worth pursuing: how do the size and importance of $\langle n_1 \rangle_B$ grow as the temperature is decreased? We will find a remarkable behavior: as a characteristic temperature is passed on the way to absolute zero, the importance of $\langle n_1 \rangle_B$ increases sharply.

To investigate the behavior of $\langle n_1 \rangle_B$, we must work with the explicit expressions for the $\langle n_s \rangle_B$'s and with their connection through the equation determining the parameter α. For mathematical convenience, let us temporarily use an energy scale with a different value for the zero of energy, one merely shifted so that the single-particle state φ_1 has an energy $\varepsilon_1' = 0$ on the new (primed) scale. Nothing in the physics can depend on where we choose to place the zero of energy, and with the new scale we conveniently sidestep the singular behavior of $\beta \varepsilon_1 = \varepsilon_1/kT$ in the $T \to 0$ limit. Single-particle energies on the new and old scale are related as

$$\varepsilon_s' = \varepsilon_s - \varepsilon_1 = \varepsilon_s - \frac{3\pi^2 \hbar^2}{2m V^{2/3}}. \tag{10.5.5}$$

Now we turn to explicit expressions for the expectation values of the boson occupation numbers. They are, from equation (9.3.27),

$$\langle n_s \rangle_B = \frac{1}{e^{\alpha' + \beta \varepsilon_s'} + 1}. \tag{10.5.6}$$

The single-particle energy ε_s' and the parameter α' are those associated with the new energy scale; hence a prime appears on each.

The parameter α' is to be determined from the equation

$$N = \sum_s \langle n_s \rangle_B = \langle n_1 \rangle_B + \sum_{s>1} \langle n_s \rangle_B$$

$$= \frac{1}{e^{\alpha'} - 1} + \sum_{s>1} \frac{1}{e^{\alpha' + \beta \varepsilon_s'} - 1}. \tag{10.5.7}$$

The term $\langle n_1 \rangle_B$ has been singled out from the sum because our interest lies primarily with its behavior. The parameter α' is a function of the temperature, as well as of N and the single-particle energies. Its value for $T = 0$ is rapidly established, for we know from equation (10.5.4) that only $\langle n_1 \rangle_B$ is non-zero at a temperature of absolute zero. Writing $\alpha'(0)$ for α' at $T=0$, we have

$$N = \frac{1}{e^{\alpha'(0)} - 1} + 0 \qquad \text{at } T = 0$$

and so

$$\alpha'(0) = \ln\left(1 + \frac{1}{N}\right) \simeq \frac{1}{N} \qquad \text{when } N \gg 1. \tag{10.5.8}$$

Thus the value of α' at absolute zero is positive but exceedingly small. At higher temperatures α' must increase in value, for then the sum over $s > 1$ contributes, requiring $\langle n_1 \rangle_B$ to be less than precisely N.

To make progress with the analysis for $T > 0$, we must bring the sum over $s > 1$ to a more convenient form. First we reexpress that sum by an integral with a density of states factor for a spinless particle:

$$\omega(\varepsilon') = \frac{2\pi(2m)^{3/2}V}{h^3} (\varepsilon')^{1/2}.$$

This step yields

$$N = \frac{1}{e^{\alpha'} - 1} + \int_{\varepsilon'=0}^{\infty} \frac{1}{e^{\alpha'+\beta\varepsilon'} - 1} \frac{2\pi(2m)^{3/2}V}{h^3} (\varepsilon')^{1/2} \, d\varepsilon'.$$

The lower limit of integration may safely be taken as $\varepsilon' = 0$, for the density of states factor vanishes as ε' goes to zero, and hence we do not inadvertently count $\langle n_1 \rangle_B$ twice. To display more clearly some of the temperature dependence of the integral, let us make a change of integration variable to

$$x = \beta\varepsilon' = \frac{\varepsilon'}{kT}.$$

After extracting constant factors from under the integral sign, we arrive at

$$N = \frac{1}{e^{\alpha'} - 1} + \frac{2\pi(2m)^{3/2}V(kT)^{3/2}}{h^3} \int_{x=0}^{\infty} \frac{x^{1/2}}{e^{\alpha'+x} - 1} \, dx. \tag{10.5.9}$$

The remaining integral, which we abbreviate as

$$I(\alpha') \equiv \int_{x=0}^{\infty} \frac{x^{1/2}}{e^{\alpha'+x} - 1} \, dx, \tag{10.5.10}$$

depends on the temperature only because the parameter α' depends on the temperature. As the temperature decreases and thus α' decreases toward $O(1/N)$, the value of the integral increases, for the denominator becomes smaller at each x in the range of integration. The integral does, however, remain finite even in the mathematical limit $\alpha' \to 0$. (A quick demonstration that $I(0)$ is greater than 1 but less than 3.2 is outlined in problem 10.4.) Already these properties of $I(\alpha')$ provide us with sufficient evidence to infer a qualitative result for the temperature dependence of $\langle n_1 \rangle_B$.

Briefly, we know that the integral $I(\alpha')$, for $\alpha' > 0$, is *less* than the *finite* value $I(0)$ that it attains in the mathematical limit $\alpha' \to 0$. This may be codified as

$$I(\alpha') < I(0) = \begin{pmatrix} \text{finite} \\ \text{number} \end{pmatrix} \qquad \text{when } \alpha' > 0. \qquad (10.5.11)$$

After comparing equations (10.5.7, 9) and using this relation, we may assert the inequality

$$\sum_{s > 1} \langle n_s \rangle_B < \frac{2\pi(2m)^{3/2}V(kT)^{3/2}I(0)}{h^3}. \qquad (10.5.12)$$

As the temperature is decreased from some high value, there comes a point at which the righthand side of the inequality (10.5.12) is numerically equal to precisely N. Let us designate that temperature by T_c, that is, define T_c by the relation

$$\frac{2\pi(2m)^{3/2}V(kT_c)^{3/2}I(0)}{h^3} = N. \qquad (10.5.13)$$

(Later we will calculate T_c for helium and find it to be about 3 K.) When the temperature is further decreased, the righthand side of the inequality will become merely a fraction of N. For instance, for $T = 0.9T_c$ the righthand side will be $(0.9)^{3/2}N$, which is $0.85N$. Since the sum plus $\langle n_1 \rangle_B$ must together yield N, we can say that for $T = 0.9T_c$ the value of $\langle n_1 \rangle_B$ must be greater than $0.15N$. We may infer that already at a temperature appreciably above absolute zero, $\langle n_1 \rangle_B$ will be of order N. The temperature need *not* be reduced to virtually zero before $\langle n_1 \rangle_B$ and the single-particle ground state φ_1 become exceedingly important. But it is the sharpness of the rise to importance—an aspect yet to be investigated—that is most striking.

Let us note in passing that the relation defining T_c may profitably be rearranged as

$$\frac{N}{V}\left(\frac{h}{\sqrt{2\pi mkT_c}}\right)^3 = \frac{2}{\sqrt{\pi}}I(0).$$

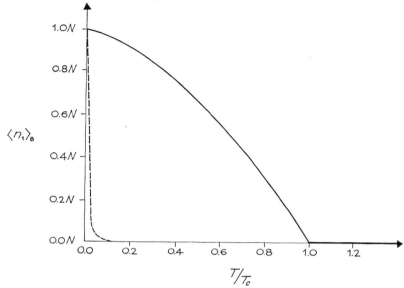

FIGURE 10.5.1
The solid curve gives $\langle n_1 \rangle_B$ as a function of temperature for a system of spinless bosons with $N \gg 1$. The values at, and in the immediate vicinity of, $T = T_c$ are of order $N^{2/3}$. On the graph these appear as virtually zero, for, relative to N, a number of order $N^{2/3}$ is exceedingly small.
    The behavior of $\langle n_1 \rangle_B$ might have been different. The dramatic rise in importance might have occurred only when kT was of the order of the energy difference between the single-particle ground state and the first excited single-particle state. That temperature is of order $T_c/N^{2/3}$, much lower than T_c. So the behavior *might* have been like that indicated by the dashed curve (which should really be squashed against the vertical axis). The "might be" curve was roughly calculated by using, for *all* temperatures, the approximate expression for $\langle n_1 \rangle_B$ that is valid only in the high-temperature classical limit.

Since we have noted parenthetically that $I(0)$ is of order unity, this tells us that T_c is of the order of that temperature at which we may expect a classical analysis to begin to fail. We should not be overly disturbed if the boson system shows a behavior at temperatures below T_c that does not conform with our intuition (developed largely from experience with classical physics).

To analyze the situation in more detail, we should return from inequalities to equalities. The introduction of the temperature T_c enables us to tidy up the equality in equation (10.5.9) as

$$N = \frac{1}{e^{\alpha'} - 1} + N \left(\frac{T}{T_c} \right)^{3/2} \frac{I(\alpha')}{I(0)} \tag{10.5.14}$$

and then to extract $\langle n_1 \rangle_B$ as

$$\langle n_1 \rangle_B = N \left[1 - \left(\frac{T}{T_c} \right)^{3/2} \frac{I(\alpha')}{I(0)} \right]. \tag{10.5.15}$$

Since the ratio $I(\alpha')/I(0)$ is less than one for $\alpha' > 0$, this expression shows us more explicitly that, when the temperature is any significant amount below T_c, the value of $\langle n_1 \rangle_{\text{B}}$ will be of order N.

The conclusion about the order of magnitude permits us to derive a simple approximate expression for $\langle n_1 \rangle_{\text{B}}$ at temperatures below T_c. Knowing that $\langle n_1 \rangle_{\text{B}}$ is then of order N, we may infer that α' is very close to zero for this temperature range:

$$\langle n_1 \rangle_{\text{B}} = \frac{1}{e^{\alpha'} - 1} = O(N)$$

implies

$$\alpha' = \ln\left(1 + \frac{1}{O(N)}\right) \simeq \frac{1}{O(N)} \simeq 0.$$

When α' is close to zero, $I(\alpha')$ is close to $I(0)$. So we may write, to good approximation,

$$\langle n_1 \rangle \simeq N\left[1 - \left(\frac{T}{T_c}\right)^{3/2}\right] \qquad \text{when } T < T_c. \qquad (10.5.16)$$

The behavior of $\langle n_1 \rangle_{\text{B}}$ as a function of temperature is displayed in figure 10.5.1, with the curve for the temperature domain $T < T_c$ being that given in equation (10.5.16). For finite N and V there is no mathematical discontinuity in the slope at T_c. The behavior is, however, "physically discontinuous," in that $\langle n_1 \rangle_{\text{B}}$ grows from order $N^{2/3}$ at $T = T_c$ to order N at temperatures a finite but small amount below T_c. This sudden growth in the importance of $\langle n_1 \rangle_{\text{B}}$ is rather like the condensation of a vapor to form a liquid at a sharply defined temperature. From this analogy comes the name *Bose-Einstein condensation* for the present phenomenon, with T_c being called the *condensation temperature*.

A demonstration that $\langle n_1 \rangle_{\text{B}}$ is indeed only of order $N^{2/3}$ for $T = T_c$ requires merely a careful treatment of equation (10.5.14) for that temperature. For small positive α', the integral $I(\alpha')$ may be expanded (in part by numerical methods) as

$$I(\alpha') = I(0)\left[1 - \frac{2\pi^{1/2}}{2.612}(\alpha')^{1/2} + O(\alpha')\right] \qquad (10.5.17a)$$

with

$$I(0) = \frac{\pi^{1/2}}{2}(2.612). \qquad (10.5.17b)$$

So $\alpha'(T_c)$, presumed to be small, may be computed to sufficient accuracy from equation (10.5.14) by employing expansions,

$$N = \frac{1}{e^{\alpha'(T_c)} - 1} + N \left(\frac{T_c}{T_c}\right)^{3/2} \frac{I(\alpha'(T_c))}{I(0)}$$

$$\simeq \frac{1}{\alpha'(T_c)} + N \left[1 - \frac{2\pi^{1/2}}{2.612} (\alpha'(T_c))^{1/2}\right],$$

and then solving for $\alpha'(T_c)$ as

$$\alpha'(T_c) \simeq \left(\frac{2\pi^{1/2}}{2.612}\right)^{-2/3} \frac{1}{N^{2/3}}.$$

This confirms that the parameter α' is quite close to zero for $T \leqslant T_c$. Then one finds

$$\langle n_1 \rangle_B \bigg|_{T = T_c} \simeq \frac{1}{\alpha'(T_c)} \simeq \left(\frac{2\pi^{1/2}}{2.612}\right)^{2/3} N^{2/3}.$$

Although $N^{2/3}$ is large compared to 1, the comparison should be made with N. Relative to the N for a macroscopic system, $N^{2/3}$ is a vanishingly small number.

The firm conclusion, then, is that as T_c is passed on the way to lower temperature, a macroscopically significant number of bosons—a number of order N—is to be found in the lowest single-particle state. In our analysis they "condense" into the state φ_1 given in equation (10.5.3a). Associated with that state is a de Broglie wavelength of order L, the edge length of the container. Thus the "condensed" bosons possess a de Broglie wavelength of macroscopic size.

One could be more elaborate in the calculation and include the interaction, typical of a laboratory, with the earth's gravitational field. Doing so is permissible without alteration of the theory, for we have had to assume only that the particles do not interact among themselves. The construction of single-particle energy eigenstates when the energy operator includes interaction with a constant gravitational field g is no trivial exercise. For the single-particle ground state, we can, however, make an educated estimate of the spatial behavior. The gravitational field, taken to be downward in the familiar cubical volume, will make it more probable that the particle is near the bottom than near the the top. Let us denote by Δz the range from the bottom in which the particle is likely to be when in this single-particle state of *lowest* energy. To estimate Δz, we use an argument based on the Heisenberg uncertainty principle. We look for the value of Δz that will minimize the energy expression

$$\varepsilon_z \equiv \frac{(\Delta p_z)^2}{2m} + mg\,\Delta z,$$

subject to the restriction

$$\Delta z \, \Delta p_z = \hbar.$$

The energy expression is an estimate of the energy associated with the z-direction position and momentum. That energy will, of course, have its minimum in the single-particle ground state, and so minimizing the expression will give us an estimate of Δz. Admittedly, this is a rough argument; we do not worry about factors of two.

After eliminating Δp_z with the aid of the restriction, we find we are to minimize

$$\varepsilon_z = \frac{\hbar^2}{2m(\Delta z)^2} + mg \, \Delta z$$

with respect to Δz. To find the minimum we set equal to zero the first derivative with respect to Δz:

$$\frac{d\varepsilon_z}{dz} = \frac{\hbar^2}{2m} \left(-\frac{2}{(\Delta z)^3} \right) + mg = 0.$$

Solved for Δz, the equation yields

$$\Delta z = \left(\frac{\hbar^2}{m^2 g} \right)^{1/3}.$$

This provides an order-of-magnitude estimate of the height within which one is likely to find the particle when in the single-particle state of lowest energy. (The result could have been inferred from dimensional arguments alone, for the addition of a gravitational potential energy mgz to the quantum-mechanical energy operator effectively introduces the above length into the problem as a second characteristic length, the first being the edge length L of the cube.) A detailed calculation of the lowest single-particle state confirms that the vast bulk of the probability is concentrated near the bottom in a layer of a few times this characteristic height. For a mass equal to that of the helium isotope He^4 the estimate for the height yields

$$\Delta z_{He^4} = 3 \times 10^{-4} \text{ cm}.$$

Thus we may conclude that the Bose-Einstein condensation into the lowest single-particle state in the presence of a gravitational field leads to a spatial separation, the "condensed" particles being very near the bottom of the container. One should not, however, push this picture too far, for real bosons do have a non-zero size, and the condensation to the thin layer would soon lead to significant direct interaction between the particles, something not taken into account in our approximation of no mutual interaction.

Excellent approximations for the total energy and pressure in the temperature domain $T < T_c$ follow readily from the preceding analysis. Continuing for the moment with the new energy scale, we have

$$\langle E' \rangle = \sum_s \varepsilon'_s \langle n_s \rangle_B$$

$$= 0 \times \langle n_1 \rangle_B + \sum_{s>1} \varepsilon'_s \langle n_s \rangle_B,$$

since $\varepsilon'_1 = 0$ on the new scale. The sum over $s > 1$ we again write as an integral with a density of states factor:

$$\langle E' \rangle = \int_{\varepsilon'=0}^{\infty} \frac{1}{e^{\alpha' + \beta \varepsilon'} - 1} \varepsilon' \omega(\varepsilon') \, d\varepsilon'$$

$$= \frac{2\pi(2m)^{3/2} V (kT)^{5/2}}{h^3} \int_{x=0}^{\infty} \frac{x^{3/2}}{e^{\alpha' + x} - 1} \, dx.$$

A change of integration variable to $x = \beta \varepsilon'$ has been made in order to extract part of the temperature dependence. For $T < T_c$ we know that α' is very close to zero, and so we may approximate the remaining integral by its value for $\alpha' = 0$:

$$\int_{x=0}^{\infty} \frac{x^{3/2}}{e^x - 1} \, dx = \frac{3\pi^{1/2}}{4} \,(1.342).$$

Upon tidying up the constant factors, we find

$$\langle E' \rangle \simeq \frac{3}{2}(1.342) \frac{(2\pi m)^{3/2} (kT)^{5/2} V}{h^3} \qquad \text{when } T < T_c,$$

the essential point being that $\langle E' \rangle$ grows with temperature as $T^{5/2}$.

At this point we may conveniently return to the old energy scale by adding $N\varepsilon_1$ to $\langle E' \rangle$:

$$\langle E \rangle = \langle E' \rangle + N\varepsilon_1$$

$$\simeq \frac{3}{2}(1.342) \frac{(2\pi m)^{3/2} (kT)^{5/2} V}{h^3} + N\varepsilon_1.$$

We can check the expression by taking the limit as the temperature goes to absolute zero:

$$\lim_{T \to 0} \langle E \rangle = 0 + N\varepsilon_1 = E_1.$$

Quite properly, the limit gives the energy of the ground state ψ_1 on the old scale.

Since an explicit expression for $\langle E \rangle$ is at hand, we may immediately write down the pressure estimate. We have

$$\langle p \rangle = \frac{2}{3} \frac{\langle E \rangle}{V}$$

$$\simeq (1.342) \frac{(2\pi m)^{3/2}}{h^3} (kT)^{5/2} + \frac{2}{3} \frac{N\varepsilon_1}{V} \qquad \text{when } T < T_c.$$

From this emerges another remarkable property of the Bose-Einstein gas: at temperatures below T_c, the pressure is virtually independent of the volume. The first term is entirely independent of V and the second is macroscopically negligible, for ε_1 is typically so very small. A pressure independent of the volume is typical of a system when it is partially condensed and hence is composed of gaseous and liquid phases in equilibrium. (Bear in mind, though, that our system is and remains a gas with no interparticle forces.)

> For a simple but good reason, the calculation of $\langle p \rangle$ was deferred until we had gotten back to the old energy scale. The relation $\langle p \rangle = \frac{2}{3}\langle E \rangle / V$ holds only if the potential energy inside the container is zero (and if two other conditions hold: we are dealing nonrelativistically with particles of non-zero rest mass, and we neglect the mutual interactions). Scrutinizing the steps that led from the general equation (9.6.1.) to equation (9.6.3) will substantiate this claim. On the primed energy scale there is, in effect, a constant potential energy of value $-\varepsilon_1$ per particle within the box. One can, of course, compute $\langle p \rangle$ in the presence of such a potential energy, but one has to fuss with algebraic details.

A word or two should be said about calculating expectation values in the temperature domain above T_c. In that domain $\langle n_1 \rangle_{\text{B}}$ is of order $N^{2/3}$ or less, and hence relatively unimportant. For calculating macroscopic properties such as total energy and pressure, one may neglect it with impunity. Then the parameter α' may be calculated from equation (10.5.9) by dropping the first term and working solely with the second. Because the integral in that second term is rather intractable, expansions and numerical methods remain necessary elements in the analysis. Since the resultant expressions for α', $\langle E \rangle$, and $\langle p \rangle$ are anything but transparent (except in the very high-temperature limit, already dealt with in Section 9.6), there seems no point in laboriously working them out here. A fine compendium is to be found in Fritz London's *Superfluids*, Vol. II, cited in the references.

We should now make the promised comparison with some experimental properties of the spinless helium isotope He^4. At sufficiently low temperatures, gaseous helium condenses to form a liquid phase as well. The temperature at which condensation sets in depends on the pressure, though the process is possible only below a critical temperature of 5.2 K. Under atmospheric pressure condensation occurs at 4.2 K. The condensation to a liquid is unquestionably due to the real forces between helium atoms. That the temperature at which the liquid phase forms is so low is (partially) a consequence of the weakness of the force between two helium atoms. The two electrons in each atom are very tightly bound to form a quite spherically symmetrical atom. Consequently it is difficult to induce by distortion an electric dipole moment, the result being that the attractive part of the interatomic force is extremely weak. That force can generate a condensed phase only when the temperature is quite low and hence the typical kinetic energy quite small.

At temperatures of order 5 K and below, one may certainly not neglect the forces between real helium atoms. Consequently, how can our analysis of a gas of not mutually interacting bosons be relevant to helium at low temperatures? The answer comes when we look at the behavior of liquid helium as the temperature is decreased below that for the onset of condensation, equilibrium between the liquid and the vapor above it being maintained. Under these conditions the liquid itself undergoes a most remarkable phase transition at a temperature of 2.17 K. The behavior of the liquid heat capacity is shown in figure 10.5.2, the shape leading to the name *lambda transition*, with the corresponding temperature being called the *lambda point*: $T_\lambda = 2.17$ K for liquid He^4. Below the lambda point, liquid helium exhibits properties that have led to the term *superfluid*: in particular, the liquid can flow through fine capillaries with imperceptible resistance and can flow as a film up over the lip of a beaker in apparent defiance of gravity.

Let us inquire, very tentatively, about the value of T_c for bosons whose mass is that of a helium atom and whose density is that of liquid helium at the latter's lambda point. Upon solving for T_c from the definition in equation (10.5.13) and using the numerical value of $I(0)$ from equation (10.5.17b), we have

$$T_c = \frac{h^2}{2\pi mk}\left(\frac{N}{V}\right)^{2/3}\frac{1}{(2.612)^{2/3}}. \tag{10.5.18}$$

The experimental mass density of liquid helium at its lambda point is

$$\left(\frac{mN}{V}\right)_{\lambda\text{ point}} = 0.146 \text{ gm/cm}^3,$$

352

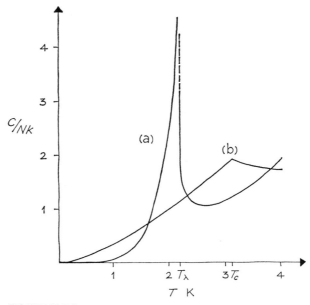

FIGURE 10.5.2

(a) The experimentally determined heat capacity of liquid helium (He^4 isotope) when the liquid is in constant equilibrium with its vapor. The heat capacity at constant volume for the liquid has a similar λ-like form. (b) The theoretical heat capacity of a perfect, spinless boson gas, held at constant volume.

about one-seventh that of water. This datum and the known values for the remaining constants yield a numerical value for T_c:

$$T_c = 3.13 \text{ K}. \tag{10.5.19}$$

Thus the temperature T_c at which the gas of noninteracting bosons shows a strikingly nonclassical behavior is quite close to the temperature T_λ at which liquid He^4 begins to exhibit its remarkable properties. This provocatively suggests a connection between the transition in liquid helium and the Bose-Einstein condensation of a perfect gas.

A plot of the heat capacity (at constant volume) of the perfect boson gas is also shown in figure 10.5.2. At the condensation temperature T_c, there is an abrupt change of slope—it is more abrupt the larger N is—and the general shape of the curve bears at least a family resemblance to the one for liquid He^4.

Perhaps the strongest argument for a connection is by contrast with a quite similar system that does now show a comparable transition. Nature has kindly provided us with a second helium isotope, He^3. Like He^4 but lacking one neutron, the isotope He^3 has (as a composite particle) an intrinsic angular momentum of $\frac{1}{2}\hbar$. Thus the isotope He^3 is a fermion, and antisymmetric wave functions are required to describe a system of such atoms. One can expect the forces between a pair of He^3 atoms to be almost identical to those between a pair of He^4 atoms, for the interatomic forces are determined primarily by the distribution of electrons around the nuclei. If wave function symmetry were of no concern, one would expect virtually no differences in macroscopic properties between the two isotopes (except such differences as might arise from the difference in atomic mass).

At low temperature, He^3 does condense to form a liquid, the maximum such temperature being 3.3 K, with He^3 condensing at 3.2 K under atmospheric pressure. There is good theoretical reason to believe that the lower temperature values for He^3 are primarily attributable to the smaller mass. The crucial question is this: does liquid He^3 show a lambda transition? For measurements down to about 5×10^{-3} K, there is no evidence for a lambda transition in liquid He^3—despite interatomic forces virtually identical to those in He^4 and a comparable atomic mass. The conclusion seems inescapable that the lambda transition in He^4, and the attendant remarkable properties, are intrinsically linked with the totally symmetric wave functions for the boson isotope. That the lambda transition is not *identical* to the Bose-Einstein condensation of a perfect gas is hardly surprising, for there are significant interatomic forces in the liquid (as the mere existence of the liquid phase demonstrates). But one is probably safe in saying that one is seeing similar macroscopic manifestations of symmetric wave functions, one in the real world of liquid He^4, the other in an idealization.

10.6. BLACK-BODY RADIATION

The history of quantum mechanics begins with Planck's analysis of black-body radiation; so it is eminently appropriate that we discuss the statistical description of electromagnetic radiation in thermal equilibrium. We will analyze the radiation in an enclosure of macroscopic size when the material walls forming the cavity are maintained at a constant temperature T. The non-zero temperature of the walls ensures that the atoms do typically possess energy that they may radiate. Taking a particulate view of the radiation, we may say that the atoms in the walls emit and absorb photons erratically. We can expect that, given time, an equilibrium will be established, there being then no general increase or decrease in the total energy of the enclosed radiation. It is this equilibrium situation that we seek to describe statistically, taking the radiation to be composed of a host of photons.

There is adequate evidence, both experimental and theoretical, that a photon has an intrinsic angular momentum—a spin—of $1\hbar$. We are dealing, therefore, with a system of identical, indistinguishable bosons. To an exceedingly good approximation, we may regard the photons in the enclosure as not interacting with one another.* There is, however, a drastic interaction with the atoms of the walls: photons may be removed from the enclosure by absorption or added by emission. Hence the number of photons in the enclosure is not fixed, even at thermal equilibrium (though we may reasonably expect that the relative fluctuations in the total number will be small).

The radiation within the enclosure may be described with energy eigenstates for the entire radiation system. We may ascribe a temperature to the radiation system by saying that the temperature of the radiation is the temperature of the material walls with which the radiation is in equilibrium, and this is the temperature that a gas thermometer would measure if permitted to interact with the radiation. The canonical distribution will give the probability that a specific state of the radiation system provides the appropriate description of the system.

Since the photons do not interact among themselves, the energy eigenstates ψ_j of the entire radiation system may be constructed out of single-photon states φ_s. Occupation numbers again provide a convenient scheme for computing expectation values, in particular, those of the total energy and of the radiation pressure. We have, as before, a unique correspondence between a set of occupation numbers $n_s(j)$ and a state ψ_j of the (radiation) system. The major departure is the need to consider states of the system with various total numbers of photons, indeed, the gamut from $N = 0$ to $N = \infty$ (though the extremes play a small role). The erratic absorption and emission of photons by the atomic walls, even at thermal equilibrium, compels us to this. We do not know how many photons are present.

It would be well to separate the analysis into its two disjoint elements. First we determine $\langle n_s \rangle$ for bosons when there is no restriction on the number of bosons present. Then we analyze the single-photon states themselves. Such a separation will enable us to distinguish the effects of two characteristics of the radiation system: (1) that the number of photons is not fixed, and (2) that the photons have spin $1\hbar$ and no rest mass.

---

* Quantum electrodynamics does predict a photon-photon interaction, which leads, for example, to the scattering of light by light. The effect is minute, however, and (to the best of my knowledge) has not yet been detected.

The Expectation Values
of the Occupation Numbers

We set out to calculate $\langle n_s \rangle$ for a system of identical bosons (not mutually interacting) when there is no restriction on the number of bosons present. The boson property—symmetric wave functions—implies that the sth occupation number may have the values

$$n_s = 0, 1, 2, 3, \ldots . \tag{10.6.1}$$

The absence of a restriction on the number of bosons present means that *any* set of non-negative integers, for instance,

$$n_1 = 5, n_2 = 1, n_3 = 0, n_4 = 7, \text{ all others zero,}$$

specifies a state of the radiation system that we will need to include in a summation over all states ψ_j. The sum of the non-negative integers need not have a specific fixed value, for there is no specified number of photons in the system. In short, there is no "fixed N" restriction like equation (9.3.20b) of the previous boson analysis.

Shortly we will turn a sum over all states ψ_j into a sum over all sets of occupation numbers. Let us examine the correspondence now with a low-dimensional example. Suppose there are only two single-particle states, φ_1 and φ_2. The occupation number n_1 may have for its value any non-negative integer; the same holds for n_2. By pairing each possible value of n_1 with each possible value of n_2 we generate all possible states ψ_j of this radiation system. Figure 10.6.1 gives a start on this indefinitely long process. A sum over all

(0, 0)	(1, 0)	(2, 0)	(3, 0)	. . .
(0, 1)	(1, 1)	(2, 1)	.	.
(0, 2)	(1, 2)	(2, 2)	.	.
(0, 3)	(1, 3)	.	.	.
.	.	.	.	.
.	.	.	.	.
.	.	.	.	.

FIGURE 10.6.1
The entries give values of the pair (n_1, n_2), and thus each entry corresponds to a state ψ_j of the (low dimensional) radiation system.

ψ_j is the same as a sum over all pairs. The sum over all pairs may be done by holding one element (n_1, say) of the pair fixed at a certain integer and summing over all values of the other element (n_2), then choosing another fixed integer and summing over all values of the other element, and so on.

This amounts to going through figure 10.6.1 column by column. We may symbolize the import of this paragraph by writing

$$\sum_j \rightarrow \sum_{n_1=0}^{\infty} \sum_{n_2=0}^{\infty}$$

as the prescription for turning a sum over states ψ_j into a sum over sets of occupation numbers in this low-dimensional case.

Now for the calculation. We write, as always,

$$\langle n_s \rangle = \sum_j n_s(j) P_j = \frac{\sum_j n_s(j) e^{-\beta E_j}}{\left(\sum_i e^{-\beta E_i} \right)}. \tag{10.6.2}$$

The sum with j goes over all energy eigenstates of the radiation system, each such state having a definite number of photons but with the number itself differing among the various states. Given the unique connection between states ψ_j and sets of occupation numbers $n_s(j)$, we may transcribe the sums above into sums over sets of occupation numbers. The generalization of our previous low-dimensional example yields

$$\langle n_s \rangle = \frac{\displaystyle\sum_{n_1=0}^{\infty} \cdots \sum_{n_s=0}^{\infty} \cdots n_s \, e^{-\beta(n_1\varepsilon_1 + \ldots + n_s\varepsilon_s + \ldots)}}{\displaystyle\sum_{n_1=0}^{\infty} \cdots \sum_{n_s=0}^{\infty} \cdots e^{-\beta(n_1\varepsilon_1 + \ldots + n_s\varepsilon_s + \ldots)}}.$$

The dots indicate "more of the same but with different subscripts." Just as we did back in Section 4.6 with a similar sum, we may rearrange the full sum as a product of factors:

$$\langle n_s \rangle = \frac{\left(\displaystyle\sum_{n_1=0}^{\infty} e^{-\beta n_1\varepsilon_1} \right) \cdots \left(\displaystyle\sum_{n_s=0}^{\infty} n_s e^{-\beta n_s\varepsilon_s} \right) \cdots}{\left(\displaystyle\sum_{n_1=0}^{\infty} e^{-\beta n_1\varepsilon_1} \right) \cdots \left(\displaystyle\sum_{n_s=0}^{\infty} e^{-\beta n_s\varepsilon_s} \right) \cdots}. \tag{10.6.3}$$

All factors except those in n_s cancel, top and bottom. So we find

$$\langle n_s \rangle = \frac{\displaystyle\sum_{n_s=0}^{\infty} n_s e^{-\beta n_s\varepsilon_s}}{\displaystyle\sum_{n_s=0}^{\infty} e^{-\beta n_s\varepsilon_s}}. \tag{10.6.4}$$

The sum in the denominator is a geometric series. It may be summed as

$$\sum_{n_s=0}^{\infty} e^{-\beta n_s\varepsilon_s} = \sum_{n_s=0}^{\infty} (e^{-\beta\varepsilon_s})^{n_s} = \frac{1}{1 - e^{-\beta\varepsilon_s}},$$

for we have $\exp(-\beta\varepsilon_s) < 1$. (Any real photon has a positive energy. If we take the zero of energy to correspond to the absence of all photons, then each ε_s is greater than zero.) The numerator may be evaluated in closed form by first rewriting it as a partial derivative:

$$\sum_{n_s=0}^{\infty} n_s e^{-\beta n_s \varepsilon_s} = -\frac{1}{\beta}\frac{\partial}{\partial\varepsilon_s}\sum_{n_s=0}^{\infty} e^{-\beta n_s \varepsilon_s}$$

$$= -\frac{1}{\beta}\frac{\partial}{\partial\varepsilon_s}\left(\frac{1}{1-e^{-\beta\varepsilon_s}}\right) = \frac{e^{-\beta\varepsilon_s}}{(1-e^{-\beta\varepsilon_s})^2}.$$

Upon inserting these into equation (10.6.4) and tidying up, we find the *exact* result

$$\langle n_s \rangle = \frac{1}{e^{\beta\varepsilon_s}-1} \tag{10.6.5}$$

for a system of identical bosons when there is *no restriction on the numbers of bosons present.*

A comparison with the result, in equation (9.3.27), when there *is* a restriction on the number of bosons present (there being N of them) shows great similarity in the expressions. We may look upon the parameter α as being an adjustable parameter by which we ensure that the equation

$$\sum_{s}\langle n_s \rangle_{\rm B} = \sum \frac{1}{e^{\alpha+\beta\varepsilon_s}-1} = N$$

is satisfied. If there is no restriction on the number of bosons present, the parameter α is unnecessary; a value of zero for α is then "natural."

Single-Photon States

Now we analyze the single-photon states themselves. That demands a subtle mixing of the particle and wave aspects of electromagnetic radiation. The single-photon states are to be associated with the admissible standing electromagnetic waves within the enclosure, a photon energy ε_s being equated to $h\nu_s$, where ν_s is the frequency of the standing wave.

From Maxwell's equations for the coupled electric and magnetic fields, one may derive the wave equation,

$$\left(\frac{\partial^2}{\partial x^2}+\frac{\partial^2}{\partial y^2}+\frac{\partial^2}{\partial z^2}\right)\mathbf{E} = \frac{1}{c^2}\frac{\partial^2\mathbf{E}}{\partial t^2}, \tag{10.6.6}$$

for the electric field $\mathbf{E}(x, y, z, t)$, with c being the speed of light in vacuum. That, however, is not the only equation to which $\mathbf{E}$ is subject. Since the enclosure itself is devoid of charge, the electric field must have a zero divergence:

$$\frac{\partial E_x}{\partial x} + \frac{\partial E_y}{\partial y} + \frac{\partial E_z}{\partial z} = 0. \tag{10.6.7}$$

The admissible standing waves are determined by solutions of these two equations, subject to appropriate boundary conditions at the walls of the enclosure.

Our major concern is with the number of single-photon states per energy or frequency interval. The density of photon states is the quantity we need in order to replace sums by integrals when computing expectation values of the total energy or pressure. Fortunately, when the volume of the enclosure is macroscopic, the density of states as a function of frequency does not depend significantly on the shape of the enclosure or on the boundary conditions. The boundary conditions affect the behavior of the admissible standing waves primarily in the immediate vicinity of the walls. For all except the very low-energy photon states, there will be many wavelengths across the enclosure; whether the boundary conditions require a node or a crest at the walls will be insignificant. The spatial behavior of the very low-energy photon states (and the precise values of their energies) does depend nonnegligibly on the boundary conditions, but precisely because they are very low-energy states, those states make an insignificant contribution to total energy or pressure.

Having recognized that there is no significant dependence on enclosure shape and specific boundary conditions, let us take the familiar cube, each

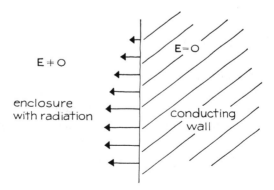

FIGURE 10.6.2
The situation where radiation and conducting wall meet.

side being of length L. For boundary conditions, let us adopt those appropriate for metallic walls in the limit of infinite conductivity: the electric field must be perpendicular to the walls at the enclosure surface. In that limit, there may be no electric field inside the conductor. A perpendicular component of an electric field may be excluded from the interior of the conductor by a surface layer of charge; so such a field component (immediately outside) is permitted. A tangential component, however, could not be cut off and excluded from the interior in such a manner; so it must be zero just outside the walls as well as within them. The situation is sketched in figure 10.6.2.

The solutions of equation (10.6.6) that meet the boundary conditions are, in component form,

$$E_x = e_x \cos\left(\frac{\pi n_x x}{L}\right)\sin\left(\frac{\pi n_y y}{L}\right)\sin\left(\frac{\pi n_z z}{L}\right)\cos(2\pi vt + \tilde{\varphi}),$$

$$E_y = e_y \sin\left(\frac{\pi n_x x}{L}\right)\cos\left(\frac{\pi n_y y}{L}\right)\sin\left(\frac{\pi n_z z}{L}\right)\cos(2\pi vt + \tilde{\varphi}),$$

$$E_z = e_z \sin\left(\frac{\pi n_x x}{L}\right)\sin\left(\frac{\pi n_y y}{L}\right)\cos\left(\frac{\pi n_z z}{L}\right)\cos(2\pi vt + \tilde{\varphi}),$$

with $\tilde{\varphi}$ being an adjustable phase angle. The wave equation implies that the integers n_x, n_y, n_z, are related to the frequency v by

$$\frac{\pi^2}{L^2}(n_x^2 + n_y^2 + n_z^2) = \frac{(2\pi v)^2}{c^2}. \tag{10.6.8}$$

The coefficients e_x, e_y, e_z, indicate the polarization of the standing electromagnetic wave. They are not totally free, for we must require that equation (10.6.7) be satisfied simultaneously with equation (10.6.6). Substitution of the tentative solution into the zero-divergence equation leads to a condition on the polarization coefficients:

$$e_x n_x + e_y n_y + e_z n_z = 0. \tag{10.6.9}$$

The expression is, effectively, the scalar product of a polarization vector $\mathbf{e}$ (having components e_x, e_y, e_z) with a vector whose components are n_x, n_y, n_z. (If we were using traveling waves rather than standing waves, the latter vector would determine the direction of propagation of the traveling wave.) The zero value for the scalar product implies that the polarization vector $\mathbf{e}$ must lie in the plane perpendicular to the direction defined by n_x, n_y, n_z. So there are typically two linearly independent polarization vectors compatible with a given set of the integers n_x, n_y, n_z.

A bit more must be said about the integers n_x, n_y, n_z. Changing the sign of one or more of them does not lead to intrinsically new solutions. Any apparently new solution is at most a linear combination of those solutions with nonnegative values for the integers. So we need consider only non-negative values in computing a density of states. Furthermore, zero values may be neglected. If two or three of the n_x, n_y, n_z, are zero, there is no electric field whatsoever. A set of values for the integers with one integer being zero is admissible, but such sets are insignificant relative to those without zeros when one is computing a density of states. Thus we may safely take the n_x, n_y, n_z, to range independently over the positive integers only.

To compute the density of single-photon states as a function of frequency, we proceed much as we did in section 9.5. We first compute the number of single-photon states with frequency less than or equal to v—this is written $\Omega(v)$—and then differentiate to determine $\omega(v)$, the number of single-photon states per unit frequency interval. The relation taking the place of equation (9.5.4) is, from equation (10.6.8),

$$R \equiv \sqrt{n_x^2 + n_y^2 + n_z^2} = \frac{2Lv}{c}.$$

The analog of equation (9.5.5) is then

$$\Omega(v) = 2 \times \frac{1}{8} \times \frac{4\pi R^3}{3} = \frac{8\pi}{3c^3} L^3 v^3.$$

The extra factor of 2 appears because of the two possible (linearly independent) polarizations for given values of the integers n_x, n_y, n_z. The result for the density follows on differentiation:

$$\omega(v) = \frac{d\Omega(v)}{dv} = \frac{8\pi}{3c^3} L^3 3v^2 = \frac{8\pi V}{c^3} v^2,$$

and so we may write

$$\begin{pmatrix} \text{the number of single-pho-} \\ \text{ton states with frequency} \\ \text{between } v \text{ and } v + dv \end{pmatrix} = \omega(v)\, dv = \frac{8\pi V}{c^3} v^2\, dv. \quad (10.6.10)$$

The preceding is a rather intricate analysis; so it would be well to exhibit a simpler, though less justifiable, approach to the same result. We regard the photons as very real particles (traveling with the speed of light) and compute $\Omega(v)$, the number of single-photon states with frequency less than or equal to v, by integrating $2\, d^3x\, d^3p/h^3$ over the appropriate domain of a phase-space.

The factor of 2 takes into account the two possible polarizations of the electromagnetic field. (Since a photon has spin $1\hbar$, one might think that the factor should be $(2S + 1) = (2 \times 1 + 1) = 3$. The reduction to merely a factor of 2 arises from the Maxwell equations, specifically, from the requirement that in charge-free space the divergence of the electric field be zero. This reduces, from 3 to 2, the possible linearly independent polarizations of the electric field of a traveling wave.) Upon relating the magnitude of a photon's momentum to the frequency by

$$|\mathbf{p}| = \frac{h}{(\text{wave length})} = \frac{h\nu}{c},$$

we may compute $\Omega(\nu)$ by integrating with d^3x over the volume V of the enclosure and by integrating with d^3p up to a maximum momentum magnitude determined by the fixed value of the frequency in $\Omega(\nu)$:

$$\Omega(\nu) = \int_{\substack{\text{specified} \\ \text{limits}}} \frac{2\,d^3x\,d^3p}{h^3} = \frac{2V}{h^3} \int_{p=0}^{p=h\nu/c} 4\pi p^2\,dp$$

$$= \frac{2V}{h^3} \frac{4\pi}{3} \left(\frac{h\nu}{c}\right)^3 = \frac{8\pi V}{3c^3} \nu^3.$$

The shape of the enclosure is irrelevant, only the volume entering. Differentiation with respect to the frequency gives us the density of single-photon states:

$$\omega(\nu) = \frac{d\Omega(\nu)}{d\nu} = \frac{8\pi V}{c^3} \nu^2.$$

That two different routes lead to the same result is always encouraging.

Black-Body Calculations

Now we can readily estimate the total energy of the radiation system, given that the enclosure walls are at a temperature T kelvin. The expression for $\langle E \rangle$, as a sum over single-photon states, is this:

$$\langle E \rangle = \sum_s \varepsilon_s \langle n_s \rangle = \sum_s \varepsilon_s \frac{1}{e^{\beta\varepsilon_s} - 1}. \tag{10.6.11}$$

In order to handle the mathematics, we transcribe the sum into an integral over frequencies with a density of single-photon states:

$$\langle E \rangle = \int_{\nu=0}^{\infty} h\nu \frac{1}{e^{\beta h\nu} - 1} \omega(\nu)\,d\nu. \tag{10.6.12}$$

The lower limit of integration has been taken to be zero, although the lowest frequency in the sum is finite and of order c/L. This simplification is admissible because the integrand itself goes to zero as v goes to zero. The upper limit is quite properly infinity. After substituting the explicit density-of-states factor, we have

$$\langle E \rangle = \int_{v=0}^{\infty} hv \, \frac{1}{e^{\beta hv} - 1} \, \frac{8\pi V}{c^3} \, v^2 \, dv.$$

The integrand, when divided by the volume V, has a useful interpretation. We may regard that function as the estimated energy per unit volume and per unit frequency interval:

$$\begin{pmatrix} \text{the estimated spatial} \\ \text{energy density per unit} \\ \text{frequency interval} \end{pmatrix} = \frac{8\pi h}{c^3} \frac{v^3}{e^{hv/kT} - 1}. \qquad (10.6.13)$$

The expression is graphed in figure 10.6.3 as a function of frequency for two values of the temperature. As one might have expected, the higher the temperature, the higher the frequency at which the maximum occurs.

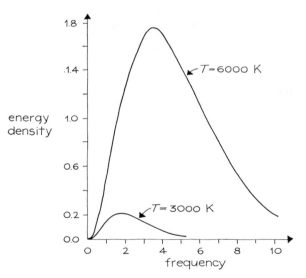

FIGURE 10.6.3

The estimated energy per unit volume and per unit frequency interval, in units of 10^{-14} erg-sec/cm³, is plotted against frequency, in units of 10^{14} cycles per second.

The temperature dependence of $\langle E \rangle$ can be made more evident with a change of integration variable to $x = h\nu/kT$. The variable change leads to the expression

$$\langle E \rangle = \left(\frac{8\pi V k^4}{c^3 h^3}\right) T^4 \int_{x=0}^{\infty} \frac{x^3}{e^x - 1}\, dx. \tag{10.6.14}$$

The remaining integral is simply a numerical constant, and so we may infer that $\langle E \rangle$ is proportional to the fourth power of the temperature, a rather strong temperature dependence. The integral is not a trivial one but may be evaluated analytically as

$$\int_{x=0}^{\infty} \frac{x^3}{e^x - 1}\, dx = \frac{\pi^4}{15}.$$

Hence the final result for the radiation energy is

$$\langle E \rangle = \left(\frac{8\pi^5 k^4}{15 c^3 h^3}\right) V T^4.$$

The radiation pressure follows readily. First we derive a convenient form from the general expression:

$$\langle p \rangle = \sum_j \left(-\frac{\partial E_j}{\partial V}\right) P_j = \sum_j -\frac{\partial}{\partial V}\left(\sum_s \varepsilon_s n_s(j)\right) P_j$$

$$= \sum_s \left(-\frac{\partial \varepsilon_s}{\partial V}\right)\langle n_s \rangle.$$

To compute the indicated partial derivative, we need the explicit dependence of ε_s on the volume V. Remembering that we deal with a cube of edge length L, for which $L = V^{1/3}$, we have, from equation (10.6.8),

$$\varepsilon_s = h\nu_s = \frac{1}{V^{1/3}}\frac{hc}{2}(n_x^2 + n_y^2 + n_z^2)_s^{1/2}.$$

The subscript s indicates that the frequency and the values of n_x, n_y, n_z are those corresponding to the sth single-photon energy. Differentiation then yields

$$\frac{\partial \varepsilon_s}{\partial V} = -\frac{1}{3}\frac{1}{V^{4/3}}\frac{hc}{2}(n_x^2 + n_y^2 + n_z^2)_s = -\frac{1}{3}\frac{\varepsilon_s}{V}.$$

After inserting this into the convenient form for $\langle p \rangle$, we arrive at

$$\langle p \rangle = \sum_s + \frac{1}{3} \frac{\varepsilon_s}{V} \langle n_s \rangle$$

$$= \frac{1}{3} \frac{\langle E \rangle}{V} \quad \text{for photons.} \qquad (10.6.15)$$

Two points are worth noting:

1. We find a factor of $\frac{1}{3}$ as the proportionality between $\langle p \rangle$ and $\langle E \rangle / V$, rather than the $\frac{2}{3}$ to which we are accustomed for particles with a non-zero rest mass.

2. Since $\langle E \rangle$ itself is proportional to V, the radiation pressure is independent of the volume, being a function of the temperature only.

Only a short comment is needed on the second point. That the radiation pressure is independent of the volume becomes less disturbing when we remember that we are not dealing with a fixed number of particles. The atoms in the walls of the enclosure can supply photons to meet any demand. Indeed, a calculation of the estimated number of photons present shows it to be proportional to the volume.

The first point is more provocative, and tracing out the origin of the $\frac{1}{3}$, $\frac{2}{3}$, difference with a simple model is worth the time it takes. Let us consider a particle in a cube. We assume that the particle makes elastic collisions with the walls and that these produce the pressure. At this stage in the analysis the particle may be either a photon or a typical gas molecule. For the pressure on the wall at $x = L$, we need to compute the momentum transferred per impact and the number of impacts per unit time. The former is twice the magnitude of the x-component of momentum, $2|p_x|$, the 2 arising because the momentum component is reversed during the collision. The number of collisions with the chosen wall in unit time will be $|v_x|/2L$. The product of these gives the momentum transferred per unit time to the wall at $x = L$. Upon dividing by the wall area L^2, we find

$$2|p_x| \times \frac{|v_x|}{2L} \times \frac{1}{L^2} = \frac{|p_x||v_x|}{V}$$

for the contribution this particle makes to the pressure.

The preceding supposes that we know the magnitudes $|p_x|$ and $|v_x|$, but really we don't, and so the next step is to form an expectation value estimate of the contribution. We start by averaging over all possible directions of the momentum and velocity, with the particle's energy held fixed. Since the momentum and velocity are parallel, we may write

$$\begin{pmatrix} \text{the expectation value of the} \\ \text{pressure contributed by one} \\ \text{particle with fixed energy} \end{pmatrix} = \frac{\langle |p_x| \, |v_x| \rangle}{V} = \frac{\langle p_x v_x \rangle}{V}$$

(10.6.16)

$$= \frac{1}{3} \frac{\langle \mathbf{p} \cdot \mathbf{v} \rangle}{V} = \frac{1}{3} \frac{|\mathbf{p}| \, |\mathbf{v}|}{V}.$$

The step to the scalar product follows by a symmetry argument, namely

$$\langle p_x v_x \rangle = \langle p_y v_y \rangle = \langle p_z v_z \rangle$$

and hence

$$\langle p_x v_x \rangle = \tfrac{1}{3} \langle \mathbf{p} \cdot \mathbf{v} \rangle.$$

The expectation value of $\mathbf{p} \cdot \mathbf{v}$ is simply the product of the magnitudes, for the expectation value is formed by averaging over all directions at *fixed* energy.

Equation (10.6.16) holds both for a photon and for a gas molecule. A difference arises, however, when one relates the magnitudes of momentum and velocity to the fixed energy ε. For a photon, one has

$$|\mathbf{v}| = c, \qquad |\mathbf{p}| = \frac{h\nu}{c} = \frac{\varepsilon}{c},$$

and hence

$$\begin{pmatrix} \text{the expectation value of the} \\ \text{pressure contributed by one} \\ \text{photon with energy } \varepsilon \end{pmatrix} = \frac{1}{3} \frac{(\varepsilon/c)c}{V} = \frac{1}{3} \frac{\varepsilon}{V}.$$

For a gas molecule, with its non-zero rest mass m, one has

$$|\mathbf{v}| = \sqrt{\frac{2\varepsilon}{m}}, \qquad |\mathbf{p}| = \sqrt{2m\varepsilon},$$

and hence

$$\begin{pmatrix} \text{the expectation value of the} \\ \text{pressure contribted by one} \\ \text{molecule with energy } \varepsilon \end{pmatrix} = \frac{1}{3} \frac{\sqrt{2m\varepsilon}\sqrt{2\varepsilon/m}}{V} = \frac{2}{3} \frac{\varepsilon}{V}.$$

The heart of the $\tfrac{1}{3}$, $\tfrac{2}{3}$ difference lies in the connections between the fixed energy and the magnitudes of momentum and velocity, connections that differ between photon and molecule. One should bear in mind that for the molecule we have made a nonrelativistic calculation, using for the energy ε the non-relativistic kinetic energy.

Let us continue for a moment and see what emerges if we use a relativistic energy ε_{rel} for the molecule,

$$\varepsilon_{\cdot el} = \frac{mc^2}{\sqrt{1 - \dfrac{v^2}{c^2}}},$$

with the momentum now being expressed in proper relativistic fashion as

$$\mathbf{p} = \frac{m\mathbf{v}}{\sqrt{1 - \dfrac{v^2}{c^2}}}.$$

Upon reexpressing the velocity and momentum magnitudes of equation (10.6.16) in terms of the relativistic energy, we find

$$\begin{pmatrix} \text{the expectation value of the} \\ \text{pressure contributed by one} \\ \text{molecule with fixed rela-} \\ \text{tivistic energy } \varepsilon_{rel} \end{pmatrix} = \frac{1}{3} \frac{\varepsilon_{rel}}{V} \left[1 - \left(\frac{mc^2}{\varepsilon_{rel}} \right)^2 \right].$$

In the limit of very high energy, so that the relativistic energy is much larger than the rest mass times c^2, the expression approaches that for a photon. In the relevant frame of reference, the particle is traveling close to the speed of light, and so should—in some ways—behave like a photon.

REFERENCES for Chapter 10

Extensive applications of the free-electron model are given by Charles Kittel in his *Introduction to Solid State Physics* (New York: Wiley, 3d ed., 1966). In addition, Kittel goes on to treat the interaction of the electrons with the positive ion cores of the metal lattice. A quite useful annotated bibliography of papers—original and review—on the "Ordinary Electronic Properties of Metals" is provided by D. N. Langenberg, *American Journal of Physics*, **36**, 777 (1968).

The Bose-Einstein condensation and its relevance to liquid helium are discussed by Fritz London in Volume II of his *Superfluids* (New York: Wiley, 1954). The discussion of relevance is authoritative, for London pioneered that approach to understanding superfluidity. In addition, London describes the "two fluid" model for superfluid helium, a model developed by Tisza and himself. All told, the book is a classic.

Further mathematical details about the sums and integrals (and their expansions) that appear when one analyzes the Bose-Einstein condensation are given by John E. Robinson, *Physical Review*, **83**, 678 (1951).

A derivation of the Bose-Einstein distribution function, with a laudable emphasis on rigor, is given by A. R. Fraser in a pair of articles: *Philosophical Magazine*, **42**, 156 and 165 (1951). Fraser treats very carefully the form of the distribution function below the condensation temperature. None of the physical conclusions we reached are altered, but they are given a more firm theoretical foundation. Remember that we did make an approximation in deriving both $\langle n_s \rangle_B$ and $\langle n_s \rangle_F$. Further justification for that approximation is provided in a short paper by F. Ansbacher and P. T. Landsberg, *Physical Review*, **96**, 1707 (1954).

A survey of the experimental properties of helium and a discussion of the numerous theoretical approaches are provided by J. Wilks, *The Properties of Liquid and Solid Helium* (London: Oxford University Press, 1967).

PROBLEMS

10.1. We consider a "gas" of free electrons in a volume V at a temperature of absolute zero. Calculate the total energy of the system, expressing it in terms of the electron mass m, Planck's constant h, the volume V, and the number of electrons N. Note that for fixed volume V, the total energy is *not* proportional to the number of electrons present. Since, by hypothesis, there are no forces between the electrons, how can this be? Doesn't it indicate that the entire analysis is rotten at the core?

10.2. There are semirealistic two-dimensional problems, as in the case of gas atoms stuck on the plane surface of a metal. Let us imagine a system of N particles, each of mass m, constrained to two-dimensional motion on a plane of area A. The particles have spin $\frac{1}{2}\hbar$ and their mutual interactions are to be neglected. Calculate the energy of the system's ground state and the heat capacity at low temperature. Answers should be explicit, that is, given in terms of $m, A, h,$ and so on.

10.3. This is a crude yet relevant model for the thermionic emission current. Suppose we have a classical monatomic gas (N atoms, mass m) in a cube of volume V at the usual temperature T, the atoms not interacting among themselves. The walls are "thin," in the sense that the wall force on an atom (acting perpendicular to the wall) is represented by a finite jump ε^* in potential energy between inside and outside the box. Thus some very energetic atoms may be able to get through the "thin" wall. Use *classical* reasoning to determine the *temperature dependence* of the number of atoms that get through the wall per second per unit wall area. (Actually, we will have to be satisfied with the temperature dependence of the expectation value estimate of the number of atoms that ....) The over-all numerical coefficient is not desired; so you need to do in detail only one easy integral, at most. How significant is the difference between the classical and the quantum-mechanical result?

10.4. **On $I(\alpha)$.** A good deal of the argument for the Bose-Einstein condensation depends on the integral $I(\alpha)$ being finite in the mathematical limit $\alpha \to 0$. The construction of a finite upper bound for $I(0)$ would both show that $I(0)$ is finite and provide an estimate of its numerical value. For the construction we split the full range of integration into two pieces, from zero to one and from one to infinity.

(a) Show that the following inequality holds and evaluate the finite righthand side:

$$\int_0^1 \frac{x^{1/2}}{e^x - 1}\, dx < \int_0^1 \frac{x^{1/2}}{x}\, dx.$$

(Hint: the inequality $e^x = 1 + x + O(x^2) > 1 + x$ for $x > 0$ is all that one needs.)

(b) Again, show that the following inequality holds and evaluate the finite righthand side:

$$\int_1^\infty \frac{x^{1/2}}{e^x - 1}\, dx < \int_1^\infty \frac{x}{e^x(1 - e^{-1})}\, dx$$

(c) Combine the inequalities in (a) and (b) to show that $I(0)$ is less than $2e/(e-1)$, which is less than 3.2.

(d) A good lower bound would be useful, too. Confirm the inequality

$$\int_0^\infty \frac{x^{1/2}}{e^x - 1} > \int_0^\infty \frac{1}{e^x}\, dx$$

by examining the two integrands in the relevant range of x, evaluate the righthand side, and thus find that $I(0)$ is greater than 1.

   You have just proved that $I(0)$ lies between 1 and 3.2. These limits are sufficient for establishing all the essential physics.

10.5. The text claimed that the second term in the expression for $\langle p \rangle$ below the Bose-Einstein condensation temperature is "macroscopically negligible." Take reasonable laboratory values for m, N/V, V, and T (with $T < T_c$) and compare the sizes of the two terms. Compare each with atmospheric pressure as well (atmospheric pressure $\simeq 10^6$ dynes/cm^2).

10.6. Assume that the radiation within the sun is adequately described as black-body radiation at some appropriate temperature. Calculate the radiation pressure (expressing it in units of atmospheric pressure, taken to be 10^6 dynes/cm^2) near the solar surface ($T \simeq 6{,}000$ K) and at the solar center ($T \simeq 10^7$ K). For a comparison, you might estimate the water pressure at the ocean floor. Impressed?

10.7. **Wien's displacement law.** If, for black-body radiation, the estimated spatial energy density per unit frequency interval is plotted against frequency, the curve will have a single peak.

(a) Show that the frequency at the peak is proportional to the temperature and hence that the wavelength at the peak is inversely proportional to the temperature. The proportionality is one aspect of Wien's displacement law, "displacement" because the wavelength for the peak is displaced when the temperature is changed.

(b) Estimate (to one significant figure only) the numerical value of the proportionality constant in Wien's law, for either frequency or wavelength.

10.8. **The Debye model.** This problem invites you to learn about the Debye model for the lattice vibrations of a solid with the aid of some plausibility arguments. The discussion of the Einstein model in Section 7.4 forms the starting point of the development. There we explicitly stated that we would calculate the position-dependent potential energy for any given atom on the assumption that all other atoms were at their equilibrium sites in the lattice structure. This assumption decoupled the vibrational motion of the various atoms from one another. Now we make less restrictive assumptions and take into account the displacements of the neighboring atoms from their equilibrium positions.

For the moment we can think of the atoms as coupled together by Hooke's law springs between neighboring pairs of atoms. (There is an amusing consistency here, for Hooke's law comes precisely from the actual behavior of interatomic forces in a solid.) In one dimension an instantaneous picture might look like this figure. The sinusoidal curve indicates the amount of

displacement from the equilibrium sites, each of which is indicated by a short vertical line.

Because the force exerted on a given atom by its neighbors now depends on *their* displacement from equilibrium as well as its own, the theory admits the possibility of *coherent* vibrations, neighboring atoms vibrating with definite amplitude and phase relations between them. We have opened the door to sound waves in the solid. The sketch displays a portion of such a wave, one wavelength's worth, to be precise. The frequency of vibration will be related to the wavelength by the speed of sound c_s in the solid. The heart of the Debye model consists of an astute association of two notions: coherently vibrating atoms and sound waves.

In the Einstein model we had N uncoupled three-dimensional harmonic oscillators, all of the same frequency. Now we deal with coupled oscillators. From the physical picture we know that coherent vibrations of various wavelengths and hence various frequencies will occur in the solid. A sound wave exhibits an oscillatory behavior rather like that of a one-dimensional harmonic oscillator. If we think of a state of the solid as described by a set of sound waves of various frequencies, amplitudes, and directions of propagation and polarization, it is plausible that the total energy will be equal to that of a set of harmonic oscillators, provided the oscillators have frequencies corresponding to those of the sound waves. One mentally associates a harmonic oscillator with each kind of sound wave, the energy of the oscillator being related to the amplitude of that kind of sound wave. This is a large jump by analogy and plausibility, but it can be justified. (If one expands the potential energy expression of the entire solid in a Taylor's series in terms of individual displacements from equilibrium sites, stops with the quadratic terms, and performs a normal mode analysis—that is, looks for the natural frequencies of vibration of the system—one finds that the quantum-mechanical energy eigenvalues of the *entire solid* are precisely those of a *set* of harmonic oscillators with the natural frequencies.) Proceeding on faith if necessary, let us represent the possible energy eigenvalues of the solid as those of a set of one-dimensional harmonic oscillators with a variety of frequencies. This is the first of two important steps.

(a) Show that the partition function of a one-dimensional harmonic oscillator of frequency ν_1, treated quantum-mechanically, is $[1 - \exp(-\beta h\nu_1)]^{-1}$, provided we arrange the energy scale so that the ground state has zero energy (a matter of convenience having no effect on the physics). Reference to Section 6.6 may be helpful.

(b) Show that if we have a system of two oscillators, of frequencies ν_1 and ν_2, so that the system has energy eigenvalues $(n_1 h\nu_1 + n_2 h\nu_2)$, with n_1 and n_2 being nonnegative integers, then the partition function is just the *product* of two functions like that in part (a).

We can generalize from part (b) and say that the partition function for $3N$ oscillators will be a product of $3N$ such functions. The logarithm of the total partition function will be a sum of $3N$ individual logarithmic terms.

Now we must ask for the frequencies of the oscillators used in this representation of the solid. Sound waves of long wavelength imply low frequency, and so we can go very close to the limit of zero frequency. Arbitrarily short wavelengths are not present, however, because the lattice spacing itself sets a natural limit to the shortness of a wavelength. There is, consequently, a finite upper limit to the frequencies. At this point we take the second of the two important steps. We follow Debye and count the number of oscillators that we should adopt for a frequency range ν to $\nu + d\nu$ by counting the number of possible standing sound waves with frequency in that range. We determine the maximum frequency by requiring that the total number of oscillators adopted equal $3N$, the same as the number in the Einstein model.

(c) Present an argument indicating why we may take over for the Debye model the counting result from Section 10.6, and write $(3/2)(8\pi V\nu^2/c_s{}^3)\, d\nu$ for the number of oscillators to be adopted for the range ν to $\nu + d\nu$. The initial factor of $3/2$ takes into account the three possible polarizations of a sound wave, two transverse and one longitudinal; the latter is absent in electromagnetic waves.

d. Determine the upper limit ν_D of the frequencies to be considered, expressing it in terms of $(N/V)^{1/3}$ and c_s. This entails integrating the expression of part (c) from $\nu = 0$ to $\nu = \nu_D$, setting that equal to $3N$, and then solving for ν_D. Does the magnitude of the cutoff wavelength agree with the previous comment?

(e) Show that the logarithm of the partition function in the Debye model may be written as

$$\ln Z_D = 9N \left(\frac{T}{\Theta_D}\right)^3 \int_0^{\Theta_D/T} \ln\left[\frac{1}{1 - e^{-x}}\right] x^2 \, dx$$

with the use of the dimensionless variable $x = h\nu/kT$. The characteristic Debye temperature Θ_D is an abbreviation for $h\nu_D/k$.

(f) Estimate a typical value for Θ_D on the basis of the lattice spacing and the speed of sound in a solid.

(g) To gain some confidence in this scheme, let us take the case of a temperature T much larger than Θ_D. Then the variable x is always much less than unity in the integral, and $1 - e^{-x}$ may be approximated as x:

$$\int_0^{\Theta_D/T} \ln(1 - e^{-x})^{-1} x^2 \, dx \simeq \int_0^{\Theta_D/T} (-\ln x) \, x^2 \, dx.$$

Use a single integration by parts to evaluate the integral. Then compute the expectation value of the energy, and find it to be equal to $3NkT$, precisely the result that the classical equipartition theorem would give directly.

(h) Low temperature is of much more interest. When $T \ll \Theta_D$ holds, the upper limit of integration may be replaced by $+\infty$:

$$\int_0^{\Theta_D/T} \ln(1 - e^{-x})^{-1} x^2 \, dx \simeq \int_0^{\infty} \ln(1 - e^{-x})^{-1} x^2 \, dx = \frac{\pi^4}{45}.$$

Calculate the expectation value of the energy, the heat capacity at constant volume, and the information-theory entropy in this low-temperature domain. The temperature dependence of the heat capacity is the famous "Debye T^3 law." It is quite well-supported by experiment.

10.9. Calculate the temperature dependence of the heat capacity (at constant volume) of a Bose-Einstein perfect gas at temperatures below the condensation temperature. Measurements on liquid helium below the λ point yield a T^3 dependence for the heat capacity (when the liquid remains in equilibrium with its vapor). The observed cubic dependence is like that in a solid, where the energy is (roughly) in the form of sound waves. Is there a favorable comparison? If not, what might be the reason for the disagreement?

10.10. A hypothetical system contains $N = 10^{20}$ fermions with negligible mutual interactions. The single-particle energy eigenstates are nondegenerate and have energies given by $\varepsilon_s = s\tilde{\varepsilon}$, with $\tilde{\varepsilon} = 10^{-31}$ ergs. The system is in equilibrium at a temperature $T = 300$ K. Is a nearly classical approximation adequate for calculating the energy? Or must one use the more elaborate "low-temperature" techniques? Calculate (approximately) the expectation value of the energy and then the heat capacity at the given temperature, expressing your answer in terms of N, $\tilde{\varepsilon}$, k, and T.

SOME CONSIDERATIONS ON TIME DEPENDENCE

Up to this point we have been estimating the properties of systems known to be in equilibrium in an unchanging environment. Now we turn to a limited consideration of time-dependent problems, specifically, how a change in an external parameter affects a system initially in equilibrium. In the end we will find that the canonical probability distribution again plays the dominant role.

Theoretical justification for that role is difficult to obtain. Therein lies not only a challenge but also scope for differences of opinion. The reader deserves to be forewarned that the approach of this chapter expresses only one consistent view of the matter. References for alternative views are given at the end of the chapter. Lest there be misunderstanding: there is general agreement that the canonical distribution may, and should, be used as in this chapter; the disagreements are on how to justify that use.

11.1. THE EFFECTS OF A CHANGE
IN AN EXTERNAL PARAMETER

As a prelude, we should review several features of our description of a system that is known to be in equilibrium in an unchanging environment and at a specified temperature. We noted that the given information was insufficient for selecting a single quantum-mechanical state to describe the system. Indeed, the paucity of information forced us to consider a set of states and to assign to each state a probability that, given our meager information, the state provided the appropriate description of the system. Because the system was known to be in equilibrium, we chose the energy eigenstates; only for them will predictions from quantum mechanics be independent of time. A good bit of labor in Chapter 4 led to the probability distribution that, following Gibbs, we called the canonical probability distribution. All our estimates of physical quantities have been based on that probability distribution.

To agree with convention, we made a specific choice of the arbitrary positive constant K in the measure of missing information. We chose K to be equal to Boltzmann's constant k and called the missing information, when $K = k$, the information-theory entropy S_I:

$$S_I \equiv MI \mid_{K=k} = -k \sum_j P(A_j \mid B) \ln P(A_j \mid B).$$

The expression for S_I has been written with the more general probability $P(A_j \mid B)$ to emphasize that the information-theory entropy is defined for any set of probabilities referring to the description of a physical system, subject to one restriction: the set of inferences A_j, each inference being of the form, "The system is appropriately described by the quantum-mechanical state labeled by the letter j," must form a set of inferences that are exhaustive and mutually exclusive on the basis of quantum-mechanics and the knowledge about the system. The states referred to here need not be energy eigenstates.

So much for our statistical description when the system is known to be in equilibrium. The object of this section is to investigate the changes induced in that description when the system's environment is altered by a change in one of the external parameters. For the sake of illustration and of definiteness, let us begin with the homely situation portrayed in figure 11.1.1. A great many molecules are confined in a cylinder with a moveable piston. The gas forms the physical system of interest. Except for our ability to move the piston in any way we may wish to prescribe, the gas plus cylinder and piston are isolated. For some period of time up to and including the time t_1, the system

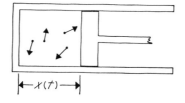

FIGURE 11.1.1
A sketch of the homely illustration.

has been in equilibrium with the piston held fixed at the distance $x(t_1)$ from the inside end of the cylinder. We are given this information, the number of molecules present, the behavior of the intermolecular forces, and so on, as well as the measured temperature. So at the time t_1 we would certainly describe the system statistically with the canonical probability distribution. Now comes the new element: between the time t_1 and a later time t_2 the piston moves in a manner known to us. Either we allow the gas to expand and push the piston or we, as an outside agent, move the piston in some prescribed fashion. The piston comes to rest at the time t_2 at some distance $x(t_2)$ from the end of the cylinder. The value of $x(t_2)$ may be larger than $x(t_1)$, smaller, or the same. The essential element is that the piston has moved in a manner known to us between times t_1 and t_2, and then remains at rest indefinitely.

Our problem is to describe the system statistically, both while the piston is moving and thereafter. If the piston is pulled out slowly, one is likely to feel that the system remains "close to equilibrium"—the notion is more intuitive than precise—and that something like a canonical probability distribution ought to be applicable throughout the period of piston motion and thereafter. If, however, the piston is pulled out rapidly, the gas will respond in a turbulent manner foreign to equilibrium behavior. To continue using a distribution of the canonical type would seem inappropriate. But, however reasonable, these views are presently a matter of conjecture. We should really seek our answers in a combination of probability theory and quantum mechanics.

First we must note how the piston's motion affects the quantum-mechanical analysis of the system. The position of the piston, together with the cross-sectional area of the cylinder, determines the volume in which the gas is contained. This means that the position must enter into U_{box}, the potential energy associated with the forces exerted by the walls. Indeed, if we take the cross section to be rectangular (for ease of expression) with edge lengths L_y and L_z, then the potential energy for molecule i arising from the walls is

$$U_{\text{box}}(x_i, y_i, z_i) = \begin{cases} 0 & \text{if } 0 < x_i < x(t), 0 < y_i < L_y, 0 < z_i < L_z, \\ +\infty & \text{otherwise.} \end{cases}$$

The relevant point is that the motion of the piston leads to a *time-dependent* potential energy. This, in turn, means that the quantum-mechanical energy operator for the entire system of molecules becomes time-dependent.

At this point we can state the generality of the situation that we will analyze quantum-mechanically. We start with a system that is known to be in equilibrium for some period of time up to and including a time t_1. The system's energy operator is independent of time for that period. Then, between time t_1 and the later time t_2, one of the external parameters specifying the environment is changed by a finite amount (though possibly returned to the original value). Such a parameter might be the position of a piston or the magnitude of an external magnetic field or the direction of an external electric field; the possibilities are numerous. The manner in which the external parameter changes is known to us, either because we prescribe the change or because we directly observe it. Since the parameter appears in the system's energy operator, the operator will change with time in the interval t_1 to t_2. We do specify that after the time t_2 the parameter remains constant at the value it attained then.

Though we will use language appropriate to the gas and movable piston problem, we are really investigating a rather general situation; it is outlined in table 11.1.1. Given that general situation—an eminently realistic one—we

TABLE 11.1.1
An outline of the general situation under consideration.

Time	Changes in the system	Energy operator
Up to t_1	The system is known to be in equilibrium at a measured temperature and is described by the canonical probability distribution.	Time-independent; may be denoted by $\mathcal{K}(t_1)$.
$t_1 < t < t_2$	One of the external parameters is changed in a manner known to us.	A function of time denoted by $\mathcal{K}(t)$.
$t_2 \leqslant t$	The external parameter remains fixed at the value it attained at time t_2. This value may or may not equal the value at time t_1.	Again time-independent; may be denoted by $\mathcal{K}(t_2)$. Only if the parameter returns to its initial value will $\mathcal{K}(t_2)$ equal $\mathcal{K}(t_1)$.

would like answers to a few pertinent questions. What happens to the quantum-mechanical states used to describe the system? What happens to the statistical description with probabilities? What happens to the information-theory entropy? We will answer these questions in order.

At the time t_1 the quantum-mechanical states that we use are the system's energy eigenstates. Let us denote them by $\psi_j(t_1)$, with the time t_1 explicitly displayed. As energy eigenstates, they satisfy the equation

$$\mathcal{3C}(t_1)\psi_j(t_1) = E_j(x(t_1))\psi_j(t_1). \tag{11.1.1}$$

The energy eigenvalue has been written as $E_j(x(t_1))$ to show that it depends on the value of the external parameter, here the position $x(t_1)$ of the piston. After all, the piston's position does affect the size of the container and hence the wave functions and their associated energies.

Now for the development of those states with time. The behavior of a quantum-mechanical state $\psi(t)$ with time is determined by the time-dependent Schrödinger equation,

$$i\hbar \frac{\partial \psi(t)}{\partial t} = \mathcal{3C}(t)\psi(t), \tag{11.1.2}$$

with $\mathcal{3C}(t)$ being the energy operator. Because the Schrödinger equation contains only a first derivative with respect to time, a state $\psi(t)$ is determined for all time if one knows $\mathcal{3C}(t)$ and knows the state at some initial time. This is the quantum-mechanical analog of the classical conclusion that if one knows the force laws for all future time and knows the positions and momenta of all particles at one instant of time, then one knows (in principle) the entire future of the classical system.

Perhaps the easiest way to see the quantum-mechanical conclusion is by examining a Taylor's series expansion for $\psi(t)$ as a function of time. Let us call the initial time the time t_1 and specify that we know $\mathcal{3C}(t)$ for all $t \geqslant t_1$. The formal Taylor's series might be written as

$$\psi(t) = \sum_{n=0}^{\infty} \frac{a_n}{n!} (t - t_1)^n.$$

The coefficients a_n are independent of time but will depend on the position variables of the particles. We do not bother to write those explicitly in either the coefficients or in $\psi(t)$. The central question is this: what information do we need in order to determine the coefficients?

When $t = t_1$, only the coefficient a_0 survives in the sum. So we can see that a_0 is determined by $\psi(t_1)$,

$$a_0 = \psi(t_1),$$

and that we must be given $\psi(t_1)$. The next coefficient, a_1, is the partial time derivative of $\psi(t)$, evaluated at t_1:

$$\left. \frac{\partial \psi(t)}{\partial t} \right|_{t=t_1} = \left. \sum_{n=0}^{\infty} \frac{a_n}{n!} n(t - t_1)^{n-1} \right|_{t=t_1} = a_1.$$

But the Schrödinger equation, evaluated at $t = t_1$, enables us to compute a_1 in terms of $\psi(t_1)$ and $\mathcal{H}(t_1)$:

$$i\hbar \left. \frac{\partial \psi(t)}{\partial t} \right|_{t=t_1} = \mathcal{H}(t)\psi(t) \Big|_{t=t_1}$$

implies

$$a_1 = \frac{1}{i\hbar} \mathcal{H}(t_1)\psi(t_1).$$

The coefficient a_2 is the second partial time derivative of $\psi(t)$, again evaluated at the initial time. If we differentiate the Schrödinger equation with respect to time and then set t equal to t_1, we find

$$i\hbar \left. \frac{\partial^2 \psi(t)}{\partial t^2} \right|_{t=t_1} = \left[\mathcal{H}(t) \frac{\partial \psi(t)}{\partial t} + \frac{\partial \mathcal{H}(t)}{\partial t} \psi(t) \right]_{t=t_1}.$$

In the first term on the righthand side, we may express the derivative in terms of $\psi(t_1)$ and $\mathcal{H}(t_1)$ with our previous result. Thus the coefficient a_2 is also determined by a knowledge of $\psi(t_1)$ and $\mathcal{H}(t)$:

$$a_2 = \frac{1}{i\hbar} \left[\mathcal{H}(t_1) \left(\frac{1}{i\hbar} \mathcal{H}(t_1)\psi(t_1) \right) + \frac{\partial \mathcal{H}(t)}{\partial t} \bigg|_{t=t_1} \psi(t_1) \right].$$

By successively differentiating the Schrödinger equation with respect to time and then evaluating at the initial time, one can express all the coefficients in terms of $\psi(t_1)$ and quantities arising from $\mathcal{H}(t)$. This justifies the claim that $\psi(t)$ is determined for all future times if one knows $\psi(t)$ at some initial time and knows $\mathcal{H}(t)$.

The physical situation that we envisage provides us with precisely the kind of knowledge assumed in the preceding paragraphs. We know the energy operator $\mathcal{H}(t)$ as a function of time—we specified that—and we know the states of interest at time t_1: they are the states $\psi_j(t_1)$. So, although the computations may be hopeless in practice, we do in principle know, for each $\psi_j(t_1)$, the state into which it evolves with time. Let us express this symbolically as

$$\psi_j(t_1) \xrightarrow[\substack{\text{deterministically} \\ \text{by the Schrödinger} \\ \text{equation}}]{} \psi_j(t) \quad \text{for} \quad t \geqslant t_1.$$

At $t = t_1$ the state $\psi_j(t)$ is merely $\psi_j(t_1)$, the state from which we start, but we know, in principle, the evolving state at all times greater than t_1 as well.

Much of our interest will be in $\psi_j(t)$ at time t_2 or later, when the energy operator has ceased to change because the piston has ceased to move. Here we should note an essential point: we have no guarantee that the state $\psi_j(t_2)$ will be an energy eigenstate. That state is completely determined by $\psi_j(t_1)$ and the Schrödinger equation. Whether $\psi_j(t_2)$ satisfies an equation like

$$\mathcal{H}(t_2)\psi_j(t_2) = (\text{constant}) \times \psi_j(t_2),$$

with the constant being an energy eigenvalue, is an open question. Later we will learn that the answer depends on whether the external parameter changed slowly or rapidly.

Now we turn to the matter of probabilities and the statistical description of the system. At the time t_1 we deal with probabilities that individual states $\psi_j(t_1)$ provide the appropriate description. Suppose, for example, that the canonical distribution ascribes a probability of 0.2 to the state $\psi_1(t_1)$ at the time t_1. That is, the number of 0.2 expresses numerically the degree of our rational belief that the system, at time t_1, is appropriately described by the state $\psi_1(t_1)$. If the system truly is appropriately described by $\psi_1(t_1)$ at that initial time, then at a later time it will truly be appropriately described by $\psi_1(t)$, the state into which $\psi_1(t_1)$ develops with time. Consistency requires that we ascribe a probability of 0.2 to the inference that, at time t, the state $\psi_1(t)$ appropriately describes the system. *The numerical values of the probabilities remain unchanged and attached to the states as the latter evolve in time.*

Let us express this seemingly simple yet profound conclusion in a symbolic fashion. We denote by $A_j(t)$ the inference that the system, at time t, is appropriately described by the state $\psi_j(t)$. The hypothesis B contains the knowledge summarized in table 11.1.1 and in equations (11.1.1, 2). Thus the hypothesis amounts to our knowledge about the system for the period up to and including time t_1, the applicability of the canonical probability distribution *at time t_1*, the manner in which the external parameter is changed, and some purely quantum-mechanical principles. The probability that the inference $A_j(t)$ is correct is denoted by $P(A_j(t)\,|\,B)$. At the time t_1 we have, with the canonical distribution,

$$P(A_j(t_1)\,|\,B) = \frac{e^{-\beta E_j(x(t_1))}}{\sum_i e^{-\beta E_i(x(t_1))}}. \tag{11.1.3}$$

The conclusion of the preceding paragraph may now be expressed as the numerical relation

$$P(A_j(t)\,|\,B) = P(A_j(t_1)\,|\,B)t \qquad \text{when} \geqslant t_1 \tag{11.1.4}$$

for each j. As the state $\psi_j(t)$ evolves with time according to the Schrödinger equation, the numerical value of the probability that we associate with the state remains unchanged. The justification for this is nothing more—*or less*—than consistency.

A simple classical analog may help to clarify the logic of the preceding argument. Suppose we know that a ball is rolling along a smooth table with a velocity v_0 to the right. Suppose further that at time t_1 we assessed the probability of the ball's being between coordinates 4 and 5 as 0.2. The future position of the ball is deterministically governed by Newton's laws. There are no horizontal forces, and so the ball moves with constant velocity to the right. If the ball truly is between coordinates 4 and 5 at time t_1, then the ball will be between coordinates $4 + v_0(t - t_1)$ and $5 + v_0(t - t_1)$ at time t. Consistency demands that we assign the same probability (0.2) to the latter interval at time t as we assigned to the former interval at the earlier time.

Now we examine the effect on the information-theory entropy. The formal expression is

$$S_I(t) = -k \sum_j P(A_j(t)\,|\,B) \ln P(A_j(t)\,|\,B). \qquad (11.1.5)$$

The relevant question is this: how does the numerical value of $S_I(t)$ change with time? The answer is that it does not change at all:

$$S_I(t) = S_I(t_1). \qquad (11.1.6)$$

A glance back at equation (11.1.4) will provide the proof.

Is the result surprising? Perhaps, but a little reflection indicates that we should have anticipated it. The piston has moved and the gas has responded. We have, however, specified that we know the manner in which the piston moved. This, together with the Schrödinger equation, permits us to calculate the response of the gas. We are able to follow the gas in time with no more, and no less, uncertainty about the "correct" quantum-mechanical state for the gas than existed at time t_1. We neither lose nor gain information.

Let us note here that neither the numerical equality of probabilities, expressed in equation (11.1.4), nor the numerical constancy of $S_I(t)$ in time depends on the rate of change with time of the external parameter. These relations hold both for slow and for rapid changes, for nowhere have we had to say anything about the rate. It is true, however, that the rate has a large influence on the states themselves and hence on the response of the system. We will find it advantageous to split the subsequent analysis into two parts: one dealing with a very slow change in the external parameter, the other with a rapid change.

11.2.   SLOW CHANGE IN AN EXTERNAL PARAMETER

If the piston in our example is pulled out very slowly, the gas will remain, in some sense, close to equilibrium. Certainly no wild, turbulent motions will appear in the gas. It is plausible, therefore, that a probability distribution rather like the canonical distribution will be appropriate both while the piston is moving and thereafter. One may be called on to adjust the temperature, for the gas will cool during the expansion (since it does work on the moving piston and thus loses energy to the external world), but that would be acceptable.

This expectation is largely borne out by a quantum-mechanical analysis. Already we know the states $\psi_j(t)$, at least implicitly in terms of $\psi_j(t_1)$ and $\mathcal{H}(t)$. Furthermore, we know each inference $A_j(t)$ that we should consider, and we know the numerical value of the probability $P(A_j(t)\,|\,B)$. There remain, however, two crucial questions that must be answered.

1. Are the states $\psi_j(t)$ energy eigenstates of the energy operator $\mathcal{H}(t)$ for $t > t_1$?

2. If the first question is answered affirmatively, is the probability distribution $P(A_j(t)\,|\,B)$ of the *form* of the canonical distribution? More precisely, does the probability distribution have the familiar simple exponential dependence on the energy eigenvalue?

Only if both questions can be answered "yes" will the probability distribution $P(A_j(t)\,|\,B)$ emerge as a canonical distribution with a temperature changing with time.

The best answer available for the first question seems to be the answer provided by Ehrenfest's principle. That quantum-mechanical theorem may be stated in the following fashion.

EHRENFEST'S PRINCIPLE:

If the energy eigenstates of $\mathcal{H}(t)$ are nondegenerate for times $t \geqslant t_1$, if $\psi_j(t_1)$ is an energy eigenstate of $\mathcal{H}(t_1)$, if $\psi_j(t)$ is the state evolved from $\psi_j(t_1)$ according to the Schrödinger equation, and if the external parameter changes very slowly, then $\psi_j(t)$, for each time $t > t_1$, is very nearly an energy eigenstate of $\mathcal{H}(t)$ at the corresponding time. In the mathematical limit of a finite change in the external parameter occurring over an infinite time interval, "is very nearly" becomes "is."

Some explanation of this theorem is certainly in order. A proof, however, would take us well beyond the bounds set for this book; references for a proof are given at the end of this chapter.

The stipulation that the energy eigenstates of $\mathcal{K}(t)$ be nondegenerate for times $t \geqslant t_1$ is a technical detail required in the proof. Though many of the idealized systems with which we have dealt do not meet this condition, there is good reason to believe that the real systems of the laboratory do have nondegenerate energy eigenstates and hence that the theorem is applicable. It appears to be the *neglect* of small interactions among the constituents of a real system that leads to degeneracy in the idealizations. (Moreover, the theorem can tolerate a limited amount of temporary degeneracy without having the proof collapse.)

The three "if"'s following the nondegeneracy "if" merely repeat the conditions that we have been discussing throughout this section. The import of the theorem is that, for a very slow change of the external parameter, we make no significant error if we take $\psi_j(t)$ to be an energy eigenstate of $\mathcal{K}(t)$. The system will smoothly readjust, provided the external parameter changes slowly enough.

There is, of course, the question, what is "slowly enough"? In our example a reasonably safe guess would be that the time for the piston's motion to increase the volume by a few per cent must be very long relative to the time it takes a typical molecule to move from piston to opposite wall and back. Probably a time that is long relative to the time it would take a sound wave to traverse the volume would suffice. Such intuitive estimates are generally adequate and are far easier to make than estimates based on the details of the proof of Ehrenfest's principle

The reply to the first question is, to all intents and purposes, affirmative. For realistic systems, Ehrenfest's principle assures us that $\psi_j(t)$ is very nearly an energy eigenstate of $\mathcal{K}(t)$, provided the external parameter changes slowly. The net change in the value of the parameter is allowed to be finite, even large —otherwise the principle would be irrelevant—but the alteration must occur slowly. Then the set of inferences $A_j(t)$ does, effectively, refer to energy eigenstates, as does the set of probabilities $P(A_j(t)\,|\,B)$. But this is not yet a demonstration that the probability distribution is like a canonical distribution. We must address ourselves to the second question.

It is both a blessing and a curse that the Schrödinger equation determines the probability distribution at later times in terms of the distribution at time t_1. We cannot impose a canonical form on $P(A_j(t)\,|\,B)$. Whether it does or does not have such a form is already determined. The numerical equality in equation (11.1.4) and the property that the distribution at time t_1 is a canonical distribution determine the numerical value of $P(A_j(t)\,|\,B)$:

$$P(A_j(t)\,|\,B) = \frac{1}{Z(t_1)} \exp\left[-\frac{1}{kT(t_1)} E_j(x(t_1))\right]. \qquad (11.2.1)$$

Here $Z(t_1)$ denotes the normalization factor—the partition function—at time t_1, and $T(t_1)$ is the known temperature of the equilibrium situation at time t_1.

We would like to know whether $P(A_j(t)\,|\,B)$ has the *form* of a canonical distribution:

$$P(A_j(t)\,|\,B) \overset{?}{=} \frac{1}{Z(t)} \exp\left[-\frac{1}{kT(t)} E_j(x(t)) \right].$$

The energy eigenvalue $E_j(x(t))$ associated with $\psi_j(t)$ will generally be unequal to the value $E_j(x(t_1))$ appearing in the exponent in equation (11.2.1). Only if a proportionality,

$$E_j(x(t)) = f(t) \times E_j(x(t_1)) \tag{11.2.2}$$

holds for all j, with $f(t)$ a time-dependent proportionality factor, will $P(A_j(t)\,|\,B)$ have the canonical form. Then one may write the exponential in equation (11.2.1) as

$$\exp\left[-\frac{1}{kT(t_1)} E_j(x(t_1)) \right] = \exp\left[-\frac{1}{kf(t)T(t_1)} E_j(x(t)) \right].$$

This makes evident that $P(A_j(t)\,|\,B)$ will then be a canonical distribution with a temperature varying slowly with time.

The question of canonical form reduces now to the question of whether the proportionality proposed in equation (11.2.2) does actually hold. There is, unfortunately, no general theorem to which one may appeal. For some idealized systems, the proportionality does hold, but not for others. For a realistic gas with some interaction between the molecules, one may show that the proportionality holds at least approximately.* We have come so far that it is disappointing to admit even a limited defeat. Yet, to the best of my knowledge, there is no theoretical proof that the proportionality holds, even approximately, for all realistic physical systems. There is, however, no disproof either.

Where do we go from here? In individual cases one may be able to demonstrate that the proportionality holds, at least approximately, but a comprehensive treatment is what we desire. To be sure, there is no absolute need for further assumptions. The numerical value of the probability $P(A_j(t)\,|\,B)$ really is determined by the information at time t_1, as equation (11.1.4) firmly asserts. We are, in principle, in a position to calculate expectation values,

---

* Amnon Katz, pp. 84–90 of the book cited in the references for Chapter 4.

anticipated deviations, and the like, at times $t \geqslant t_1$. But the computations would be far more difficult than those that may be done with the canonical distribution and the associated partition-function technique. Indeed, they would be prohibitively difficult.

In the present " state of the art " of statistical mechanics, we are condemned to taking an utterly pragmatic approach. The bald assumption that the probabilities $P(A_j(t)|B)$ do have the canonical form leads to statistical estimates that are in excellent agreement with observation. So let us take the following stand.

PRAGMATIC APPROACH :

When an external parameter $x(t)$ of a physical system (in equilibrium at the initial time t_1) is *slowly* changed, we will frankly assume that the probability $P(A_j(t)|B)$ has the canonical form,

$$P(A_j(t)|B) = \frac{1}{Z(t)} \exp\left[-\frac{1}{kT(t)} E_j(x(t))\right],$$

and we will determine the temperature $T(t)$ from the proven constancy of the information-theory entropy of the evolving probability distribution,

$$S_I(t) = S_I(t_1).$$

This is a nakedly empirical treatment. Yet it is eminently reasonable, for common sense tells us that if the change in external parameter is very slow, the system will remain close to equilibrium, and so the probability distribution that we use for equilibrium itself should be a good approximation.* Since the canonical distribution depends on the temperature, we need some means of determining the change in temperature accompanying the change in external parameter. The constancy of the information-theory entropy, which we proved for the probability distribution that evolves rigorously from the initial distribution, provides a consistent (and convenient) method. This we will see at first hand in the applications, where we will note that this method is equivalent to using energy conservation to choose the new temperature. Once the temperature has been determined, the statistical situation is reduced to manipulations with a fully determined canonical distribution.

---

* That such is the case in a purely classical analysis was demonstrated—under plausible assumptions—by L. Rosenfeld, *Proceedings, Nederlandse Akademie van Wetenschappen*, **45**, 970 (1942).

Here is a way to visualize the content of this formidable section. The lefthand drawing in figure 11.2.1 represents the statistical situation at the initial time. To each state $\psi_j(t_1)$ corresponds a dot at the end of a vertical line. The height of the dot is equal to the probability $P(A_j(t_1) \mid B)$, and the horizontal position is determined by the energy $E_j(x(t_1))$ of the state. There is no reason for the differences in energy between neighboring states to be equal, and so the vertical lines are not uniformly spaced. Since the initial probability distribution is a canonical distribution, the dots do, however, fall along an exponential curve.

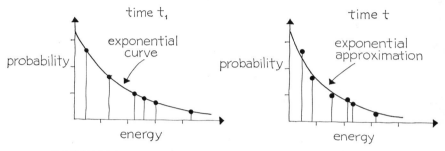

FIGURE 11.2.1
The two statistical situations. Only six of the many states are depicted.

The righthand drawing represents the statistical situation at a later time t. The external parameter has slowly changed in the interval t_1 to t, but, by Ehrenfest's principle, each state $\psi_j(t)$ is still an energy eigenstate, and so we may represent it by a unique dot on the graph. The energy $E_j(x(t))$ will generally differ from the initial energy, and so the dot and vertical line will be shifted either to the right, for higher energy, or to the left. By equation (11.2.1) the probability $P(A_j(t) \mid B)$ is numerically equal to the initial probability, and so the height of each dot remains constant as the dot shifts to the right or left. Provided there is never any degeneracy, no two vertical lines will ever cross one another.

The second question at the beginning of this section asks, in effect, whether the dots at time t fall along an exponential curve. Unfortunately, we can't give a general answer to that question. What we can do is conjecture that the pattern of dots will not differ much from an exponential curve. If this is indeed so, then we can adequately approximate the actual probabilities by a canonical probability distribution, provided we choose the new temperature so that the exponential gives as good a fit as possible to the actual pattern of dots.

The numerical value of S_I depends on the heights of the dots—on the probabilities—but not on the horizontal spacing. At any specific energy, the height of the exponential approximation depends on the temperature chosen. If we choose the temperature by demanding that S_I for the canonical approximation be equal to S_I for the exact dots, then we have reasonable assurance that the exponential curve will be placed at a good height.

One can say a bit more about the approximation. When one is estimating the system's macroscopic properties, only states with energy near $\langle E \rangle$ should be important. This conjecture is certainly plausible; if valid, then the smallness of ΔE relative to $\langle E \rangle$ suggests that the region "near $\langle E \rangle$" will be very small. Provided there is never degeneracy, and hence no crossing of one line by another while the probability distribution evolves, the heights of the dots will decrease monotonically as one goes to states with higher energy. Thus it is also plausible that a decaying exponential—the canonical form—plus the "right" choice of temperature will provide a good approximation to the actual probability distribution in the crucial region near $\langle E \rangle$, even if not everywhere.

11.3. APPLICATIONS WHEN THE CHANGE IS SLOW

In this section we commence applying the canonical distribution to situations where an external parameter changes slowly. Certainly we should finish analyzing a gas in a container of variable size, and so that is first on the agenda. Then we will deal, by a somewhat different technique, with black-body radiation in a similar container. The application to magnetic cooling warrants a separate section, which follows this one.

The Expansion of a Monatomic Gas

In order to apply the canonical distribution to the slow expansion of a gas, we need to relate the temperature to the volume. To do so, we use the constancy of the information-theory entropy. The form associated with the canonical distribution,

$$S_I(t) = k\left(\beta\langle E \rangle + \ln Z\right),\qquad(11.3.1)$$

continues to hold, though we must remember that the quantities on the righthand side are now functions of the time t. When we analyze black-body radiation, we will work directly with S_I. Here the most convenient approach uses the constancy of S_I to derive a differential equation relating temperature and volume. Let us differentiate both sides with respect to time:

$$\frac{dS_I(t)}{dt} = k\left(\frac{d\beta}{dt}\langle E \rangle + \beta\frac{d\langle E \rangle}{dt} + \frac{\partial \ln Z}{\partial \beta}\frac{d\beta}{dt} + \frac{\partial \ln Z}{\partial V}\frac{dV}{dt}\right).$$

The partition function depends explicitly on β and implicitly on the volume V (because the energy eigenvalues depend on the volume), and so there are two contributions from the change of ln Z with time. Nonetheless, the righthand side may be greatly simplified. The first and third terms cancel exactly, since the partial derivative of ln Z with respect to β yields $-\langle E \rangle$. The partial derivative in the fourth term is simply β times the estimated pressure. Because the information-theory entropy is constant in time, the derivative on the lefthand side is zero, and so we arrive at the tidy result,

$$0 = k\beta\left(\frac{d\langle E \rangle}{dt} + \langle p \rangle \frac{dV}{dt}\right). \tag{11.3.2}$$

The next steps hinge on the explicit forms of $\langle E \rangle$ and $\langle p \rangle$. Just to see how the calculations go, let us take first a perfect classical monatomic gas. Then we have, for the first term,

$$\frac{d\langle E \rangle}{dt} = \frac{d}{dt}(\tfrac{3}{2}NkT) = \tfrac{3}{2}Nk\,\frac{dT}{dt}.$$

Upon inserting this and the value of $\langle p \rangle$, we find a relatively simple differential equation:

$$0 = k\beta\left(\tfrac{3}{2}Nk\,\frac{dT}{dt} + \frac{NkT}{V}\frac{dV}{dt}\right).$$

To prepare this for integration, we rearrange it as

$$\frac{1}{T}\frac{dT}{dt} = -\frac{2}{3}\frac{1}{V}\frac{dV}{dt}.$$

Integration with respect to time from time t_1 will give logarithms:

$$\ln\frac{T(t)}{T(t_1)} = -\tfrac{2}{3}\ln\frac{V(t)}{V(t_1)}.$$

Solved for $T(t)$, this yields

$$T(t) = T(t_1)\left(\frac{V(t_1)}{V(t)}\right)^{2/3}$$

for a perfect classical monatomic gas.

If the volume increases slowly between times t_1 and t, the righthand side decreases, implying a reduction in the temperature. This is the anticipated behavior, for the gas does work against the piston during the expansion. Thus it loses energy to the external world, and we expect it to cool.

Now let us be more sophisticated and include, to some approximation, the interactions between the atoms of the monatomic gas. The logarithm of the partition function will provide us with the expectation values of energy and pressure that we need. If we are satisfied with the first-order corrections produced by the interatomic forces, we may take over results from Section 8.5. From equation (8.5.7), we have

$$\ln Z = \ln\left[\frac{1}{N!}\left(\frac{2\pi m}{h^2 \beta}\right)^{\frac{3}{2}N}\right] + N \ln V - \frac{N^2}{V} B_2(T).$$

The contribution from the interatomic forces has been rewritten tidily as a multiple of the second virial coefficient. From Section 9.7 we know that a factor of $1/N!$ should appear in the argument of the first logarithm to give the "classical" compensation for the indistinguishability of the atoms; that has been inserted. As we shall see, the presence or absence of that factor has no effect on the differential equation we derive.

The pressure follows from $\ln Z$ as

$$\langle p \rangle = \frac{1}{\beta}\frac{\partial \ln Z}{\partial V} = \frac{1}{\beta}\left(\frac{N}{V} + \frac{N^2}{V^2} B_2(T)\right).$$

A similar differentiation gives the estimated energy:

$$\langle E \rangle = -\frac{\partial \ln Z}{\partial \beta} = -\left(-\tfrac{3}{2}N\frac{1}{\beta} - \frac{N^2}{V}\frac{dB_2(T)}{d\beta}\right)$$

$$= \tfrac{3}{2}NkT - \frac{N^2 kT^2}{V}\frac{dB_2(T)}{dT}.$$

The interatomic forces make $\langle E \rangle$ dependent on the particle density N/V as well as on the temperature.

To relate the temperature to the volume, we return to equation (11.3.2), the equation that we derived from the constancy of S_I. Since the time t never appears explicitly in that equation, we may work with a differential form:

$$0 = d\langle E \rangle + \langle p \rangle \, dV. \qquad (11.3.3)$$

In loose mathematical language, one arrives at this by multiplying through by dt in equation (11.3.2). The differential of $\langle E \rangle$ is the change associated with the differential of V. This minor alteration of procedure is introduced not because the differential form is any better—if anything, it is less clear in meaning—but because it is the conventional form.

We should, however, pause here and note that equation (11.3.3) expresses the conservation of energy in the volume expansion: the change in the system's energy plus the work done on the external world sum to zero. This vindicates the claim of consistency for our method of choosing the new temperature. To choose the temperature in our canonical distribution by imposing constancy for S_I is equivalent to choosing it by imposing energy conservation. In different words: if we construct (by grouping states) a probability distribution specifically for the energy, the peak will occur at the location demanded by energy conservation.

In computing $d\langle E \rangle$ we must take into account the dependence on both temperature and volume; so we must write

$$\left[\left(\frac{\partial \langle E \rangle}{\partial T} \right)_V dT + \left(\frac{\partial \langle E \rangle}{\partial V} \right)_T dV \right] + \langle p \rangle \, dV = 0. \qquad (11.3.4)$$

The subscripts on the parentheses indicate the variable that is held fixed during the partial differentiation. This equation will provide us with the desired relation between temperature and volume for a slow change in the volume.

If we took a general form for the second virial coefficient, we would soon be lost in an algebraic morass. Let us take merely the leading terms that arise from the repulsive and attractive forces. Upon referring to equation (8.6.2) and using the a, b, terminology of the van der Waals equation, we take $B_2(T)$ to have the form

$$B_2(T) = b - \frac{a}{kT}.$$

The term in b represents a hard-core repulsive force, and the term in a is the leading term arising from the attractive forces in the Sutherland model.

With this form for the second virial coefficient, the estimated energy becomes

$$\langle E \rangle = \tfrac{3}{2} NkT - \frac{N^2 kT^2}{V} \frac{d}{dT} \left(b - \frac{a}{kT} \right)$$

$$= \tfrac{3}{2} NkT - N \left(\frac{aN}{V} \right).$$

Because the repulsive part of the Sutherland model rises infinitely steeply, the repulsive forces make no contribution to this low order of approximation. To the same order, the contribution of the attractive forces depends only on the density N/V and the total number N of atoms. Quite reasonably, the higher the density, the larger the effect, for then the atoms feel the short-ranged attractive forces more frequently. Higher-order terms in the attractive part of $B_2(T)$ would produce temperature-dependent contributions to $\langle E \rangle$.

The two partial derivatives of $\langle E \rangle$ follow readily:

$$\left(\frac{\partial \langle E \rangle}{\partial T}\right)_V = \tfrac{3}{2}Nk,$$

$$\left(\frac{\partial \langle E \rangle}{\partial V}\right)_T = +\frac{aN^2}{V^2}.$$

Upon inserting these and the explicit expression for $\langle p \rangle$ into equation (11.3.4), we arrive at the equation

$$\left[\tfrac{3}{2}Nk\,dT + \frac{aN^2}{V^2}\,dV\right] + kT\left[\frac{N}{V} + \frac{N^2}{V^2}\left(b - \frac{a}{kT}\right)\right]dV = 0.$$

The terms in a cancel, a fortuitous occurrence arising from our low order of approximation. Rearrangement in preparation for integration leads to

$$\frac{1}{T}\,dT = -\frac{2}{3}\left(\frac{1}{V} + \frac{Nb}{V^2}\right)dV.$$

This equation is to be integrated term by term. On the righthand side the integration is from the initial volume V_1, corresponding to $V(t_1)$ in the previous notation, to V, corresponding to $V(t)$. The integration on the left goes from the initial temperature T_1, corresponding to $T(t_1)$, to a temperature T, the temperature associated with $V(t)$ and previously called $T(t)$. Fortunately, the integrations are all easy:

$$\ln \frac{T}{T_1} = -\tfrac{2}{3}\ln \frac{V}{V_1} + \tfrac{2}{3}Nb\left(\frac{1}{V} - \frac{1}{V_1}\right).$$

Solved for T, this yields

$$T = T_1 \left(\frac{V_1}{V}\right)^{2/3} \exp\left[\tfrac{2}{3}Nb\left(\frac{V_1 - V}{V_1 V}\right)\right]$$

for a simple imperfect gas.

When the volume is expanded, the temperature drops more rapidly than the calculation with neglect of interatomic forces predicted. The repulsive forces, whose presence is indicated by the positive coefficient b, increase the pressure over that of a gas without those forces at the same temperature and density. Consequently, the work done by the gas against the piston is greater when b is non-zero. This means more energy loss, and so a lower final temperature is natural.

Black-Body Radiation

The "gas" we now examine is the gas of photons comprising the electromagnetic radiation within an enclosure. The changes in the radiation attendant on a slow change in volume are the object of the investigation. Though the expressions that we will derive are analytically simple, there is a subtle point associated with the effects of the container walls. Both in equilibrium and during a slow change of volume, photons are continually absorbed and emitted by the walls. In principle, one should take "the system" to consist of both photon gas and walls. One may, however, stipulate or assume a negligible net exchange of energy between the photon gas and the walls due to emission and absorption of photons. (Some such net exchange is sure to occur, for the temperature of the walls will follow the temperature of the photon gas, and so it is negligible, not zero, net exchange that one may reasonably specify.) Then one may regard the photon gas as the system, though one will still permit it to do work on a movable piston and in this macroscopic manner exchange energy with the external world. There is a good measure of consistency here, for the expectation value of the number of photons present (of all frequencies and polarizations) has the same dependence on temperature and volume as does the information-theory entropy. If one regards the photon gas as an isolated system—except for a pressure interaction with the piston—one is guaranteed the constancy of S_I. That, in turn, implies a constancy of the estimated number of photons during an expansion. Were the photons to change significantly in number, one would justifiably have doubts about viewing the photon gas as an isolated system (again, except for the pressure interaction).

As before, we use the constancy of S_I during an expansion to relate temperature and volume. We could use that constancy to derive a differential equation and then integrate it; explicit expressions for $\langle E \rangle$ and $\langle p \rangle$ are already at hand in Section 10.6. More instructive, however, will be a direct calculation of S_I as a function of temperature and volume. From that we will be able to recognize at a glance the desired temperature-volume relation.

To compute S_I we will use the partition function for the photon gas, a gas of identical bosons with no restriction on the number of particles present.

Most of the hard work has already been done in Section 10.6. The denominator on the righthand side of equation (10.6.2) is precisely the partition function we want. Upon tracing that denominator along to equation (10.6.3), we find an intermediate result:

$$Z = \left(\sum_{n_1=0}^{\infty} e^{-\beta n_1 \epsilon_1} \right) \left(\sum_{n_2=0}^{\infty} e^{-\beta n_2 \epsilon_2} \right) \cdots .$$

Each factor is itself a geometric series; a typical factor was evaluated in the analysis following equation (10.6.4). The upshot is that the partition function is expressible as a product over all single-photon states:

$$Z = \prod_{s=1}^{\infty} (1 - e^{-\beta \epsilon_s})^{-1} .$$

For S_I we will want $\ln Z$ and $\langle E \rangle$. Since the last is derivable from $\ln Z$ by appropriate differentiation, we concentrate on bringing $\ln Z$ into convenient form. We may write the logarithm of Z as

$$\ln Z = \ln \prod_s (1 - e^{-\beta \epsilon_s})^{-1} = \sum_s \ln(1 - e^{-\beta \epsilon_s})^{-1} .$$

Provided the volume is macroscopic, we may replace the summation over single-photon states by an integration with the factor $\omega(v)$, the density of single-photon states:

$$\ln Z = \int_0^{\infty} \ln(1 - e^{-\beta h v})^{-1} \frac{8\pi V}{c^3} v^2 \, dv$$

$$= \frac{8\pi V}{c^3} (\beta h)^{-3} \int_0^{\infty} \ln(1 - e^{-x})^{-1} x^2 \, dx$$

$$\equiv \beta^{-3} \frac{8\pi V}{c^3 h^3} I .$$

The explicit form for $\omega(v)$ is that given in equation (10.6.10). In the step to the second line, the temperature dependence has been extracted from the integral by the variable change $x = \beta h v$. Thus the remaining integral is merely a numerical constant, denoted in the last line by I.

The estimated energy follows by partial differentiation of $\ln Z$:

$$\langle E \rangle = - \frac{\partial \ln Z}{\partial \beta} = +3\beta^{-4} \frac{8\pi V}{c^3 h^3} I .$$

To find S_I, we need only put together the pieces:

$$S_I = \left(k\beta \langle E \rangle + \ln Z \right) = \left(\frac{32\pi k^4 I}{c^3 h^3} \right) T^3 V .$$

If we wished to, we could determine the numerical value of the integral I by comparing the present expression for $\langle E \rangle$ with that given in Section 10.6, but doing so is really unnecessary. The salient point is that S_I for a photon gas in equilibrium is equal to a numerical constant times $T^3 V$. The information-theory entropy is constant in an expansion (or compression) of the volume. When the volume changes slowly, so that continued use of the canonical distribution is permissible, we may infer from S_I the connection between temperature and volume:

$$T^3 V = T_1{}^3 V_1,$$

which implies

$$T = T_1 \left(\frac{V_1}{V} \right)^{1/3}$$

for a photon gas. In an expansion the photon gas cools. Indeed, the relation is quite *similar* to that for a perfect gas.

11.4.   MAGNETIC COOLING

In 1926 Debye and Giauque independently proposed that one could cool a paramagnetic system, initially in equilibrium in a strong external magnetic field but otherwise isolated, merely by slowly reducing that external field to zero. The scheme does, in fact, work admirably, and for years provided an indispensable means of achieving very low temperatures, of the order of one thousandth of a kelvin. Let us first develop a qualitative microscopic understanding of why the cooling occurs, and then use our analytic machinery—primarily the constancy of S_I in such a process—to discuss magnetic cooling quantitatively.

Let us suppose that the paramagnetic system consists of atoms each having a magnetic moment μ parallel to a net angular momentum of $\frac{1}{2}\hbar$. We will refer to the latter as the spin. The atoms form a crystal, and that crystalline paramagnetic system is initially in equilibrium at some known temperature in a strong external magnetic field. The interaction of the paramagnets with the external field makes a spin orientation that is parallel to the field an energetically favorable orientation of low potential energy. If we were able to look in microscopic detail at the crystal at some instant of time, we would find most—though not all—of the magnetic moments lined up parallel to the field. This microscopic situation is illustrated in figure 11.4.1.

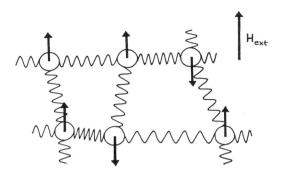

FIGURE 11.4.1

A sketch of the microscopic situation within the
paramagnetic crystal at one instant of time. Each
arrow represents a magnetic moment μ. The
majority are aligned along the external magnetic
field H_{ext}. The springs represent the electrostatic
forces between the over-all neutral atoms. Properly,
there should be a spring connecting each atom with
every other atom; it is easier to bear this in mind
than to cope with a hopelessly cluttered diagram.

We must bear in mind that in a real crystal the atoms do not sit quiescently
at their equilibrium positions in the lattice structure. Each atom typically
possesses some kinetic energy. An atom may not, however, wander about
freely. Even for atoms that are over-all neutral, there are electrostatic interactions between the atoms, and those interactions tend strongly to restrain
an atom to the vicinity of its equilibrium position in the crystalline lattice.
Those interactions are represented by springs—some extended, some com-
pressed—between the atoms in the sketch. So we may picture each atom as
vibrating (in three dimensions) about its equilibrium position. Looking at
the crystal as a whole, we may conveniently refer to the totality of those
individual vibrations as the *lattice vibrations*.

Thus far we have described the interaction of the paramagnets with the
external field, and we have described the electrostatic interactions of the
atoms among themselves and hence the lattice vibrations. There are further
interactions crucial to a microscopic understanding of the system: the
magnetic interaction between pairs of magnetic moments, and the resultant
coupling between the magnetic moments and the lattice vibrations. In figure
11.4.2 two of the paramagnetic atoms from the preceding sketch are shown,
along with portions of the magnetic field produced by one of the two paramagnets. The magnetic moment μ_1 produces a small but crucial magnetic
field in the vicinity of the second moment μ_2. That field is not parallel

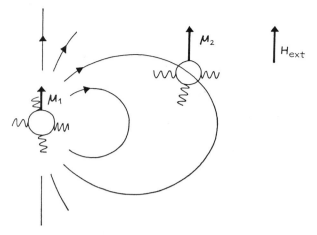

FIGURE 11.4.2
Indicated here is the magnetic interaction between a pair of
paramagnetic atoms. The curves are lines drawn tangent to
a portion of the magnetic field produced by the magnetic
moment μ_1. This small field gives rise to both a torque and
a net force on the magnetic moment μ_2. There will be an
analogous effect on μ_1 due to μ_2.

to the external magnetic field. It exerts a torque on the second moment that
may flip the spin of the second atom. Moreover, the field due to μ_1 is not
homogeneous in the immediate neighbourhood of μ_2 (nor elsewhere). In an
inhomogeneous magnetic field, a magnetic moment experiences a *net force*
in addition to a possible torque. The direction and magnitude of that force
depend on the orientation of the magnetic moment with respect to the
inhomogeneous field. Thus the magnetic moment μ_1 exerts a net force on the
atom with moment μ_2. So long as the spins retain their respective orientations,
we may regard that net force as effectively constant. But now suppose that
the spin associated with μ_2 flips to the opposite orientation. The net force
then changes sign. The effect of this is rather like giving a kick to the atom
whose spin flipped. The kick may either increase or decrease the kinetic
energy of the vibrating atom. Which it will be depends on the direction of
the kick relative to the instantaneous vibrational motion of the atom. There
is, of course, an analogous effect on the atom with μ_1 due to the magnetic
moment μ_2.

Let us summarize the essential points of the preceding paragraph. There
is a small but nonnegligible magnetic interaction between pairs of magnetic
moments. The interaction may induce a spin to flip, at least if there is energy
available for the change of orientation. Moreover, the magnetic interaction

couples the magnetic moments to the lattice vibrations. When a spin flips, the energy associated with the lattice vibrations may simultaneously be increased or decreased.

The foregoing provides all the elements that we need for a qualitative understanding of paramagnetic cooling. As the external magnetic field is slowly reduced in magnitude, the magnetic interaction of the paramagnets among themselves becomes better able to flip a spin from an orientation parallel to the external field to the opposite orientation. (In a strong external field we expect much net alignment; in zero field, we expect no net magnetic moment for the system.) With the external field still being far from zero, the antiparallel orientation is an orientation of high potential energy relative to that of parallel orientation. If the flip is to occur, some energy must be made available to the flipping spin. The source of that energy will be the energy of the lattice vibrations (and the energy of the mutual interactions of the spins). Thus, as the spins flip in the slowly decreasing external field, energy is extracted from the lattice vibrations. In consequence, the atoms vibrate less rapidly. Since we naturally correlate diminished vibration with reduced temperature, we have here a picture of why the temperature of the crystal decreases as the external magnetic field is slowly reduced to zero.

Getting to low temperatures by magnetic cooling is necessarily a two-step process: before one can slowly remove an external field, that field must first be applied. To ensure a net cooling, the environment for the magnetization and demagnetization processes must differ significantly. One must turn on the magnetic field in an environment in which the energy given *to* the lattice by the spin flips leading to alignment may be transferred away as heat.

The general experimental arrangement is indicated schematically in figure 11.4.3. The paramagnetic sample is suspended in an inner vessel surrounded by liquid helium. The liquid helium will provide the initial, already low temperature of a few kelvin or so. Before the external field is turned on, the inner vessel is filled with gaseous helium. The function of the gas is to provide a "thermal contact" between the paramagnetic sample and the liquid helium. As the magnetic field is turned on, the magnetic moments tend to line up parallel to the field. When these spin flips occur in a non-zero external field, the magnetic moments go to orientations of low potential energy; the energy released thereby goes, in part, to the lattice vibrations. Thus there is an inherent tendency for the lattice to warm up as the magnetic field is turned on. With the helium present, however, the energy released by the spin flips can be transferred from the lattice to the gas and thence to the liquid helium. In this way one avoids increasing the temperature of the paramagnetic system.

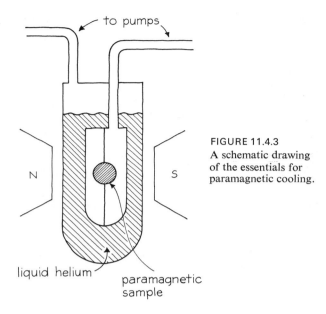

FIGURE 11.4.3
A schematic drawing
of the essentials for
paramagnetic cooling.

After the sample has been magnetized, the helium gas is pumped off. This breaks the thermal contact between the sample and the liquid helium. The paramagnetic system is now isolated but at the same temperature as the liquid helium. The demagnetization to achieve a still lower temperature may now begin.

The discussion earlier gave us a qualitative picture of why the paramagnetic system cools as the external field is reduced to zero. Provided that the reduction to zero field is made slowly, we may analyze the process quantitatively by invoking the constancy of S_I and the legitimacy of using a canonical distribution. Our primary aim will be to calculate, in order of magnitude, the temperature that may be reached by demagnetization from liquid-helium temperatures.

> The magnetic cooling process just described is often called cooling by "adiabatic demagnetization." Although the adjective *adiabatic* has two distinct meanings, fortunately both apply here. The first meaning implies that there is *no energy transfer as heat* to the system while the external magnetic field is reduced to zero (the "demagnetization" process). The second meaning implies that the external field is reduced to zero *very slowly*. In the laboratory, one always strives for an adiabatic reduction in the first sense, and, typically, also in the second sense. The first meaning of adiabatic may be familiar from thermodynamics; the second is associated with Ehrenfest's principle, which is sometimes called Ehrenfest's adiabatic principle.

An exact calculation, from the canonical distribution, of S_I for the paramagnetic crystal is hopelessly difficult. Some reasonable approximations are in order. Though the coupling of the magnetic moments and the lattice vibrations is essential to an understanding of the cooling process, we may (successfully) approximate the information-theory entropy for the system by the sum of two terms, one for the spins alone and the other for the lattice:

$$S_I \simeq S_{I\,\text{spins}} + S_{I\,\text{lattice}}. \tag{11.4.1}$$

Each magnetic moment μ is, in the situation we envisage, associated with a spin of $\frac{1}{2}\hbar$. In Chapter 5, and again in Chapter 7, we analyzed a system of *ideal* paramagnets, that is, paramagnets which interact with an external field but for which mutual interactions are neglected. Here we should take into account, even if only crudely, the mutual interactions. As a means to this end, let us represent the mutual interactions by an *effective* magnetic field H_{eff}, as though the net influence on any one magnetic moment due to all the others were equivalent to the production of a small magnetic field H_{eff}. This field is superimposed on the external magnetic field, which we now write as H_{ext}. Then the potential energy associated with a single magnetic moment μ is given by

$$(-\boldsymbol{\mu} \cdot \mathbf{H}_{\text{ext}}) + (-\boldsymbol{\mu} \cdot \mathbf{H}_{\text{eff}}) = -\boldsymbol{\mu} \cdot \mathbf{H} \tag{11.4.2a}$$

with

$$\mathbf{H} \equiv \mathbf{H}_{\text{ext}} + \mathbf{H}_{\text{eff}}. \tag{11.4.2b}$$

This is, admittedly, a crude method for handling the mutual interactions, but the complexity of the actual situation forces us to make do with it.

The virtue of this approach to the mutual interactions lies in the great simplification it provides for the approximate calculation of $S_{I\,\text{spins}}$. From the canonical distribution applied to the spins alone, we have

$$S_{I\,\text{spins}} = k\big(\beta\langle E\rangle_{\text{spins}} + \ln Z_{\text{spins}}\big). \tag{11.4.3}$$

Since we approximate the true energy operator for the N spins by

$$\sum_{i=1}^{N} -\boldsymbol{\mu}_i \cdot (\mathbf{H}_{\text{ext}} + \mathbf{H}_{\text{eff}}) = -\sum_{i=1}^{N} \boldsymbol{\mu}_i \cdot \mathbf{H},$$

and since we deal with a system of paramagnets with spin $\frac{1}{2}\hbar$, we may take over results from Chapter 5. The analysis in that chapter preceded the introduction of the partition function; nonetheless, we can extract Z_{spins} from the equations there. The denominator on the righthand side in equation

(5.1.5) is the desired partition function; it appears in an algebraically different form in equation (5.1.6) and is evaluated in equation (5.1.7). With the aid of those equations we may write

$$Z_{\text{spins}} = (e^{\beta\mu H} + e^{-\beta\mu H})^N.$$

For the energy we take the estimated total moment from equation (5.1.9) and write

$$\langle E \rangle_{\text{spins}} = -\langle \mathbf{M} \rangle \cdot \mathbf{H} = -N\mu H \frac{e^{\beta\mu H} - e^{-\beta\mu H}}{e^{\beta\mu H} + e^{-\beta\mu H}}.$$

(We could also have differentiated the logarithm of the preceding partition function.) In the interests of brevity, let us write these elements in terms of hyperbolic functions when we insert them in the general expression for $S_{I \text{ spins}}$:

$$S_{I \text{ spins}} = Nk\left[-\frac{\mu H}{kT} \tanh \frac{\mu H}{kT} + \ln\left(2 \cosh \frac{\mu H}{kT}\right) \right]. \qquad (11.4.4)$$

Noteworthy in this approximation is the property that $S_{I \text{ spins}}$ depends on H and T through the ratio H/T only. Later we will use this.

About S_I for the lattice, little need be said, beyond that it is already quite small at liquid-helium temperature and, of course, decreases with a further reduction of the temperature. Whether one uses the Einstein model or the Debye model for an approximate description of the motion of the atoms is irrelevant to the practical conclusion. The Debye theory of lattice vibrations yields, for low temperatures,

$$S_{I \text{ lattice}} \simeq \frac{4\pi^4}{5} Nk\left(\frac{T}{\Theta_D}\right)^3, \qquad (11.4.5)$$

where Θ_D is the characteristic temperature of the Debye theory. The value of Θ_D is typically of the order of several hundred kelvin, and so at temperatures of a few kelvin and below, $S_{I \text{ lattice}}$ is quite small. It is, in fact, so small that we will not need to make any quantitative use of it.

What truly matters is the size of $S_{I \text{ lattice}}$ relative to $S_{I \text{ spins}}$. For the situations of concern to us, the latter is of order Nk, since the quantity in brackets in equation (11.4.4) will be of order unity. Thus the information-theory entropy of the lattice is smaller than that of the spins by roughly the factor $(T/\Theta_D)^3$, which is quite small.

Now we can readily estimate the temperature reached by demagnetization. At the initial temperature T_i, the magnetic field is $H_{ext} + H_{eff}$, and so for S_I we have

$$(S_I)_i \simeq S_{I\,spins}\left(\frac{H_{ext} + H_{eff}}{T_i}\right) + S_{I\,lattice}(T_i).$$

Because the external field is slowly reduced to zero, we continue to use a canonical probability distribution. Then the expression after demagnetization, with final temperature T_f, is

$$(S_I)_f \simeq S_{I\,spins}\left(\frac{H_{eff}}{T_f}\right) + S_{I\,lattice}(T_f).$$

To determine T_f, we invoke the constancy of S_I and equate the final and initial expressions. If we drop the lattice contribution as being quantitatively negligible, we have

$$S_{I\,spins}\left(\frac{H_{eff}}{T_f}\right) \simeq S_{I\,spins}\left(\frac{H_{ext} + H_{eff}}{T_i}\right). \qquad (11.4.6)$$

Since the same function appears on the two sides of the (approximate) equality, we may conclude that the arguments are (approximately) equal. We infer that the final temperature is given by

$$T_f \simeq \frac{H_{eff}}{H_{ext} + H_{eff}}\, T_i. \qquad (11.4.7)$$

Typical values for the effective field range between 50 and 400 gauss. (Here is a fast semiquantitative check: if we take the distance between lattice sites to be 5 Ångströms and take a magnetic moment of one Bohr magneton, then one paramagnet produces a magnetic field of order 80 gauss at the neighboring lattice sites.) Thus the effective field is always small relative to the external fields of 10 to 50 thousand gauss commonly employed. We are assured a significant temperature reduction.

To illustrate the effectiveness of magnetic cooling, let us take some modest values:

$$H_{eff} = 100 \text{ gauss};$$
$$H_{ext} = 10,000 \text{ gauss};$$
$$T_i = 1 \text{ K}.$$

The low value for T_i can be achieved by pumping on the liquid helium and thereby cooling it by evaporation. The practical limit of this evaporative cooling is about 1 kelvin. Upon inserting these figures into the preceding relation, we find

$$T_f \simeq 0.01 \text{ K},$$

an impressive reduction.

Lest there be confusion about H_{eff}, one should point out that, as the external field is reduced to zero, the magnetic moment of the paramagnetic crystal typically goes to zero. The coupling with the lattice vibrations is responsible for this. The effective field H_{eff} does not (usually) support a residual macroscopic net alignment. Rather, it is primarily a crude but mathematically tractable means of handling the interactions among the magnetic moments. The major reason for insisting on the inclusion of H_{eff} is this: one avoids generating the erroneous impression that magnetic cooling can take a paramagnetic sample all the way down to absolute zero. The microscopic origin of this particular effective field is discussed in detail in Section 12.4.

11.5. AN EXTENSION OF THE TEMPERATURE SCALE

According to our theoretical analysis, magnetic cooling can achieve temperatures of order 0.01 K and less. The operations are performed routinely in laboratories all over the world. Nonetheless, one must be cautious in talking about a temperature of 0.01 K. No one has ever measured so low a temperature with a gas thermometer. The incredibly small gas pressure that would exist as such a low temperature makes it unlikely that anyone ever will perform such a measurement. If there is to be a precise practical meaning to a temperature of 0.01 K, an extension of the scale provided by a gas thermometer is needed.

Let us review for a moment. Thus far the temperature scale has been defined in terms of measurement with a gas thermometer. We chose a unit (the kelvin) and specified, arbitrarily but conveniently, the numerical value of the temperature of water at its triple point; then we took the gas law for sufficiently dilute gases,

$$\frac{p_{\text{obs}} V}{Nk} = T,$$

as a reasonable measure of temperature. Boltzmann's constant k has a value of about 1.38×10^{-16} ergs/K as an experimental consequence. Whenever it is feasible to bring a gas thermometer into contact with a system in equilibrium, we have a means of measuring the system's temperature on an intuitively reasonable scale.

Since we cannot always employ a gas thermometer, there is a definite need to extend the temperature scale. The extremes of low-temperature physics and of solar physics can hardly be handled with a gas thermometer. Moreover, the prescription of an operational procedure would certainly have to be different in these two cases. An apparatus that might provide a suitable scale for low-temperature physics would melt if one were to carry out seriously a proposal to put it in contact with the sun.

If we look for an extension of the temperature scale that will encompass, in one fell swoop, the entire range of qualitatively possible temperatures, it will necessarily be somewhat theoretical. The most we can expect is this: for any given temperature range and class of physical phenomena, the extension will provide an unambiguous basis for an experimental procedure that will measure temperature. The extension should, of course, lead to the gas-thermometer temperature when that scale is applicable. One is looking for an extension, not a revision.

The source, in statistical mechanics, of such a general extension lies in the canonical probability distribution. To recapitulate, that is the probability distribution for a system in equilibrium when our information consists merely of temperature, environment, and constituents. The distribution was derived with the explicit use of the gas thermometer and the scale of temperature that such a thermometer provides. Moreover, the analysis in Section 7.5 showed that the canonical distribution was consistent with the general view of temperature advanced in that section.

Let us recall that estimates, from the canonical distribution, of the macroscopic properties of macroscopic systems are typically very sharp estimates. For example, in Section 5.2 we found that the probability distribution for the total magnetic moment of a macroscopic paramagnetic system exhibits an incredibly sharp peak at (or exceedingly close to) the expectation value $\langle M_z \rangle$.

These properties of macroscopic systems and of the canonical probability distribution suggest that we use the canonical distribution itself to extend the temperature scale to domains where a gas thermometer is impractical. For a macroscopic system known to be in equilibrium, we compare observations with the estimates arising from the canonical probability distribution, choose the value of " T " appearing in the distribution to fit the observations, and take that " T " to be the numerical value of the system's temperature. The extension is a rather natural one, and by proceeding in this fashion we are assured of agreement when a measurement by a gas thermometer is practical. Moreover, we place ourselves in a position to generate specific experimental procedures as they are needed.

> The definition of a general temperature scale in thermodynamics has a very different starting point. The present "statistical mechanical" extension is, however, effectively equivalent to the Kelvin definition and to definitions based on the second law of thermodynamics. The effective equivalence arises because of two elements: (1) our use of macroscopic systems, and (2) the close connection between the information-theory entropy, *when evaluated for the canonical probability distribution*, and the thermodynamic entropy in thermodynamically reversible processes. A bit more is said about this connection in Appendix D.

An immediate illustration may be worthwhile. High temperatures can be measured, without contact, with the aid of Planck's distribution law for black-body radiation. The intensity of electromagnetic radiation in a narrow range of frequencies around frequency v is proportional to

$$[\exp(hv/kT) - 1]^{-1},$$

a property that one may infer from equation (10.6.13). Filters in an optical pyrometer enable one to select some such frequency range (often in the red). A comparison of two radiation intensities, one from a source at a known temperature (for calibration of the instrument), the other from the hot system with unknown temperature, and then an appeal to Planck's relation, permit a determination of the unknown high temperature. The value of the calibration temperature can be determined with a gas thermometer, for they have been successfully used to temperatures of order 1,800 K. (In practice, corrections for nonideal conditions are necessary, but this is the essence of the technique.) Although the radiation does presumably come in the form of single photons, the observations entail the reception of huge numbers of photons. In this sense the measurement is macroscopic, and one need not worry about the fact that the Planck distribution refers to a statistical expectation value.

The statistical-mechanical extension of the temperature scale provides us with an extension all the way down to absolute zero. There is, now, a well-defined meaning for the temperature of 0.01 K that we calculated in Section 11.4 as being attainable by magnetic cooling. The calculations of that section do, moreover, provide an illustration of the practical impossibility of cooling to absolute zero in a finite number of steps. The argument is brief.

Equation (11.4.7) informs us that the final temperature T_f attained by slow demagnetization is proportional to the initial temperature T_i. Since we must start at a non-zero value for T_i, we cannot achieve a zero value for T_f in a single step. Even if we imagine the cooling process repeated many times, with the starting temperature in each instance being the final temperature of the preceding cycle, the initial temperature of each cycle will always be non-zero, and hence the final temperature will also be non-zero. No finite number of steps will take the system to absolute zero.

Although absolute zero may be unattainable, that does not, remarkably, preclude the production of negative absolute temperatures. To be sure, this is possible only in special circumstances, but systems of paramagnetic nuclear spins have been brought to (effective) equilibrium at negative temperatures. It is here that the natural extension of the temperature scale becomes all important. A gas thermometer is incapable of being at a negative absolute temperature and hence is incapable of measuring such a temperature, but statistical mechanics can give meaning to the concept.

Our previous work on paramagnetism can illustrate how the notion of a negative temperature can arise. The curve in figure 11.5.1 displays $\langle M_z \rangle$ as a function of temperature. The external field H_{ext} points along the positive z-axis, and the interactions among the paramagnets are crudely represented

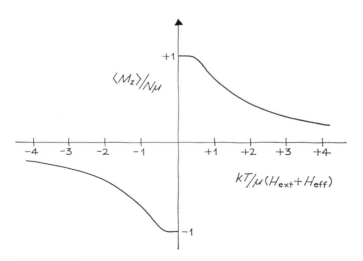

FIGURE 11.5.1
A display of $\langle M_z \rangle / N\mu$ for a system of N paramagnets with spin $\frac{1}{2}\hbar$ and individual magnetic moment μ. The calculated magnetic moment approaches zero both for T going to $+\infty$ and for T going to $-\infty$.

by a field H_{eff}, taken to have the same direction. The theoretical expression, calculated with the canonical distribution, is that of equation (5.1.9) and is reproduced here:

$$\langle M_z \rangle = N\mu \tanh\left[\frac{\mu}{kT}(H_{ext} + H_{eff})\right].$$ (11.5.1)

At positive temperatures the estimated magnetic moment is parallel to the external field. Nothing prevents one from inserting a negative value for T; then the estimated moment points antiparallel. If one can prepare a paramagnetic system (in effective equilibrium) such that the observed total moment points antiparallel, then the natural extension of the temperature scale says that the system is at a negative absolute temperature.

Let's now investigate the circumstances under which a negative absolute temperature can be attained. This means, in part, answering the question: when may we take T to be negative in the canonical probability distribution without running into trouble? In that distribution,

$$P_j = \frac{e^{-E_j/kT}}{\sum\limits_{i=1}^{n} e^{-E_i/kT}};$$

the summation in the denominator will not converge when T is negative if the system has an infinite number of energy eigenstates with increasing energy eigenvalues. Such is certainly true of the states for a gas, and indeed, for any real physical system, because there is no upper bound to the kinetic energy of internal motions. Nonetheless, a real system may sometimes be analyzed into effectively isolated parts. To jump ahead for a moment, a lithium fluoride crystal may be. The crystal as a whole may be regarded as a system of nuclear spins and a system of vibrating atoms, with a relatively weak interaction between the two. To be sure, the nuclei are part of the atoms; the phrase "system of nuclear spins" refers to the spins alone and to their orientations. The nuclear spin system has only a finite number of energy eigenstates, and those have finite energies. A negative T in a canonical distribution for the spin system alone leads to no trouble, whereas it definitely would for the lattice vibrations.

Let us note that if we imagine a negative temperature, the states with high energy will be more probable than those with low energy, exactly the reverse of the situation for ordinary positive temperature. In a sense, the basic problem of achieving a negative temperature is to make the high-energy states more probable than the low-energy states. If the system is a gas, the foregoing is impossible. Since a gas may have unlimited internal kinetic energy, one could supply energy indefinitely in the (futile) attempt to make states successively higher in energy more probable than the preceding states.

By now we can see the conditions that must be met if a system is to be capable of being at a negative temperature.

1. The system must be macroscopic and in equilibrium, so that the very notion of temperature is applicable;

2. the possible values of the energy of the system must have a finite upper limit; and

3. the system must be isolated (effectively, at least) from systems not satisfying the second condition.

The upper energy bound is necessary if a finite amount of energy—all that one ever controls or observes—is to be able to produce a negative temperature. The effective isolation is necessary because systems in contact have a common temperature at equilibrium. If two systems interact significantly, a negative temperature is possible for either only if a negative temperature is possible for both.

The notion of a negative absolute temperature is no mere academic curiosity. In 1951 Purcell and Pound produced the first such temperature, using the nuclear spin system in a crystal of lithium fluoride. For the fluorine isotope F^{19} the nuclear spin is $\frac{1}{2}\hbar$; that for Li^7 is $\frac{3}{2}\hbar$. The paramagnetic nuclei are not truly "spatially fixed," for the nuclei do vibrate about their average positions in the crystal lattice. The exchange of energy between the magnetic moments and the lattice vibrations occurs on a time scale long enough so that one may conceive of the nuclear *spins* as a system effectively isolated from the lattice vibrations. This is critical, for there is no upper bound to the energy that the lattice vibrations may possess. A negative temperature equilibrium is not possible for the crystal as a whole, but the effective isolation of the nuclear spins does permit *them alone* to have a negative temperature equilibrium. There is sufficient interaction within the nuclear spin system itself to ensure that it can come to equilibrium on a time scale short relative to the time scale for significant interaction between the spins and the crystal lattice.

Thus the system of paramagnetic nuclear spins in lithium fluoride is theoretically capable of being at a negative temperature. The experimental problem is this: how does one prepare the system so that the observed total magnetic moment will be antiparallel to an external field, an indication of a negative temperature? The essence of the scheme used by Purcell and Pound was a double reversal of an external magnetic field of order 100 gauss. The initial field was reversed in a time of order 2×10^{-7} seconds. This is the crucial reversal. A second reversal, on a time scale of 10^{-3} seconds, returned the magnetic field to its original value. The crystal was then taken out of the 100 gauss field and quickly placed, for purposes of observation only, in a large external field of about 6,400 gauss. No matter how the crystal was geometrically oriented in the large field, the observed total nuclear magnetic moment always pointed in the " wrong " direction, antiparallel to the external field. The nuclear spin system could indeed be said to be at a negative absolute temperature.

This is all well and good, but one would like some microscopic picture of how the negative temperature was achieved. The interactions of the magnetic moments among themselves are crucial. In a crystal a magnetic moment interacts not only with an external magnetic field but also with the neighboring magnetic moments. Following convention, we call the energy associated with the latter interaction the *spin-spin energy*. Initially the spin system is at a positive temperature, with the nuclear magnetic moments predominantly parallel to the external field, an orientation of negative potential energy. The first field reversal is so fast that the spins cannot respond, cannot change their orientation during the reversal. Consequently, the spins find themselves pointed predominantly antiparallel to the reversed field, an orientation of positive potential energy. In a short time some of this energy is transferred, by mutual interactions, to spin-spin energy. The energy associated with the mutual interactions becomes relatively large. In fact, that energy becomes larger than what one would predict at any positive temperature.

Since the interaction between the spin system and the lattice vibrations is weak, the spin system holds onto that relatively large amount of energy, even as the crystal is transferred for observation from the small external field to the large magnetic field. There is, in fact, so much spin-spin energy around that the magnetic moments "prefer" to line up antiparallel to the new external field, a position of positive potential energy, rather than line up in the usual parallel position, the position of negative potential energy. There is enough energy around so that a spin can easily line up in what is ordinarily an energetically unfavored orientation.

Indeed, the examination of the crystal in the large field was quite directly a test for the amount of energy possessed by the nuclear spin system. Purcell and Pound applied electromagnetic radiation at a frequency chosen so that Li^7 nuclei could flip from an orientation of low to high potential energy by the absorption of a photon. Instead of absorption they found—as they had hoped—stimulated *emission* of radiation at that frequency. The spin system had so much energy that spins predominantly flipped to low potential energy orientations and simultaneously emitted a photon. From the direct observation of stimulated emission of radiation—rather than absorption—one infers both that the total magnetic moment is antiparallel to the external field and that the nuclear spin system is at a negative temperature.

Let us now consider a different aspect of the experiment. Originally, both spin systems and lattice were at room temperature, a positive 300 K. After the double reversal and placement of the crystal in the observing field, the lattice remained unaffected at 300 K, but the spin system was found to exhibit a behavior such as one would expect statistically for a negative temperature of about -350 K. The isolation of the spin system is not perfect, and over a long time the interaction between spin system and lattice is not negligible. The interaction led to a slow, continuous transfer of energy from the spin system to the lattice. The observations showed a change in $\langle M_z \rangle$ from its negative temperature value antiparallel to the large external field, through zero value for $\langle M_z \rangle$, to a new value parallel to the field. Figure 11.5.2 shows

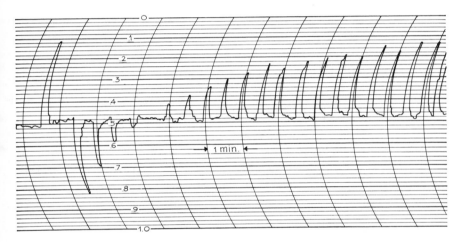

FIGURE 11.5.2
The following quotation is the caption that appeared with the curve in the paper by E. M. Purcell and R. V. Pound, *Physical Review*, **81**, 279 (1951):
"A typical record of the reversed nuclear magnetization. On the left is a deflection characteristic of the normal state at equilibrium magnetization ($T \approx 300°$K), followed by the reversed deflection ($T \approx -350°$K), decaying ($T \rightarrow -\infty$) through zero deflection ($T = \infty$) to the initial equilibrium state." By "deflection" the authors mean absorption of electromagnetic radiation and by "reversed deflection," emission.

the now-classic experimental curve that accompanied the Purcell-Pound paper. The inference is that, as the energy of the spin system decreased, the spin temperature passed from a negative temperature through infinite temperature (at $\langle M_z \rangle = 0$) to a commonplace positive temperature. Such a transition for the temperature is, perhaps, another surprise, but it does follow from the display in figure 11.5.1.

Positive and negative temperatures are *not* joined across absolute zero. The practical unattainability of absolute zero remains a valid empirical conclusion. (The assertion may be extended to the statement that one cannot attain absolute zero from *either* positive *or* negative temperature in a finite number of operations.) A system passes from negative to positive temperature, or vice versa, through the "back door" of infinite temperature. As far as the exponentials in the canonical distribution are concerned, the effect is the same for both $T = +\infty$ and $T = -\infty$. Each leads to a perfectly flat probability distribution, for the parameter β is then zero.

Although there is no need to distinguish between plus and minus infinity for the temperature, there is a vast difference between approaching absolute zero from positive and from negative temperatures. The physical difference is that for $T \to +0$, the system settles into its ground state with certainty for the inference, while for $T \to -0$, the system is pushed into its highest energy state, again with certainty for the inference in the limit. In both limits the information-theory entropy goes to zero (provided there is no degeneracy in the lowest and highest energy levels).

11.6. RAPID CHANGE IN
AN EXTERNAL PARAMETER

Now we investigate a rapid change in an external parameter, the second of the two cases into which we agreed (at the end of Section 11.1) to split the analysis. To continue with the illustrative gas system, let us suppose that the piston is pulled out extremely rapidly, specifically, much faster than the speed of sound in the originally quiescent gas. Very few molecules will have sufficiently high speed to be able to follow the piston's motion, and so for a brief time interval there will be virtually no molecules in the new volume made available by the displacement of the piston. This is far from being an equilibrium situation. In short, when an external parameter changes rapidly, the system does *not* remain close to equilibrium, and there is no reason to expect a probability distribution of canonical form to apply during the change of external parameter.

But one surely expects that after the external parameter has ceased to change, the system will begin to settle down to a new equilibrium. The length of time required for the settling down will depend on the system and the extent to which its previous equilibrium was disturbed, but everyday experience indicates that macroscopic systems do, in time, settle down; so, empirically, something like a canonical probability distribution should eventually be appropriate for describing the system statistically.

The physical situation is clear enough. The difficulty comes in trying to show that the rigorous analysis given in Section 11.1 yields results in agreement with one's expectations. While the external parameter is changing, each quantum-mechanical state will evolve according to the time-dependent Schrödinger equation. A state denoted by $\psi_j(t_1)$ at time t_1 evolves into the state $\psi_j(t_2)$ at time t_2. If the change in the parameter is rapid, the latter state will not, in general, be an eigenstate of the energy operator at time t_2. Consequently, the probability distribution at time t_2 will not be of the canonical type, but we have no reason to expect that type immediately after the parameter has ceased to change. Surely some time must elapse during which the system settles down.

Nonetheless, one does find cause to be worried here. If the state $\psi_j(t_2)$ is not an energy eigenstate, the state that evolves from it for $t > t_2$ will never be one, and so the probability distribution given by $P(A_j(t)|B)$ will never become precisely a canonical distribution. The reasoning is based on the superposition principle of quantum mechanics (outlined in Section 12.1) and goes like this. If $\psi_j(t_2)$ is not an energy eigenstate, it may be looked upon as a superposition of many different energy eigenstates of $\mathcal{H}(t_2)$, each such state appearing with a definite numerical coefficient. As $\psi_j(t)$ evolves for $t > t_2$ according to the Schrödinger equation, with the energy operator being time-independent and equal to $\mathcal{H}(t_2)$ for $t > t_2$, the numerical values of the coefficients will not change. The state $\psi_j(t)$ itself changes because the phase relations between the states in the superposition change, but $\psi_j(t)$ remains a superposition of different energy eigenstates. The state never evolves into an energy eigenstate for $t \geqslant t_2$.

Thus the rigorous integration of the Schrödinger equation tells us that, for the kind of system we are considering, the exact probability distribution never evolves into a canonical distribution as the system settles down. One may expect that estimates of macroscopic properties, calculated with the evolved probability distribution, will, with time, approach those calculated with a canonical distribution, but one may not expect much more.

The conclusions of the preceding paragraph are surprising and a bit disturbing. They are, however, reasonably well-established by modern theoretical work on rigorous treatments of the approach to equilibrium. To state those results precisely, some of the formal quantum-mechanical apparatus to be developed in Chapter 12 is needed, but all the essentials can be indicated here.

Let us first remind ourselves of two points in the context: (1) we are dealing with a physical system of finite spatial extent; and (2) for times later than time t_2, when the external parameter ceases to change, the evolution of the quantum-mechanical states is governed by a known energy operator that is *independent* of time. Infinite systems exist only as limits and then only in the mind of the theorist. The import of the second point is that the system is "isolated," interaction with the external world being represented solely by external parameters having fixed values.

Now we turn to the crucial theorems. The first is that the probability distribution for a finite isolated system is an almost periodic function of time.* As the probability distribution evolves in time, it does not quite repeat itself. Yet if one picks an instant of time and specifies a small margin of difference in probability distributions, then one can always find a finite time interval such that the probability distribution at that later time differs from the distribution at the first instant by less than the margin of difference. Indeed, return to the first probability distribution (within the margin of difference) will occur arbitrarily often, though not necessarily at equally spaced intervals of time. This theorem implies that the evolving probability distribution will never settle down.

Before we allow the theorem to dismay us completely, we should ask for the time-scale of such a recurrence time. When the system is macroscopic, the time-scale is tremendously long. Boltzmann considered a one cubic-centimeter sample of gas at roughly room temperature and density and estimated the time for recurrence (within a modest error margin) of a given arrangement of molecular positions and velocities. (That is not a recurrence time for a probability distribution, but it will give us an idea of the order of magnitude.) He wrote:

> Though the number N/b [Boltzmann's deliberate underestimation of the recurrence time] is enormous, one can obtain some idea of its magnitude by noting that it has many trillions of digits. For comparison, suppose that every star visible with the best telescope has as many planets as does the sun, and on each planet live as many men as are on the earth, and each of these men lives a trillion years; then the total number of seconds that they all live will still have less than 50 digits.†

---

* I. C. Percival, *Journal of Mathematical Physics*, **2**, 235 (1961); P. Bocchieri and A. Loinger, *Physical Review*, **107**, 337 (1957).

† S. G. Brush, *Kinetic Theory* (New York: Pergamon, 1966), II, 228; this book contains a translation of Boltzmann's article, *Annalen der Physik*, **57**, 773, (1896). Boltzmann's theory of the approach to equilibrium had been sharply criticized by Zermelo, who had invoked Poincaré's recurrence theorem: a *classical* system, if bounded and isolated, will return in a finite length of time to any configuration of molecular positions and momenta that it once had—if not exactly, then at least to within any non-zero margin of error one wishes to specify.

The typical recurrence time is so fantastically long (compared, even, with the estimated age of the earth) that it has no direct significance in the laboratory. That the probability distribution for a finite isolated system is almost periodic in time is a major point of principle, but it can have little to do with the observed approach by a system to (effective, at least) equilibrium in the laboratory.

The approach to equilibrium by an infinite physical system with finite density is the subject of the second major point. Such a system is, of course, only an idealization, a mathematical limit, but we will find that it is relevant. In the limit of infinite volume and an infinite number of particles, the recurrence time becomes infinite. (Crudely, a particle cannot bounce repeatedly between walls indefinitely far apart.) Thus a strict approach to equilibrium is theoretically possible. For physically reasonable initial conditions, the work of Prigogine and his collaborators* indicates that in such a limit the estimated macroscopic properties approach, in time, the values that would emerge from a calculation with the canonical probability distribution and an appropriately chosen temperature. Microscopic properties dependent on only a few particles at a time do the same. For example, the velocity probability distribution for a single particle approaches, with time, the form given by applying the canonical probability distribution to the system.

The results of the preceding three paragraphs enable us to make a reasonable conjecture about the approach to equilibrium in the laboratory. The recurrence time is so long that the almost-periodicity of the probability distribution is irrelevant in practice. The empirical behavior should be much more like that for an infinite system. So the probability distribution should, for reasonable times, evolve in such a way as to give estimates that are indistinguishable from those of the canonical distribution for most relevant quantities. Thus we have theoretical justification for the modest expectation that a finite isolated system will settle down (at least effectively) to behavior appropriately described with the canonical distribution. Since calculations with that distribution are likely to be simpler than those using the rigorously evolved probability distribution, there is good reason for adopting the canonical distribution when permissible.

In order to use the canonical distribution a long time after time t_2, we need to know the appropriate temperature. Direct measurement is certainly one way of determining it. Another method is possible if we know the amount of energy transferred between system and external world in the time

---

* I. Prigogine, R. Balescu, F. Henin, and P. Resibois, *Physica*, **26**, S36 (1960); P. Resibois, *Physica*, **27**, 541 (1961).

interval $(t_2 - t_1)$ during which the external parameter was changing. We can then choose the temperature by requiring that the new expectation value of the energy equal that at time t_1 minus the energy loss to the external world. (We could, instead, use $\langle E \rangle$ as computed with the evolved probability distribution, but that distribution is likely to be intractable.) The significant point is that we *cannot* determine the final temperature by using the information-theory entropy of the evolved distribution. The procedure that we used when the external parameter changed slowly is not applicable here, for the following reason. Although the evolved distribution and the new canonical distribution may make indistinguishable estimates for macroscopic properties, the two probability distributions *are* different in microscopic detail. The evolved distribution retains information about the initial situation at time t_1 and about the subsequent changes in external parameter. Consequently, the amount of missing information associated with the newly adopted canonical distribution will be greater than that of the evolved distribution.

We have at our disposal the two lemmas needed to prove the preceding contention. They are the following.

Lemma (1): The temperature appearing in the newly adopted canonical distribution should surely be chosen so that the energy estimate calculated with the canonical distribution is equal to the energy estimate calculated with the rigorously evolved distribution.

This is eminently reasonable, though perhaps difficult to achieve in a practical calculation. We may write this as the equation

$$\langle E \rangle_{\substack{\text{adopted} \\ \text{canonical} \\ \text{distribution}}} = \langle E \rangle_{\substack{\text{evolved} \\ \text{probability} \\ \text{distribution}}} .$$

Lemma (2): When the expectation value of the energy is specified, a probability distribution of the canonical form has, among all compatible probability distributions, the maximum missing information (equivalently, the maximum information-theory entropy).

This was the outcome of the analysis with Lagrange multipliers in Section 4.5.

Upon combining the two lemmas, we can immediately say that S_I for the newly adopted canonical distribution cannot be less than that of the evolved distribution. Moreover, only if those two probability distributions were precisely equal could their S_I's be equal. When the history of the physical system includes a fast change in external parameter, the two probability distributions will not be precisely equal. Hence S_I for the newly adopted distribution will be greater than that of the evolved distribution.

We can profitably go one step further. We know that S_I for the probability distribution that evolves according to the Schrödinger equation remains constant in time, a result we derived in Section 11.1. In view of it and of the conclusion of the preceding paragraph, we may state the following theorem:

> The information-theory entropy associated with the newly adopted canonical distribution (adopted for the sake of convenience in calculations) is greater than the initial S_I at time t_1 if the system has been subject to a fast change in external parameter. Only if the parameter changed very slowly will the two S_I's be equal.

The analysis tells us that we must carefully distinguish between fast and very slow changes in an external parameter. A scheme for computing the temperature that works for one type of change may not work for the other. The theorem's relevance to the second law of thermodynamics is developed in Appendix D. The crucial point to remember is this: in adopting a canonical distribution to describe the system, even long after the fast change in external parameter, we are throwing away information about the history of the system, information that is contained in the probability distribution that evolved according to the Schrödinger equation. In consequence, it should not be surprising that a difference in information-theory entropy results, for S_I is precisely the measure of missing information (with a particular choice of scale constant).

There is a more technical but still useful way of pointing out the difference. The two probability distributions refer to inferences about different sets of states, one to evolved states, the other to energy eigenstates. The hypotheses are different, one containing statements about the change of an external parameter, the other containing a bald assumption of equilibrium. Each distribution provides the "best" probability distribution, given its context. Although the two distributions describe statistically the same physical system, the "given information" is different for the two; so there is no reason to expect the same amount of "missing information," or of uncertainty.

An Illustration

To illustrate these abstract arguments, we turn again to the gas and piston. At time t_1 the gas is in equilibrium and is properly described statistically with the canonical distribution and the known temperature. The first sketch in figure 11.6.1 depicts this initial situation. The piston is then pulled out very rapidly, at a speed many times the typical speed of a gas molecule. The motion of the piston is halted, at a time t_2, when the volume available to the

414

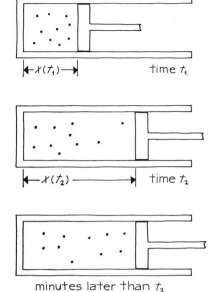

time t_1

FIGURE 11.6.1
Sketches depicting typical spatial
distributions of molecules at three
qualitatively different times in the system's
evolution.

time t_2

minutes later than t_2

gas has been doubled. The spatial distribution of the gas molecules at that time might look like the distribution in the second sketch. A few molecules possessing high speed were able to follow the piston's motion, but most of the molecules remain in the region available initially.

The physical situation at time t_2 is certainly not what we would consider to be "equilibrium." Also, the probability distribution that evolved from the canonical distribution at time t_1 is not another canonical distribution, but that is as it should be: we know the system is not in equilibrium, and so we do not expect that the evolved distribution will have the canonical form. Still, we can assert the equality

$$S_I(t_2) = S_I(t_1)$$

because the change in the probability distribution is deterministic—we know $\mathcal{H}(t)$ for $t_1 \leqslant t \leqslant t_2$—and information in the mathematical sense is neither gained nor lost.

Regardless of whether we view matters classically or quantum-mechanically, the probability distribution continues to evolve after time t_2. A nearly uniform spatial distribution of molecules in the doubled volume will become highly probable after a short time has elapsed. Minutes later will do, though the time required is likely to be no more than a small multiple of the time that a typical molecule needs to cross the new container volume. The anticipated situation is shown in the last sketch.

Now we must face the recurrence problem. The rigorous probability distribution continues to evolve with time, and in such a manner that at some time in the future the spatial distribution will be very close to the " nonequilibrium" distribution of time t_2. The time required for this " almost-recurrence" is, however, incredibly long; geologic time is short by comparison. Although an almost-recurrence to the nonequilibrium spatial distribution is unequivocally predicted, the prediction is experimentally irrelevant.

For describing statistically the effective equilibrium minutes after time t_2, we have an unambiguous probability distribution: the distribution that evolved from time t_1 through time t_2 to the present. An attempt to compute the evolution would, however, pose an extraordinarily difficult mathematical problem. For the sake of convenience, we may simply *adopt* a new canonical probability distribution, using the energy eigenstates associated with the doubled volume. Then the only question is what value of the temperature to insert.

In our example the energy of the gas has suffered only negligible change, for only a few molecules were fast enough to collide with the moving piston and thereby lose energy. To determine the temperature for the adopted canonical distribution, we can use the modest requirement

$$\langle E \rangle_{\substack{\text{adopted} \\ \text{canonical} \\ \text{distribution}}} = \langle E \rangle_{\text{at time } t_1}$$

in this specific example. To the extent that a few molecules manage to transfer energy to the moving piston, this equation is an approximation, but an excellent one. If the work done by the gas on the piston were measured, one could correct for that loss of energy.

Thus far we have not needed to say whether the gas is dilute or not, monatomic or diatomic, and the like. These questions become relevant now, for we must write down explicit expressions for $\langle E \rangle$ according to the canonical distribution, one expression at the initial volume and temperature, the other at double the volume and the unknown final temperature, and from the equality extract the value of the final temperature. For simplicity, let us take a dilute structureless monatomic gas with negligible intermolecular forces. We are idealizing, of course, but at modest temperature and dilution the idealization is not bad for gases such as helium and argon. The form of the estimated energy is then $\frac{3}{2}NkT$, independent of volume. Consequently, we infer that the temperature for the adopted canonical distribution should be numerically the same as the initial temperature. (Taking cognizance of the intermolecular forces, for instance, would alter this conclusion, leading to a different but well-determined final temperature.)

Now we look at the information-theory entropy. That for the evolved probability distribution will not have changed:

$$S_I(t > t_2) = S_I(t_2) = S_I(t_1),$$

whatever the gas. The situation with S_I for the newly adopted distribution will be significantly different. Since our gas is dilute and structureless, we may use the expression for S_I written toward the end of Section 9.7. Because the temperature of the effective equilibrium is equal to the initial temperature, the sole effect arises from the doubling of the volume, and so we find

$$\begin{pmatrix} (S_I)_{\text{c.p.d.}} \text{ of the} \\ \text{c.p.d. adopted} \\ \text{when } t \gg t_2 \end{pmatrix} = S_I(t_1) + k \ln 2^N.$$

The information-theory entropy of the adopted canonical distribution is larger than the initial S_I by the positive quantity $k \ln 2^N$. The reason for the increase lies in the *discarding* of information about subtle correlations among molecular positions and velocities, correlations present in the evolved probability distribution but absent in the adopted canonical distribution.

Why Does It Happen?

The preceding theorems are fine, but how does one understand intuitively the approach to equilibrium? Here is a way, based on classical statistical mechanics. To set the scene, let's specify that the system's energy and external parameters have fixed values. Under these mild restrictions, many different molecular configurations are possible, but most of them have macroscopic properties that are nearly the same as some single set of macroscopic properties. Such "macroscopic uniformity" is characteristic only of systems containing many, many particles, and is so important that we should confirm it with an example.

Let's mentally divide a container into halves and ask for the number of molecular configurations that show n molecules in the first half and $N - n$ molecules in the second half. Our initial move is to subdivide each half into many equal cubical volumes of molecular size; the number of such tiny cells in each half is N^*, say. Roughly, the repulsive intermolecular forces ensure that, at most, one molecule can occupy such a tiny cell. Let's agree that a spatial arrangement of the N identical molecules is specified once we specify (a) which n cells out of N^* are occupied in the first half and (b) which $N - n$ out of N^* are occupied in the second. Next we need to ask, how many different ways are there in which n cells can be occupied in the first half? If we let "spin up" correspond to "occupied," the question is mathematically equivalent to asking, how many states are there with n spins up and $N^* - n$ spins down? Therefore Section 5.1 provides the answer: $N^*!/n!(N^* - n)!$. For the second half of the container, there is a similar question and answer; we need only replace n by $N - n$. Provided we neglect the attractive part of the intermolecular forces, each of the spatial arrangements has the same potential energy. Consequently, the number of different ways in which the kinetic energy can be shared among the molecules is the same for all spatial arrangements, and so let's forget about the momenta when we talk about molecular configurations. Once we agree on this, we may write

$$\begin{pmatrix} \text{the number of molecular} \\ \text{configurations with } n \\ \text{molecules in the first} \\ \text{half, } N - n \text{ in the second} \end{pmatrix} = \frac{N^*!}{n!(N^* - n)!} \times \frac{N^*!}{(N - n)!(N^* - [N - n])!}.$$

Before we can see what's going on, something must be done with the mess of factorials. Provided $N^* \gg N$, which will be true for a gas, we may approximate just as we did for the function $\mathcal{N}(m)$ in Section 5.2. The techniques used there, plus judicious rearranging, lead to

$$\begin{pmatrix} \text{same as} \\ \text{above} \end{pmatrix} \simeq \frac{(2N^*)^N}{N!} \frac{1}{(\pi N/2)^{1/2}} \exp[-(n - \tfrac{1}{2}N)^2/(\tfrac{1}{2}N)].$$

The distribution of configurations is sharply peaked around $n = \tfrac{1}{2}N$; indeed, an overwhelming majority of configurations have n within the range $\tfrac{1}{2}N \pm 100N^{1/2}$. Let's take one extreme of this range and compute the ratio of the number of molecules in the two halves:

$$\frac{\tfrac{1}{2}N - 100N^{1/2}}{\tfrac{1}{2}N + 100N^{1/2}} \simeq 1 - 400N^{-1/2} = 1 - 4 \times 10^{-8}$$

if $N = 10^{20}$. The ratio is so close to unity that, macroscopically, the molecules will appear to be equally divided between the two halves. In summary, though there are a few configurations with almost all molecules in one half, the majority of configurations have a spatial distribution that is very nearly the same as "some single spatial distribution," which is here a uniform distribution.

Now we ask, what is the macroscopic behavior of the system as it evolves, that is, as the spatial distribution changes because of molecular motion and the momentum distribution because of intermolecular and external forces? We have just seen that when the number of molecules is large, most configurations have macroscopic properties very nearly the same as some single set of macroscopic properties. If the configuration when we start observing the system is not one of those many, evolution is likely to take the system to such a configuration shortly, not for any teleologic reason, but simply because there are so many of them. Once the system is in that set of configurations, evolution is not likely to lead out of the set—at least not for a long time. (The recurrence theorems of both classical and quantum physics predict that evolution will eventually carry the system close to its starting configuration, but only after an inconceivably long time.) There is continual microscopic change, but after a short time, if not from the beginning, the macroscopic properties are dominated by "some single set of macroscopic properties," and those we then *call* the equilibrium properties. The very existence of "thermal equilibrium" depends on a simple proposition, but one difficult to prove comprehensively: an overwhelming majority of the configurations through which a system evolves have nearly the same macroscopic properties.

REFERENCES for Chapter 11

The classic paper on Ehrenfest's principle is by M. Born and V. Fock, *Zeitschrift für Physik*, **51**, 165 (1928). A reasonably accessible derivation is given by Richard C. Tolman (Section 97b) and a sophisticated derivation by Amnon Katz (Sections 8 and 10 of his Chapter 6); both books were cited in the references for Chapter 4.

The pioneering work of E. M. Purcell and R. V. Pound on negative temperatures is reported in the *Physical Review*, **81**, 279 (1951), in a letter entitled "A Nuclear Spin System at Negative Temperature." Perhaps the best single discussion of the Purcell-Pound experiment is that given by J. H. Van Vleck in an article, "The Concept of Temperature in Magnetism," in *Nuovo Cimento Supplement*, **6**, series 10 (1957). The general concept of a negative temperature in both statistical mechanics and thermodynamics is discussed by Norman F. Ramsey, *Physical Review*, **103**, 20 (1956).

On Alternatives to Almost-Periodicity

Some physicists look to the world outside the system proper for a mechanism and a theory giving a monotonic (or nearly so) approach to equilibrium. They argue that one may not idealize a " container wall " by replacing it with a time-independent infinite potential jump; rather, they feel, one must regard the wall as made of atoms that interact in a "random" fashion with the system proper. The "random" interaction with the outside world provides, in their analysis, a mechanism that destroys the almost-periodic behavior of an ideally isolated system and generates a rather uniform relaxation to thermal equilibrium. This point of view is nicely presented by J. M. Blatt, *Progress of Theoretical Physics*, **22**, 745 (1959); his paper contains references to earlier papers on the same theme.

A lengthy discussion by Tolman (cited in the references for Chapter 4) and a review article by D. ter Haar, *Reviews of Modern Physics*, **27**, 289 (1955), provide an extensive introduction to theories based on a "coarse-graining" of the probability distribution, that is, on a grouping together of states with nearly the same properties. A theorem by Percival in the *Journal of Mathematical Physics*, **2**, 235 (1961), is particularly germane to this approach.

PROBLEMS

11.1. Use the approximate expression for $\ln Z$ derived in Section 8.5 and calculate S_i, taking the intermolecular potential energy to be that of the Sutherland model. (If you express the integral in $\ln Z$ in terms of $B_2(T)$, you can save yourself some labor.) In the contribution to S_i from the attractive part of the potential, keep only the first *nonvanishing* term in powers of $1/T$. Then compare the effects that the repulsive and attractive contributions have on the change in temperature when the volume is very slowly expanded. If you can, offer qualitative explanations of the effects.

11.2. a. Starting with the probabilities (from the canonical distribution) for states of the entire photon system, show that the expectation value $\langle N \rangle$ of the number of photons may be written as

$$\langle N \rangle \equiv \sum_J N_J P_J = \sum_s \langle n_s \rangle,$$

where N_J denotes the number of photons in state ψ_J.

b. Turn the sum over single-photon states into an integral and show that $\langle N \rangle$ has the same dependence on T and V as does S_I. So constancy of S_I implies constancy of $\langle N \rangle$.

c. Form the ratio $\langle E \rangle / \langle N \rangle$, which will be proportional to kT, and estimate the numerical value of the proportionality constant. This will give you an estimate of a "typical" photon energy in black-body radiation. What would be the typical energy (in electron volts) for a kitchen oven? For the solar surface ($T \simeq 6{,}000$ K)?

11.3. In the numerical example in Section 11.4, the information-theory entropy of the lattice was taken to have a negligible quantitative influence on the final temperature and so was fully neglected. Reconsider the example, taking $\Theta_D = 300$ K and $\mu = 10^{-20}$ ergs/gauss (about one Bohr magneton); assuming only that the influence is very small, estimate numerically the (very minor indeed) extent to which the final temperature is not as low as the previous analysis indicated.

11.4. Could the Doppler broadening of a spectral line be used to measure an unknown gas temperature? Justify your answer. If yes, derive a relation yielding the temperature in terms of the observed line width and fixed parameters. Would you prefer a dense or dilute gas? Why? (There is some experimental evidence to support the belief that one can measure temperatures of order 450,000 K in this manner.)

11.5. Let us consider the maintenance of a probability distribution of canonical form for a gas of N structureless molecules whose intermolecular forces we neglect. This is an extreme idealization, for molecules do have internal structure (even monatomic ones) and do mutually interact. The gas is initially in equilibrium at a known temperature in a *cubical* container.

a. Show (or merely argue) that the energy eigenvalues E_j of the system are proportional to $V^{-2/3}$, both for fermions and for bosons.

b. Suppose that one very slowly expands the volume *while maintaining the cubical shape*. Does the probability distribution that slowly evolves retain a canonical form? If yes, how is the instantaneous temperature related to the instantaneous volume?

c. Construct the single-particle energy eigenstates of a particle in a *rectangular* box, with sides L_x, L_y, L_z, with no equalities among pairs of sides. (The energy eigenvalues will be proportional to $n_x^2/L_x^2 + n_y^2/L_y^2 + n_z^2/L_z^2$.)

d. Consider now the same idealized physical situation, but with the rectangular box replacing the cube. Suppose that one very slowly expands the rectangular box, *again while maintaining the shape;* that is, a ratio such as L_y/L_x remain fixed. Is canonical form maintained?

e. Suppose now that one very slowly expands the volume by extending the length of *one side only* (L_x, say) of the rectangular box. Is canonical form maintained?

A disaster appears to befall us, but we must remember that we have adopted an extreme idealization of a gas. Indeed, Katz's analysis on pp. 84–90 of his book, cited in Chapter 4, demonstrates that not all is lost; already slight interactions among the molecules can rescue us. Nonetheless, perhaps the most valid conclusion is that much theoretical work remains to be done on the questions of when and how canonical form is maintained.

THE DENSITY MATRIX
TECHNIQUE

This chapter reformulates some of the preceding results in a more general quantum-mechanical language. It requires some familiarity with quantum mechanics when expressed in terms of operators and matrix elements of those operators, and so the first section outlines a few essential quantum-mechanical notions. The second section introduces the density matrix in a rather general manner; the third considers the special form associated with the canonical probability distribution. The final section uses the density matrix to analyze paramagnetism without neglecting the mutual interactions, and so indicates the power of the technique. Indeed, we will find that the analysis calls into play virtually all the techniques of calculation that we have studied.

12.1. THE SUPERPOSITION OF STATES

Quantum mechanics is full of remarkable propositions, not the least of which is the principle of the superposition of states: if a system is known to be in a specific quantum-mechanical state, we may nonetheless represent the state of the system by a linear combination of some other states. An example may be useful. Suppose we measure the spin orientation of a particle of spin $\frac{1}{2}\hbar$, and find the spin orientation to be that indicated by this arrow: ↗. The particle is in a state $\psi_\nearrow$ with spin orientation ↗. Nonetheless, we may represent that quantum-mechanical state as a linear superposition of a state $\psi_\uparrow$ with spin orientation ↑ and a state $\psi_\downarrow$ with spin orientation ↓. We must be careful to choose the coefficients correctly, but that is minor. The procedure is analogous to expressing a linearly polarized electromagnetic wave as a superposition of righthanded and lefthanded circularly polarized waves, and vice versa.

Let us now set up the mathematical pattern. The superposition principle tells us that we may represent a definite state ψ of a system as a linear combination of states ψ_j with appropriately chosen coefficients c_j:

$$\psi = \sum_j c_j \psi_j. \tag{12.1.1}$$

Contrary to our previous usage, the states ψ_j are not necessarily energy eigenstates of the system. They might, for example, be states with definite values of angular momentum, or perhaps of linear momentum, neither of which are necessarily compatible with definite values of the energy.

The states ψ_j need not be orthogonal to one another, but convenience usually dictates such a choice. Whenever we write an expansion like that in equation (12.1.1), we will take the states ψ_j to be mutually orthogonal and normalized to unity:

$$\int \psi_k^* \psi_j = \begin{cases} 1 & \text{if } k = j \\ 0 & \text{if } k \neq j \end{cases} \equiv \delta_{kj}. \tag{12.1.2}$$

For the sake of brevity, we will leave off the differentials associated with the integration. The asterisk indicates that ψ_k^* is the complex conjugate of the state ψ_k. The symbol δ_{kj} has the meaning indicated in the curly brackets; it is commonly called the *Kronecker delta*. States with the properties indicated in equation (12.1.2) are called *orthonormal*.

An expansion like the preceding is possible only if we include enough states ψ_j in the set over which we sum. If the number of states in the set is large enough that any ψ for the system may be expanded in terms of them, the set of states ψ_j is called *complete*. The choice of a specific complete set of orthonormal states is called the choice of a specific *representation*.

This is the place for a remark of some importance. In the earlier chapters we often spoke of summing over "all" the energy eigenstates of a system. The precise meaning of the statement is that we should sum over a complete orthonormal set of energy eigenstates of the system. The precision is relevant when there is degeneracy, that is, when there are two or more distinct (meaning linearly independent) energy eigenstates with the same numerical value for the energy.

Given the state ψ and the set of states ψ_j, the determination of the correct coefficients is, in principle, quite simple. To determine c_k we multiply both sides of equation (12.1.1) by $\psi_k{}^*$ and integrate:

$$\int \psi_k{}^* \psi = \sum_j c_j \int \psi_k{}^* \psi_j$$
$$= \sum_j c_j \delta_{kj} = c_k. \tag{12.1.3}$$

The step to the second line uses the mutual orthogonality and proper normalization of the states in the set. Thus the coefficient c_k may be calculated by evaluating the integral on the left. As one would hope, the coefficients are uniquely determined by ψ and the set of states used for the expansion.

The next matter on the agenda is that of quantum-mechanical expectation values. Suppose we are concerned with the total magnetic moment along the z-axis of some specific physical system. Let us write the associated quantum-mechanical operator as $\mathcal{M}$, omitting a subscript z to avoid cluttering subsequent equations. The system is in state ψ. If ψ is an eigenstate of the operator $\mathcal{M}$, matters are simple, for then measurement of the total magnetic moment along z is certain to yield the eigenvalue of $\mathcal{M}$ in state ψ. Typically, ψ is not an eigenstate of the operator. Nonetheless, measurement will always yield some one of the eigenvalues of the operator $\mathcal{M}$. In estimating the component of the total magnetic moment, we will have to be satisfied with an estimate that will probably be borne out if measurements are made on a large number of systems, each in state ψ, and the results are then averaged. We may take this to be the provisional meaning of the *quantum-mechanical expectation value*; it is written symbolically on the left in the following equation and is to be computed by evaluating the righthand side:

$$\langle M \rangle_{\text{q.m.}} = \int \psi^* \mathcal{M} \psi. \tag{12.1.4}$$

The expression on the right is sometimes derived, sometimes postulated. In a few steps we can bring it to a form such that we can see the connection with expectation values as we defined them in Section 2.8.

Let us first substitute for ψ and ψ^* a general expansion like that given in equation (12.1.1), and then interchange the orders of integration and summation:

$$\langle M \rangle_{\text{q.m.}} = \int \left(\sum_k c_k^* \psi_k^* \right) \mathcal{M} \left(\sum_j c_j \psi_j \right)$$

$$= \sum_j \sum_k c_k^* c_j \int \psi_k^* \mathcal{M} \psi_j$$

$$= \sum_j \sum_k c_k^* c_j M_{kj}. \tag{12.1.5}$$

An abbreviation,

$$M_{kj} \equiv \int \psi_k^* \mathcal{M} \psi_j, \tag{12.1.6}$$

gets us to the third line. The quantity M_{kj} is called the *matrix element* of the operator $\mathcal{M}$ between the states ψ_k and ψ_j.

Thus far the states in the set ψ_j have not been assumed to have any special properties. Let us now take each to be an eigenstate of the operator $\mathcal{M}$, with eigenvalue m_j:

$$\mathcal{M} \psi_j = m_j \psi_j. \tag{12.1.7}$$

The matrix element simplifies greatly:

$$M_{kj} = \int \psi_k^* \mathcal{M} \psi_j = \int \psi_k^* m_j \psi_j = m_j \delta_{kj}.$$

This we substitute into the general expression, equation (12.1.5), and then use the properties of the Kronecker delta to perform the sum over k:

$$\langle M \rangle_{\text{q.m.}} = \sum_j \sum_k c_k^* c_j m_j \delta_{kj}$$

$$= \sum_j c_j^* c_j m_j.$$

The product $c_j^* c_j$ is just the square of the absolute value of c_j, and so we may write:

$$\langle M \rangle_{\text{q.m.}} = \sum_j m_j |c_j|^2 \tag{12.1.8}$$

when the states ψ_j are eigenstates of $\mathcal{M}$. The form of this expression suggests a reasonable interpretation for the expansion coefficient c_j:

$$|c_j|^2 = \left| \int \psi_j^* \psi \right|^2$$

$$= \begin{pmatrix} \text{the probability, given that the} \\ \text{system is in state } \psi, \text{ that} \\ \text{measurement will yield the} \\ \text{eigenvalue } m_j \text{ of the operator } \mathcal{M} \end{pmatrix} \qquad (12.1.9)$$

Provided we accept this interpretation, the quantum-mechanical expectation value is perfectly analogous in form and meaning to all the other expectation values with which we have dealt.

The last statement warrants elaboration. Provided one accepts equation (12.1.9), one may use the notion of probability as we adopted it in Chapter 2 and may regard $\langle M \rangle_{\text{q.m.}}$ as an expectation value estimate in full accordance with the definition adopted in Section 2.8. The probabilities will provide us with estimates of frequencies of occurrence, as we found in Section 2.10. Certainly it is reasonable to test a quantum-mechanical calculation experimentally by observing frequencies of occurrence and seeing whether they are in reasonable agreement with the theoretical estimates. The meaning for the quantum-mechanical expectation value that was called provisional when introduced can indeed be replaced by the combination of equations (12.1.8) and (12.1.9), with "probability" understood as we have used it throughout this book. The sole reason for introducing the provisional meaning is that it is conventional and hence likely to be more familiar. It can, however, be replaced by a more comprehensive meaning.

Incidentally, the probabilities $|c_j|^2$ do satisfy the normalization condition for probabilities referring to inferences that are exhaustive and mutually exclusive on the given hypothesis. This follows directly from the normalization condition for ψ and from the orthonormality of the states ψ_j:

$$1 = \int \psi^* \psi = \int \left(\sum_k c_k^* \psi_k^* \right) \left(\sum_j c_j \psi_j \right)$$

$$= \sum_j \sum_k c_k^* c_j \delta_{kj} = \sum_j |c_j|^2. \qquad (12.1.10)$$

12.2. THE DENSITY MATRIX

In the typical situation in statistical mechanics, we cannot associate a unique quantum-mechanical state with the physical system. The information granted us is far too meager. Even in choosing a state to describe the system, we are forced to deal with probabilities. Quantum statistical mechanics—on which the entire development in this book has been based—requires a synthesis of probabilistic methods. Probabilities appear first when we select quantum-mechanical states to use for describing the system, and then again when we use the states to estimate the values of physical quantities.

Let us set up a rather general context. We are concerned with estimating the properties of a specific system. The information that we have does not enable us to select a single state as uniquely appropriate for describing the system. We must deal with a set of them. On the basis of our information, we select a set of states, an individual member of which is designated by ψ_α. The states in the set need not form a complete set of states, in contrast to the states ψ_j; the switch to a Greek subscript is intended to indicate this. Also, the states need not be energy eigenstates. We will, however, specify that they are orthonormal.

Let us write P_α for the probability of the inference that the state ψ_α is appropriate for describing the physical system, given our information. Further, let us specify that the inferences are exhaustive and mutually exclusive on the hypothesis. Then the probabilities must satisfy the normalization condition:

$$\sum_\alpha P_\alpha = 1. \tag{12.2.1}$$

An example is in order; we take one from nuclear physics. Protons from a cyclotron can be scattered off a succession of carbon targets in such a way that a partially polarized beam is produced. For an individual proton in the beam, one may know the momentum quite well but one knows (at most) a probability for spin up (↑) and a probability for spin down (↓). The set of states ψ_α would have two members, $\psi_\uparrow$ and $\psi_\downarrow$, states with the same momentum but different spin orientations. The characterization "partially polarized" means that the probabilities $P_\uparrow$ and $P_\downarrow$ differ, but that neither is unity. Now we return to the more general context.

We look for an estimate, one that neglects no possibilities, of the value of a physical quantity such as the total magnetic moment along the z-axis. We can construct such an estimate $\langle M \rangle$ by forming two successive expectation values:

$$\langle M \rangle = \sum_\alpha P_\alpha \int \psi_\alpha^* \mathcal{M} \psi_\alpha. \tag{12.2.2}$$

First we form the quantum-mechanical expectation value for state ψ_α in accordance with equation (12.1.4). Next we weight each such quantity with the probability that the associated state is appropriate for describing the system, and finally we sum over the states. Equation (12.2.2) is the crucial equation. The remainder of this section consists merely of developing mathematical techniques for working with it. But before we turn to those techniques, two remarks are in order.

The first remark concerns the connection with the canonical probability distribution. If the states ψ_α form a complete orthonormal set of energy eigenstates and if the probabilities have the canonical form,

$$P_\alpha = \frac{e^{-\beta E_\alpha}}{Z},$$

then $\langle M \rangle$ is precisely the expectation value according to the canonical distribution. In the past we have dealt with problems in which the quantities corresponding to the integral in equation (12.2.2) could be assessed without elaborate theory. Such is certainly true when we estimate the energy, for then the integral is just E_α. In the magnetic problems we analyzed, the energy eigenstates were simultaneously eigenstates of the z-component of the total magnetic moment; so the integral was simply the value of the moment in the state. (To be sure, the extension in Appendix C on the magnetic moment operator leads to precisely the integral in equation (12.2.2) when the states ψ_α are energy eigenstates.) The only exception was for pressure calculations, where a special argument based on energy conservation gave us the quantity to insert. The point of this remark is that we have indeed been using the proper quantum-mechanical canonical probability distribution all along.

The second remark concerns the extent to which a single measurement of a physical quantity could yield numerical agreement with an estimate like that in equation (12.2.2). Let's take magnetic moment and the canonical probability distribution. Quantum mechanics tells us that the measurement of a component of the moment can yield only an eigenvalue of the corresponding operator. The estimate $\langle M \rangle$ may not be equal, numerically, to any of the eigenvalues. (An analogous situation prevailed in the box-of-matches example that introduced expectation values in Section 2.8.) If the system is macroscopic, however, the eigenvalues are very closely spaced, and so, although the estimate may fall between a pair of eigenvalues, that is really unimportant. In statistical mechanics the expectation value can be, and generally is, an excellent estimate. Now we return to the general analysis and the development of mathematical techniques.

The first step consists of substituting for each ψ_α and ψ_α^* an expansion like that in equation (12.1.1). To avoid ambiguity, we need to place two indices on the expansion coefficients:

$$\psi_\alpha = \sum_j c_j(\alpha)\psi_j. \tag{12.2.3}$$

We neither need to nor should specialize the states ψ_j. Any complete ortho-normal set will do. Now we substitute and reorder the summations:

$$\langle M \rangle = \sum_\alpha P_\alpha \int \left(\sum_k c_k^*(\alpha)\psi_k^* \right) \mathcal{M} \left(\sum_j c_j(\alpha)\psi_j \right)$$

$$= \sum_j \sum_k \left[\sum_\alpha P_\alpha c_k^*(\alpha)c_j(\alpha) \right] M_{kj}. \qquad (12.2.4)$$

Only through the presence of the matrix element of the magnetic moment operator is the righthand side specialized to yield $\langle M \rangle$. The quantity in square brackets will appear whenever we estimate a physical quantity by the general scheme of equation (12.2.2). This leads one to introduce a symbol and a name for it. The matrix multiplication is simplest if we arrange the indices in the abbreviation as follows:

$$\rho_{jk} \equiv \sum_\alpha P_\alpha c_k^*(\alpha)c_j(\alpha). \qquad (12.2.5)$$

The ensuing matrix is called the *density matrix*. (The name arises, in part, because in one representation the diagonal elements give the spatial probability density for the particles in the system.) We may profitably look upon ρ_{jk} as the matrix element, between states ψ_j and ψ_k, of an operator ρ:

$$\rho_{jk} = \int \psi_j^* \rho \psi_k. \qquad (12.2.6)$$

The operator ρ is defined such that its matrix elements give, numerically, the righthand side of equation (12.2.5).

With the aid of the density matrix, we can tidy up equation (12.2.4), though we certainly do not change its content:

$$\langle M \rangle = \sum_j \sum_k \rho_{jk} M_{kj}$$

$$= \sum_j (\rho M)_{jj} \equiv \mathrm{Tr}\,(\rho \mathcal{M}). \qquad (12.2.7)$$

The sum over k, looked upon as matrix multiplication, yields the matrix element of the product $\rho \mathcal{M}$ between states ψ_j and ψ_j. The final sum, that over j, is a sum of the diagonal elements of that matrix. Such a sum is often called the *trace* of the matrix and abbreviated as on the far right. We have used no properties of the set of states ψ_j that are not shared by all complete orthonormal sets. To compute $\langle M \rangle$ we may evaluate the trace of the operator $\rho \mathcal{M}$ with any complete orthonormal set of states. (Problem 12.2 provides an assist in proving explicitly that the trace of an operator is independent of the representation used.) Herein lies an advantage of the density-matrix technique, for we may use the set that is easiest to work with.

Before we restrict matters and look at the operator ρ associated with the canonical distribution, we should establish three general properties of the density matrix. Each is really nothing more than a restatement, in the density-matrix language, of statistical properties that we already know about.

Normalization Condition

The trace of the density matrix has the numerical value one:

$$\text{Tr}\,\rho = 1. \tag{12.2.8}$$

To show this, we need only take the trace of both sides of the equation defining the density matrix, equation (12.2.5):

$$\sum_j \rho_{jj} = \sum_j \sum_\alpha P_\alpha c_j^*(\alpha) c_j(\alpha)$$

$$= \sum_\alpha P_\alpha \sum_j c_j^*(\alpha) c_j(\alpha)$$

$$= \sum_\alpha P_\alpha = 1.$$

The step to the third line uses equation (12.1.10), applicable because the states ψ_α are properly normalized.

Time Development

If a physical system has a definite energy operator* $\mathcal{H}$, then the time development of the operator ρ is governed by a deterministic equation:

$$i\hbar\,\frac{\partial \rho}{\partial t} = \mathcal{H}\rho - \rho\mathcal{H}. \tag{12.2.9}$$

A glance back at equations (12.2.2, 3, 5) indicates that the operator ρ is determined by the states ψ_α and the associated probabilities P_α. Each state ψ_α will evolve in time according to the Schrödinger equation:

$$i\hbar\,\frac{\partial \psi_\alpha}{\partial t} = \mathcal{H}\psi_\alpha.$$

---

* Here "definite" means that the explicit expression for $\mathcal{H}$ is known by us and that any temporal variation in $\mathcal{H}$ arises from known changes in external parameters; the term excludes a coupling to another system also described quantum-mechanically, for its influence could be known only in a statistical sense.

Each probability P_α will, however, remain constant, for there is no reason to change the assessment of the probability, given the initial information, that the evolving state ψ_α appropriately describes the system. This paragraph forms the basis for deriving the deterministic equation for ρ.

Equation (12.2.9) holds for both time-independent and time-dependent $\mathcal{H}$. In an attempt to pare the algebra down to reasonable proportions, we will work out a derivation adequate for the time-independent case and many (though not all) time-dependent cases. The fully general derivation is analogous.

The operator equation will follow readily once we have derived the corresponding equation in matrix-element form. We will need to differentiate both sides of the equation defining the density matrix. Let us prepare by computing derivatives of the expansion coefficients. For convenience, we take the states on the right in the expansion

$$\psi_\alpha = \sum_l c_l(\alpha)\psi_l$$

to be independent of time, that is, the ψ_l carry the spatial dependence of ψ_α but the $c_l(\alpha)$ carry all the time-dependence. (It is this step, made to simplify the algebra, that limits somewhat the generality of the derivation.) We substitute the expansion into the Schrödinger equation,

$$i\hbar \sum_l \frac{\partial c_l(\alpha)}{\partial t}\psi_l = \sum_l c_l(\alpha)\mathcal{H}\psi_l,$$

multiply from the left with $\psi_j{}^*$, and then integrate to get

$$i\hbar \frac{\partial c_j(\alpha)}{\partial t} = \sum_l c_l(\alpha)H_{jl}.$$

In addition to this, we will need an equation for $\partial c_k{}^*(\alpha)/\partial t$. Upon taking the complex conjugate of the above equation and replacing j by k, we arrive at

$$-i\hbar \frac{\partial c_k{}^*(\alpha)}{\partial t} = \sum_l c_l{}^*(\alpha)H_{kl}{}^*.$$

Now we compute the time derivative of ρ_{jk}:

$$i\hbar \frac{\partial}{\partial t}\rho_{jk} = i\hbar \sum_\alpha P_\alpha \frac{\partial}{\partial t}c_k{}^*(\alpha)c_j(\alpha)$$

$$= \sum_\alpha P_\alpha \sum_l [c_k{}^*(\alpha)c_l(\alpha)H_{jl} - c_l{}^*(\alpha)H_{kl}{}^*c_j(\alpha)]$$

$$= \sum_l (H_{jl}\rho_{lk} - \rho_{jl}H_{kl}{}^*)$$

$$= \sum_l (H_{jl}\rho_{lk} - \rho_{jl}H_{lk}) = (\mathcal{H}\rho - \rho\mathcal{H})_{jk}.$$

The step to the last line uses a property of the matrix elements of $\mathcal{H}$, namely $H_{kl}{}^* = H_{lk}$. Since we used time-independent states in the expansion and matrix elements, the matrix equation permits us to infer the quoted operator equation for ρ.

Equation (12.2.9) tells us that ρ will not change with time only if the operators ρ and $\mathcal{H}$ commute, that is, only if $\mathcal{H}\rho = \rho\mathcal{H}$. Having $\mathcal{H}$ independent of time is far from sufficient to guarantee that ρ is constant in time. The value of ρ at an initial time also enters into the question.

In Section 11.6 the statement was made that, when $\mathcal{H}$ is independent of time, the probability distribution which evolves with time is (if not constant) almost periodic in time. That statement may be formulated in the density-matrix language by saying that the operator ρ is almost periodic in time. In principle, the operator ρ never settles down to equilibrium if it is not initially there. In practice, the difference between the evolving operator and an operator representing the canonical distribution (and hence describing thermal equilibrium) is often negligible after a reasonable waiting time.

Information-Theory Entropy

The information-theory entropy S_I associated with the probabilities P_α may be written in terms of the density matrix as

$$S_I = -k \operatorname{Tr} (\rho \ln \rho). \tag{12.2.10}$$

The trace may, of course, be evaluated in any representation.

> A function of an operator, such as $f(\rho) = \ln \rho$, is to be understood in the following sense. To compute the effect of $f(\rho)$ operating on a state ψ, one first expands ψ, in the fashion of equation (12.1.1), as a sum over eigenstates of the operator ρ; then one multiplies each eigenstate in the sum by f(eigenvalue of ρ), and takes the ensuing sum to be the new state $f(\rho)\psi$. Often a simpler procedure is permissible: one merely substitutes the operator ρ in a formal Taylor's series expansion of the function $f(\rho)$.

To prove equation (12.2.10), we need only show that the value of the righthand side agrees numerically with the more familiar expression

$$S_I = -k \sum_\alpha P_\alpha \ln P_\alpha. \tag{12.2.11}$$

Since the states ψ_α are orthonormal, we may use them in forming a complete orthonormal set, augmenting them with other states, if necessary, to ensure that we have a complete set. In that representation all the expansion coefficients appearing in equation (12.2.5) are either 1 or 0. The density matrix is diagonal, each diagonal element being either one of the probabilities P_α or zero. Hence in that representation the trace appearing in equation (12.2.10) yields precisely the sum in equation (12.2.11). Since the numerical value of a trace is independent of the representation, the expression for S_I in terms of the density matrix is valid in general.

Incidentally, our equation for the time development of ρ can be used to prove the constancy in time of S_I, provided, of course, that the system has a definite $\mathcal{H}$. The derivation we made in Section 11.1, when the initial probability distribution was a canonical distribution, is fully as rigorous (for that special case) and shows far more clearly the reason for the constancy. Although the states ψ_α evolve in time, they do so deterministically, *and* we have no reason to change the probabilities P_α. We neither lose nor gain information about the system; so S_I must remain constant in numerical value.

Remarks on Generality

The foregoing development is adequately general for our purposes. We should note in passing, however, that the density matrix can be promoted to the status of a primary concept, becoming independent of any scheme used to introduce it. (In an article cited in the references, D. ter Haar discusses three introductory schemes, one of them equivalent to our development.) The density matrix can become the primary statistical or probabilistic element both in quantum theory and in quantum statistical mechanics. As part of the promotion process, equation (12.2.7) becomes, by definition, the prescription for computing expectation value estimates of physical quantities. Likewise, equation (12.2.10) becomes the general measure for the "missing information," in the sense of the additional information needed to achieve the optimum state of knowledge about the system. That optimum is itself limited by quantum-mechanical principles (such as the Heisenberg uncertainty principle) and corresponds to knowing "the correct" quantum-mechanical state for describing the system. There remains, of course, the problem of determining ρ, given some physical context. An illustration will appear toward the end of the next section.

12.3.   THE CANONICAL DENSITY MATRIX

It is time now to see how the canonical probability distribution appears in the framework of the density-matrix technique. To determine the operator ρ, we first compare two expressions for $\langle M \rangle$ in terms of matrix elements and then infer what the operator must be. If we go back to equation (12.2.2), take the set of states ψ_α to be a complete orthonormal set of energy eigenstates, and take the associated probabilities P_α to have the canonical form, then we may write out $\langle M \rangle$ as

$$\langle M \rangle = \sum_j \frac{e^{-\beta E_j}}{Z} \int \psi_j^* \mathcal{M} \psi_j = \sum_j \frac{e^{-\beta E_j}}{Z} M_{jj}.$$

(The change from α to j subscripts is permissible because we are dealing, in the canonical distribution, with a complete set of states; the change gives the expression a more familiar appearance.) The expression for $\langle M \rangle$ in terms of the density matrix is

$$\langle M \rangle = \sum_j \sum_k \rho_{jk} M_{kj}.$$

Comparison shows us that the canonical density matrix has the form

$$\rho_{jk} = \frac{e^{-\beta E_j}}{Z} \delta_{jk} \tag{12.3.1}$$

when the matrix element is formed with energy eigenstates.

Now we look for an operator ρ such that its matrix element between energy eigenstates ψ_j and ψ_k equals the righthand side of the preceding equation, that is, such that

$$\int \psi_j^* \rho \psi_k = \frac{e^{-\beta E_j}}{Z} \delta_{jk}.$$

The solution is

$$\rho = \frac{1}{Z} e^{-\beta \mathcal{H}}. \tag{12.3.2}$$

The exponential of the energy operator $\mathcal{H}$ is to be understood as an abbreviation for the Taylor's series expansion, that is,

$$e^{-\beta \mathcal{H}} = \sum_{n=0}^{\infty} \frac{1}{n!} (-\beta \mathcal{H})^n. \tag{12.3.3}$$

The partition function Z may first be written in terms of energy eigenstates and then expressed as a trace:

$$Z = \sum_j e^{-\beta E_j} = \sum_j \int \psi_j^* e^{-\beta \mathcal{H}} \psi_j$$

$$= \mathrm{Tr}\ e^{-\beta \mathcal{H}}. \tag{12.3.4}$$

In the first line, the last expression is the trace of $\exp(-\beta\mathcal{H})$ when formed with energy eigenstates. Since the value of a trace does not depend on one's choice of representation, Z may be evaluated as a trace with any complete orthonormal set of states, as indicated in the second line. That form for Z may seem to be only a minor mathematical generalization. When we finally struggle with the mutual interactions of paramagnetic particles, we will find that it provides a tremendous computational advance. Moreover, the trace formulation is the starting point for a general derivation of the classical partition function as a sometimes-valid limit of the quantum-mechanical one.

This is the point at which to comment on the derivation back in Section 4.5 of a probability distribution of the canonical form. We were dealing with a system in equilibrium and had presented arguments for using the energy eigenstates to describe the system. Further, we supposed that we knew enough about the actual system so that it was reasonable to force an expectation value calculation of the energy to yield a specific known value, denoted by $\tilde{E}$. Then we turned to Criterion II on the assignment of probabilities and determined the " best " probability distribution by looking for the distribution that would maximize the missing information subject to the normalization condition and the energy constraint.

The analogous procedure in terms of the density matrix would be governed by the following instructions. To choose the " best " operator ρ in accordance with Criterion II, look for the ρ that maximizes

$$S_I = -k \operatorname{Tr}(\rho \ln \rho)$$

subject to the normalization condition

$$\operatorname{Tr} \rho = 1$$

and the energy constraint

$$\operatorname{Tr}(\rho\mathcal{H}) = \tilde{E}.$$

If one follows the instructions,* there emerges the operator of equation (12.3.2), though with β to be determined by the energy constraint and not yet related to temperature, just as in section 4.5. In effect, the work of Sections 4.6 and 4.7 remains before one arrives at precisely the canonical probability distribution. The advantage of the approach with the density matrix is that

---

* A sophisticated derivation is given by Amnon Katz (cited in the references for Chapter 4), and a more accessible derivation by Gregory H. Wannier, *Statistical Physics* (New York: Wiley, 1966), pp. 88–92.

one need not make the physical argument about using energy eigenstates; Criterion II and the mathematics select them automatically.

The canonical probability distribution is designed to describe a system known to be in equilibrium in an unchanging environment. As such, it is to provide statistical estimates that are independent of time. We have now the equations necessary to verify that independence.

A physical quantity, such as the z-component of the total magnetic moment, will be represented in quantum mechanics by an operator, such as $\mathcal{M}$. Let us form the expectation value estimate according to the canonical distribution and then differentiate with respect to time:

$$\frac{d}{dt} \langle M \rangle = \frac{d}{dt} \operatorname{Tr} \rho \mathcal{M} = \frac{d}{dt} \sum_l \int \psi_l^* \rho \mathcal{M} \psi_l$$

$$= \operatorname{Tr} \left(\frac{\partial \rho}{\partial t} \mathcal{M} + \rho \frac{\partial \mathcal{M}}{\partial t} \right).$$

Moving the differentiation inside the trace symbol is permissible. If the states used to evaluate the trace are not time-independent, but rather evolve with the time-dependent Schrödinger equation, the effects of differentiating ψ_l and ψ_l^* cancel, yielding the last line in any case.

Since the operator ρ associated with the canonical distribution is a function of the operator $\mathcal{H}$ only, ρ and $\mathcal{H}$ commute, and thus ρ does not change with time; that is,

$$\mathcal{H} \frac{e^{-\beta \mathcal{H}}}{Z} - \frac{e^{-\beta \mathcal{H}}}{Z} \mathcal{H} = 0$$

implies

$$\frac{\partial \rho}{\partial t} = 0,$$

upon appeal to equation (12.2.9). In our scheme of quantum mechanics, the operator for a physical quantity such as a magnetic moment will not depend on the time: $\partial \mathcal{M}/\partial t = 0$. The upshot is then

$$\frac{d}{dt} \langle M \rangle = 0$$

for the canonical distribution. Only a non-zero result could be a surprise; nonetheless, it is gratifying to find the time-independence of the statistical estimates confirmed by the elaborate machinery.

12.4. SPIN-SPIN INTERACTIONS IN PARAMAGNETISM

Throughout this book we have exercised new theoretical machinery by using it to analyze a system of paramagnets. The decisions to do so were based partially on analytic tractability. Such pragmatism should not, however, be allowed to undermine an appreciation of how amazingly fruitful the study of paramagnetic systems has been. For decades paramagnetic crystals provided the sole means of producing temperatures below 1 K. Yet in all our work we have had to sidestep, in one way or another, a confrontation with the mutual interactions of the individual paramagnets. Either we neglected those interactions (as being relatively minor) or we incorporated them only through the device of an effective field H_{eff} (as in Section 11.4). The theoretical advance provided by the trace expression for Z finally enables us to deal microscopically with the mutual interactions. The calculation is not trivial, and approximations will again be necessary, but we will be able to calculate their influence on $\langle M_z \rangle$ in microscopic detail. Moreover, we will be able to determine the final temperature in a magnetic cooling process without resort to the effective field notion. The approach is that pioneered by Ivar Waller and J. H. Van Vleck. It is fitting that this book conclude with an analysis of magnetic dipole-dipole interactions, both because of the continuing research interest in paramagnetic systems and because virtually all of the formidable theoretical machinery that we have developed will come into play.

With one prominent exception, the assumptions about the paramagnetic system are those of Section 7.3. The notation is the same. The new element is the inclusion, in the energy operator for the system, of the mutual interactions of the paramagnets. First we will work out, in abbreviated fashion, the mutual potential energy of a pair of magnetic dipoles. Then we will sum over all distinct pairs to get the total potential energy of the mutual interactions.

The magnetic field produced by a magnetic dipole moment μ_1 falls off with distance as the inverse third power of the distance. There is also a strong angular dependence. Let us denote the vector distance between the position of μ_1 and that of a second dipole μ_2 by $\mathbf{r}_{12} = r_{12}\hat{\mathbf{r}}_{12}$, the latter vector being a unit vector pointing from the position of μ_1 to that of μ_2. Then we may write

$$\begin{pmatrix} \text{the magnetic field} \\ \text{at } \mu_2 \text{ due to } \mu_1 \end{pmatrix} = -\frac{1}{r_{12}^3}(\mu_1 - 3\mu_1 \cdot \hat{\mathbf{r}}_{12}\, \hat{\mathbf{r}}_{12}).$$

To compute the mutual potential energy of the pair of dipoles, we need only proceed as we have previously with a true external magnetic field: we take the scalar product of μ_2 with the field due to μ_1 and apply a minus sign. Doing so yields

$$\left(\begin{array}{c}\text{the } \textit{mutual} \text{ potential}\\ \text{energy of dipoles}\\ \mu_1 \text{ and } \mu_2\end{array}\right) = \frac{1}{r_{12}{}^3}(\mu_1 \cdot \mu_2 - 3\,\mu_1 \cdot \hat{\mathbf{r}}_{12}\,\mu_2 \cdot \hat{\mathbf{r}}_{12}).$$

The absence of a factor of 2 in the mutual energy is correct, though perhaps not easy to see. (Problem 12.3 outlines a direct calculation for a pair of electric dipoles, and from that one may argue for the above result by analogy.) The total potential energy of the dipole-dipole interactions is the sum over all distinct pairs of a term like the above.

The energy operator $\mathcal{H}$ for the system consists of two parts: (1) the dipole-dipole interactions, and (2) the potential energy of interaction with an external magnetic field $\mathbf{H} = H_z \hat{\mathbf{z}}$. Following closely the notation of Section 7.3, we write the magnetic dipole moment of the ith paramagnet as

$$\mu_i = g\mu_B \mathbf{J}_i,$$

with g the Landé factor, μ_B the Bohr magneton, and $\mathbf{J}_i$ the dimensionless angular momentum operator, that is, the angular momentum operator after division by $\hbar$. Then, with N spatially fixed paramagnets in the system, we have

$$\begin{aligned}\mathcal{H} &= -g\mu_B \sum_{i=1}^{N} J_{iz} H_z \\ &\quad + (g\mu_B)^2 \sum_{k<i} r_{ik}{}^{-3}(\mathbf{J}_i \cdot \mathbf{J}_k - 3\mathbf{J}_i \cdot \hat{\mathbf{r}}_{ik}\,\mathbf{J}_k \cdot \hat{\mathbf{r}}_{ik}) \\ &\equiv \mathcal{A}H_z + \mathcal{B}.\end{aligned} \qquad (12.4.1)$$

The somewhat cryptic notation $k < i$ under the second summation sign indicates a sum over both i and k but with the proviso that k be less than i, that is, $1 \leqslant k < i, 1 \leqslant i \leqslant N$. This ensures that the sum includes each distinct pair of dipoles once but only once, precisely what we want. The symbol $k < i$ can be replaced mentally by the phrase "all distinct pairs." The symbols $\mathcal{A}$ and $\mathcal{B}$ designate convenient abbreviations for the corresponding operators. The expression for $\mathcal{B}$ in terms of the $\mathbf{J}$'s leads to the designation "spin-spin interactions" used in the title of this section.

Our first interest lies in $\langle M_z \rangle$, the estimate that the canonical distribution provides of the z-component of the system's magnetic moment. The analysis in Section 7.2 and in Appendix C indicates that knowing the partition function suffices for the calculation; we may use the relation

$$\langle M_z \rangle = \frac{1}{\beta Z} \frac{\partial Z}{\partial H_z}. \qquad (12.4.2)$$

Recapitulating the essential points in the derivation may be useful. The primary result of Appendix C may be summarized as

$$\begin{pmatrix} \text{the quantum-mechanical expectation} \\ \text{value of the } z\text{-component of the} \\ \text{total magnetic moment for the} \\ \text{energy eigenstate } \psi_j \end{pmatrix} = -\frac{\partial E_j}{\partial H_z}.$$

Now we form a second expectation value, weighting each state ψ_j with the probability according to the canonical distribution:

$$\langle M_z \rangle = \sum_j \left(\frac{e^{-\beta E_j}}{Z}\right)\left(-\frac{\partial E_j}{\partial H_z}\right) = \frac{1}{\beta Z}\frac{\partial}{\partial H_z}\sum_j e^{-\beta E_j}.$$

The rearrangement, and then recognition of the remaining sum as simply Z, complete the derivation.

To calculate the energy eigenstates and eigenvalues of the operator $\mathcal{H}$ is a hopeless task. The partition function cannot be computed in the fashion that we have formerly used: directly summing $\exp(-\beta E_j)$. Yet our new machinery is so powerful that we can compute an approximation for Z that is adequate from high temperature down to low (though limited) temperature, provided the external field is not too strong. The nature of the approximations is best spelled out as we go along.

The first move consists of expressing the partition function as the trace of the appropriate operator. With the aid of equations (12.3.3, 4) we may write

$$Z = \mathrm{Tr}\ e^{-\beta \mathcal{H}}$$

$$= \mathrm{Tr}\ \sum_{n=0}^{\infty} \frac{1}{n!}(-\beta\mathcal{H})^n. \tag{12.4.3}$$

Each term $\mathcal{H}^n$, the n-fold product of the operator $\mathcal{H}$ with itself, may be written out in terms of $\mathcal{A}H_z$ and $\mathcal{B}$. Then we may rearrange the sum under the trace symbol according to powers of H_z. This may be symbolized as

$$Z = \mathrm{Tr}\ \sum_{n'=0}^{\infty} \mathcal{O}_{n'}\, H_z^{n'},$$

where $\mathcal{O}_{n'}$ is the operator, emerging from the rearrangement, associated with the n'th power of H_z. There is method in this madness. The partition function must be an *even* function of H_z, and so the trace of each operator $\mathcal{O}_{n'}$, for n' odd, must be zero. We immediately eliminate half the terms in the sum over n'. An indication of why Z must be an even function of H_z is certainly in order; it follows now.

If we replace H_z by $-H_z$, we must find that the new expectation value of the z-component of the total moment is the negative of the old value. After all, the change to $-H_z$ amounts merely to reversing the direction of the external magnetic field. Now we look at the expression for $\langle M_z \rangle$ in equation (12.4.2). The differentiation $\partial/\partial H_z$ changes the odd-even property of each term in Z. Unless Z has only even terms or only odd terms, $\langle M_z \rangle$ will not behave properly, that is, will not be an odd function of H_z. Since Z has at least one even term, namely Tr 1, arising from $(-\beta \mathcal{K})^0$, the partition function must be an even function of H_z.

We have gone about as far as we can with exact relations. Back in Section 5.1 we found that the dimensionless factor $\mu_B H_z / kT$ had the small value of about 0.18 when H_z was as large as 10^4 gauss and T as low as 4 K. So we should get an approximation for Z that is adequate for a wide range of fields and temperatures if we expand in ascending powers of the small quantity $\mu_B H_z / kT$. Let us adopt this course.

Upon going back to equation (12.4.3), expanding $\mathcal{K}^n$, and collecting the lowest order terms that are even in H_z, one finds

$$Z = \text{Tr}\,[1 - \beta\mathcal{B} + \tfrac{1}{2}\beta^2\mathcal{B}^2 + O(\beta)^3]$$
$$+ H_z{}^2\,\text{Tr}\,[\tfrac{1}{2}\beta^2\mathcal{A}^2 - \tfrac{1}{6}\beta^3(\mathcal{A}^2\mathcal{B} + \mathcal{A}\mathcal{B}\mathcal{A} + \mathcal{B}\mathcal{A}^2) + O(\beta^4)] + O(H_z{}^4).$$

$$(12.4.4)$$

No terms odd in H_z are necessary, for we know that the trace formation will make them vanish. The μ_B/kT associated with the expansion in ascending powers of $\mu_B H_z / kT$ remains hidden in the β's and $\mathcal{A}$ and $\mathcal{B}$ operators of the second square bracket. At the cost of greater clumsiness they could be pulled out. The order of the products of two $\mathcal{A}$'s and a single $\mathcal{B}$ is that which the expansion of $\mathcal{K}^3$ yields. The point worthy of special attention is this: associated with each power of H_z is an infinite series of terms, which has here been cut off at some power of β. Since β is not dimensionless, each series must really be in powers of (some energy)/kT. The appearance of the operator $\mathcal{B}$, representing the spin-spin interactions, suggests that "some energy" is associated with the mutual potential energy of the dipoles. Later we will see explicitly that we are here making a second expansion, this time in terms of (typical spin-spin energy)/kT. This does entail a second limitation on the domain in which the approximation for Z is adequate.

Now we start to evaluate the traces. There is no need to use the energy eigenstates of $\mathcal{K}$; any complete orthonormal set of states will do, for the value of a trace does not depend on the representation used. This is the tremendous advantage provided by the trace form for Z. Let us use states of the system that are eigenstates of J_{iz} for $1 \leqslant i \leqslant N$, that is, eigenstates of the z-component of each individual angular momentum. Because of the spin-spin interactions expressed by $\mathcal{B}$, these states are not energy eigenstates of $\mathcal{H}$, but that is irrelevant. This choice of representation will help us in using symmetry arguments to evaluate the traces.

Next, let us count the total number of states in the complete orthonormal set. The eigenvalues of J_{iz} range in integral steps from $-J$ to J, where the positive quantity J is determined by the total angular momentum of the paramagnetic particle and may be either an integer or half an odd integer. Thus there are $2J + 1$ possible values for the z-component of the angular moment of each paramagnet. Since there are N paramagnets, the total number of states in the complete orthonormal set is $(2J + 1)^N$.

Examples of Trace Calculations

To see how the trace calculations will go, let us work out a low-dimensional example in detail. We take two lattice sites and two particles of spin $\frac{1}{2}\hbar$ each. The spin at a lattice site is either up ($\uparrow$) or down ($\downarrow$). With four states we can construct a complete orthonormal set of states for this system of two paramagnets:

$$\psi_1 = (\uparrow, \uparrow), \qquad \psi_2 = (\downarrow, \uparrow),$$
$$\psi_3 = (\uparrow, \downarrow), \qquad \psi_4 = (\downarrow, \downarrow).$$

The first slot in the symbol (,) refers to the spin orientation of particle 1 at its site; the second slot, to particle 2. The property that these states are eigenstates of J_{iz} implies, for example, the eigenvalue equation

$$J_{1z}\psi_2 = -\tfrac{1}{2}\psi_2.$$

The trace of J_{1z} is

$$\mathrm{Tr}\, J_{1z} = \sum_{l=1}^{4} \int \psi_l^* J_{1z} \psi_l$$

$$= \tfrac{1}{2} + (-\tfrac{1}{2}) + \tfrac{1}{2} + (-\tfrac{1}{2}) = 0.$$

The zero value could have been predicted by a symmetry argument. In the set of states we are using, there is no preference for either up spin or down spin. Opposite orientations enter equally, for there is *no preferential weighting* with $\exp(-\text{energy}/kT)$ *in the trace formation*. In a useful sense, the traces are just quantum mechanics, not statistical mechanics. So up-and-down symmetry indicates that $\mathrm{Tr}\, J_{1z}$ must be zero.

Now we investigate $\operatorname{Tr} J_{1z}J_{2z}$. In our set of states the spins of the two paramagnets are not correlated. For a *single* state they are correlated, but not for the *set* of states as a whole. Since $\operatorname{Tr} J_{1z} = 0$ and likewise for J_{2z}, symmetry and the absence of correlations indicates a zero value for the trace. That is confirmed by the direct calculation:

$$\operatorname{Tr} J_{1z}J_{2z} = (\tfrac{1}{2})(\tfrac{1}{2}) + (-\tfrac{1}{2})(\tfrac{1}{2}) + (\tfrac{1}{2})(-\tfrac{1}{2}) + (-\tfrac{1}{2})(-\tfrac{1}{2}) = 0.$$

The situation is different with the trace of the unit operator:

$$\operatorname{Tr} 1 = 1 + 1 + 1 + 1 = 4 = (2 \times \tfrac{1}{2} + 1)^2.$$

The trace of the unit operator yields simply the total number of states in the complete orthonormal set, here 4 for $J = \tfrac{1}{2}$ and $N = 2$.

As the final example, we look at another non-zero trace:

$$\operatorname{Tr} J_{1z}^2 = (\tfrac{1}{2})^2 + (-\tfrac{1}{2})^2 + (\tfrac{1}{2})^2 + (-\tfrac{1}{2})^2.$$

There is another, more useful way of evaluating this trace. One relates it to the trace of the square of the particle's total angular momentum, to the trace of $\mathbf{J}_1^2$. On symmetry grounds, we may assert the relations

$$\operatorname{Tr} J_{1x}^2 = \operatorname{Tr} J_{1y}^2 = \operatorname{Tr} J_{1z}^2.$$

That the z-direction is a preferred direction in the real physical situation, because of the external magnetic field, is irrelevant in the trace formation. Our complete set of states does not "know" about H_z, nor need it know. Since

$$\mathbf{J}_1^2 = J_{1x}^2 + J_{1y}^2 + J_{1z}^2,$$

we may write

$$\operatorname{Tr} J_{1z}^2 = \tfrac{1}{3} \operatorname{Tr} \mathbf{J}_1^2.$$

The value of $\mathbf{J}_1^2$ is fixed by the angular momentum of the particle and is always $J(J + 1)$. So we may replace $\mathbf{J}_1^2$ by $J(J + 1)$ times the unit operator, arriving at

$$\operatorname{Tr} J_{1z}^2 = \tfrac{1}{3}J(J + 1)\operatorname{Tr} 1.$$

Since $J = \tfrac{1}{2}$ in our example and $\operatorname{Tr} 1 = 4$, this expression for $\operatorname{Tr} J_{1z}^2$ agrees numerically with that from the preceding explicit calculation. Now back to where we left off.

The Expectation Value $\langle M_z \rangle$

A confirmation that the present scheme yields a result for $\langle M_z \rangle$ in agreement with Section 7.3 when we drop the spin-spin interactions will be a comfort. For that confirmation we need evaluate only one of the traces in equation (12.4.4), that in $\mathcal{A}^2$:

$$\text{Tr } \mathcal{A}^2 = \text{Tr}\left(-g\mu_B \sum_i J_{iz}\right)\left(-g\mu_B \sum_k J_{kz}\right)$$

$$= (g\mu_B)^2 \sum_i \sum_k \text{Tr } J_{iz} J_{kz}$$

$$= (g\mu_B)^2 \sum_i \text{Tr } J_{iz}^2$$

$$= (g\mu_B)^2 N\tfrac{1}{3}J(J+1)\text{Tr } 1.$$

The step to the third line follows from the relation

$$\text{Tr } J_{iz} J_{kz} = 0 \qquad \text{if } k \neq i,$$

a result of symmetry and the absence of correlations in the trace formation, as we found explicitly in the sample calculations. The step to the last line follows from

$$\text{Tr } J_{iz}^2 = \tfrac{1}{3} \text{Tr } J_i^2 = \tfrac{1}{3}J(J+1)\text{Tr } 1,$$

again a result derived in the examples.

So, upon neglecting the spin-spin interactions and terms of order H_z^4, we have

$$Z_{\text{no spin-spin}} \simeq [1 + \tfrac{1}{6}NH_z^2(g\mu_B)^2\beta^2 J(J+1)]\text{Tr } 1.$$

Appeal to equation (12.4.2) yields

$$\langle M_z \rangle_{\text{no spin-spin}} \simeq \tfrac{1}{3}N(g\mu_B)^2 J(J+1)\beta H_z, \qquad (12.4.5)$$

in perfect agreement with the high-temperature limit of the calculation in Section 7.3. (Consistency requires that one drop the H_z^2 term relative to 1 when Z appears in the denominator, for we have neglected the $\partial O(H_z^4)/\partial H_z$ term in the numerator; the latter is of quadratic order in H_z relative to the leading term there.)

Of course we have not gone to all this labor just to recover an old no-spin-spin result. The traces in Z that contain the operator $\mathscr{B}$ are now on the agenda. Let us start with the easiest:

$$\text{Tr } \mathscr{B} = (g\mu_B)^2 \sum_{k<i} r_{ik}^{-3} \text{ Tr}(\mathbf{J}_i \cdot \mathbf{J}_k - 3\mathbf{J}_i \cdot \hat{\mathbf{r}}_{ik} \ \mathbf{J}_k \cdot \hat{\mathbf{r}}_{ik})$$

$$= 0.$$

Since k is always different from i in the sum, the operators $\mathbf{J}_i$ and $\mathbf{J}_k$ always refer to different paramagnets. The two spins are uncorrelated in the set of states that we are using to evaluate the trace. That, plus symmetry, imply relations such as

$$\text{Tr } J_{iz} J_{kz} = 0, \qquad \text{Tr } J_{ix} J_{kz} = 0, \qquad \text{when } k \neq i.$$

The first of these we worked out explicitly in the sample calculation. Thus each term in Tr $\mathscr{B}$ yields zero.

The other traces linear in $\mathscr{B}$ are more complicated and non-zero. For the first of three we have

$$\text{Tr } \mathscr{A}^2 = (g\mu_B)^4 \sum_l \sum_m \sum_{k<i} r_{ik}^{-3} \text{ Tr } J_{lz} J_{mz} (\mathbf{J}_i \cdot \mathbf{J}_k - 3\mathbf{J}_i \cdot \hat{\mathbf{r}}_{ik} \ \mathbf{J}_k \cdot \hat{\mathbf{r}}_{ik}).$$

Let us imagine the scalar products written out in component form. The entire expression is a sum of terms of the form

$$\text{Tr } J_{lz} J_{mz} J_{i*} J_{k*'},$$

where $*$, $*'$ stand independently for one of x, y, or z. The notation $k < i$ under the third summation sign reminds us that only combinations with $k \neq i$ will appear. Unless $l = i$ and $m = k$, or $l = k$ and $m = i$, there will be at least one particle that appears only once in the trace, and so symmetry with regard to plus-minus values of its angular momentum components will make the expression vanish. Hence we need consider only the two special cases, such as

$$\text{Tr } J_{iz} J_{kz} J_{i*} J_{k*'}.$$

Unless $* = z$ and $*' = z$, this trace, too, will vanish because of symmetry and the absence of correlations between components in orthogonal directions, for example, z and x. Continuing with this trace, we evaluate it as

$$\text{Tr } J_{iz} J_{kz} J_{iz} J_{kz} = \text{Tr } J_{iz}^2 J_{kz}^2$$

$$= \text{Tr } \tfrac{1}{3}\mathbf{J}_i^2 \tfrac{1}{3}\mathbf{J}_k^2$$

$$= \tfrac{1}{9}[J(J+1)]^2 \text{ Tr } 1.$$

The legitimacy of the last two steps, if not clear, can be checked with the aid of the low-dimensional example.

Now we go back to $\text{Tr } \mathcal{A}^2 \mathcal{B}$ and perform the sum over l and m for fixed i and k. We get contributions only for $l = i$, $m = k$ and for $l = k$, $m = i$; these are identical, yielding the factor of 2 in the expression

$$\text{Tr } \mathcal{A}^2 \mathcal{B} = \tfrac{2}{9}(g\mu_B)^4 [J(J+1)]^2 \sum_{k<i} r_{ik}{}^{-3}[1 - 3(\hat{\mathbf{r}}_{ik} \cdot \hat{\mathbf{z}})^2] \text{Tr } 1.$$

The two remaining traces linear in $\mathcal{B}$ yield the same result.

For convenience, let us abbreviate the remaining sum as

$$\Phi \equiv -\frac{V}{N(N/2)} \sum_{k<i} r_{ik}{}^{-3}[1 - 3(\hat{\mathbf{r}}_{ik} \cdot \hat{\mathbf{z}})^2]. \tag{12.4.6}$$

The sum with $k < i$ is a sum over all distinct pairs of paramagnets. Since there are $N(N-1)/2$ such distinct pairs, the division by $N(N/2)$ makes Φ effectively independent of N for the large N that we are considering. The insertion of the volume factor V makes Φ dimensionless. Later we will find that the inclusion of these factors ensures that the numerical value of Φ will typically be of order unity.

Let us take stock of our achievements to date, writing out the partition function through $O(H_z{}^2)$ as explicitly and profitably as we are presently able:

$$Z/\text{Tr } 1 = 1 + \tfrac{1}{2}\beta^2 \text{ Tr } \mathcal{B}^2/\text{Tr } 1 + O(\beta^3)$$

$$+ \tfrac{1}{6}NH_z{}^2(g\mu_B)^2 J(J+1)\beta^2 \left[1 + \tfrac{1}{3}(g\mu_B)^2 J(J+1)\frac{N}{V}\beta\Phi + O(\beta^2) \right].$$

$$\tag{12.4.7}$$

The new factorization shows that the primary expansion, that containing H_z, is effectively an expansion in ascending powers of the dimensionless quantity

$$g\mu_B\sqrt{J(J+1)} \, \frac{H_z}{kT}.$$

Since the quantity $g\mu_B\sqrt{J(J+1)}$ is the magnitude of the magnetic dipole moment of a single paramagnet, this dimensionless quantity is precisely what a more detailed analysis would have suggested in place of simply $\mu_B H_z/kT$, our earlier expression for the primary expansion parameter.

We can, moreover, now discern the characteristic dimensionless parameter of the secondary expansion. Let us extract a provocative factor from the second term in the square brackets and introduce τ as

$$\tau \equiv \frac{1}{k}(g\mu_B)^2 J(J+1)\frac{N}{V}. \qquad (12.4.8)$$

The quantity τ has the dimensions of a temperature. The factor V/N, as the volume per paramagnet in the crystal, is roughly the cube of the distance between a pair of neighboring paramagnets. Upon remembering that the dipole-dipole interaction depends on the inverse cube of the separation, we may write

$$k\tau \equiv \frac{(g\mu_B\sqrt{J(J+1)})^2}{(V/N)} = \begin{pmatrix} \text{the typical value of the} \\ \text{mutual potential energy of a} \\ \text{pair of neighboring magnetic} \\ \text{dipoles} \end{pmatrix}.$$

An expansion in ascending powers of τ/T is indeed an expansion in terms of (typical spin-spin energy)/kT.

A numerical estimate of τ is in order. In a typical paramagnetic salt, the distance between neighboring dipoles is about 10 Ångströms, implying $V/N \simeq 10^{-21}$ cm. The Landé factor g is frequently close to 2, and a reasonable value for J would be $J = 3$. (The experimental curves in figure 5.1.2 were for paramagnets with $J = \frac{5}{2}$ and $J = \frac{7}{2}$.) These values lead to the estimate

$$\tau \simeq 0.035 \text{ K}.$$

This characteristic temperature is quite low, though it is of the order of the physical temperature achieved by magnetic cooling; so the spin-spin interactions can exercise an appreciable influence.

Thus far we have worked out the traces linear in the dipole-dipole operator $\mathcal{B}$ in our basic approximation for Z. They modify the expression for $\langle M_z \rangle$, so that it now reads

$$\langle M_z \rangle \simeq \langle M_z \rangle_{\text{no spin-spin}} \left[1 + \frac{1}{3}\frac{\tau}{T}\Phi\right], \qquad (12.4.9)$$

when only terms permitted by consistency in powers of H_z are kept. Before we can draw firm conclusions about the influence of the spin-spin interactions, we must investigate the function Φ.

The numerical value of Φ depends on both the crystal structure and the shape of the paramagnetic sample. Overwhelmingly the simplest case is that of cubic crystal structure and spherical shape: then Φ is zero. The derivation, though not obvious, is nonetheless simple. In this special case, the $\hat{x}$ and $\hat{y}$ directions in the sample are no different from the $\hat{z}$ direction; that is, the spatial arrangements of paramagnets and the boundaries of the sample are the same. If we replace $\hat{z}$ in Φ by $\hat{x}$ or $\hat{y}$, we must necessarily get the same numerical value. Let us imagine doing that, adding the three expressions, and then dividing by three to get Φ. The over-all sum will now be a sum of terms with the factor

$$3 - 3[\hat{\mathbf{r}}_{ik} \cdot \hat{x})^2 + (\hat{\mathbf{r}}_{ik} \cdot \hat{y})^2 + (\hat{\mathbf{r}}_{ik} \cdot \hat{z})^2].$$

That factor is zero, for the expression in square brackets is the sum of the squares of the components of the unit vector $\hat{\mathbf{r}}_{ik}$ and hence is just unity. So for the very special case of a spherical sample of a cubic crystal, we have $\Phi = 0$.

For other shapes and crystal structures, Φ will typically be non-zero. For a long thin rod, oriented along $\hat{z}$, and cubic crystal structure, the value of $(\hat{\mathbf{r}}_{ik} \cdot \hat{z})^2$ will predominantly be close to unity, and so Φ will be positive. For a flat disc one can anticipate a negative value for Φ. Despite this diversity, the magnitude of Φ will be a small number, roughly in the range of zero to ten, provided that the sample has a reasonable shape and that the crystal structure is (at least in some average sense) cubic. The argument in support of this claim follows.

The function Φ depends on a sum over all distinct pairs of paramagnets. Let us pick some paramagnet not near the surface and imagine performing the sum over all other paramagnets. The contributions from those in a spherical volume extending to the nearest boundary will sum to zero, for the average of $(\hat{\mathbf{r}}_{ik} \cdot \hat{z})^2$ will be $\frac{1}{3}$. The nonvanishing contribution comes from the distant paramagnets. The number of those is of order N, that is, some modest fraction of N, and $\hat{\mathbf{r}}_{ik}^{-3}$ will be of order V^{-1} if the shape is not too contorted. Consequently, the contribution of the distant paramagnets to the sum will be of order NV^{-1}. In this scheme of estimating Φ, the number of "first" paramagnets that we must pick is $N/2$. So, with a little carelessness about those "first" paramagnets near the boundary, we find that the full summation yields a value of order $NV^{-1}N/2$. That cancels in order of magnitude with the initial factor in the definition of Φ. The function Φ will indeed be of order unity for a sample of reasonable shape and, on the average, cubic crystal structure.

At last we can assess the influence of the spin-spin interactions on the estimated total moment. Since Φ is of order unity, the combination $\tau\Phi/3$ will typically be of order 0.05 K. At a temperature T of 4 K, the value we considered back in section 5.1, there will be an effect of order 1 per cent. This provides ample justification for neglecting the complicated spin-spin interactions at that early stage in the development of statistical mechanics.

At lower temperatures, the inclusion of the spin-spin interactions becomes essential. Whether they increase or reduce the estimated moment relative to the idealized no-spin-spin value depends on the sign of Φ and hence on the shape of the sample. Magnetic cooling can produce temperatures as low as the order of τ itself; then the mutual interactions become crucial. By the time such temperatures are reached, however, one needs terms in higher powers of τ/T as well, and indeed the usefulness of the expansion itself becomes questionable.

Before we go on, let us try to get a more intuitive picture of the spin-spin influence on $\langle M_z \rangle$. The $\tau\Phi$ modification of $\langle M_z \rangle$ comes from trace terms with two $\mathcal{A}H_z$ operators and a single $\mathcal{B}$ operator; it depends linearly on the mutual interactions. This, as well as intuition, suggests the following picture: the external magnetic field tends to align the dipoles along $\hat{\mathbf{z}}$; the partial alignment influences the effect of the mutual interactions; the altered mutual interactions either enhance the alignment tendency ($\Phi > 0$) or diminish it ($\Phi < 0$).

The peculiar form of the sum defining Φ actually supports this intuitive picture in a striking fashion. Upon going back to the beginning of this section and using the expression for the magnetic field produced by a single dipole $\boldsymbol{\mu}_1$, we may write

$$\begin{pmatrix} \text{the component of the magnetic} \\ \text{field along } \hat{\mathbf{z}}, \text{ at the position} \\ \text{of dipole } \boldsymbol{\mu}_2, \text{ due to dipole } \boldsymbol{\mu}_1 \\ \text{when aligned along } \hat{\mathbf{z}} \end{pmatrix} = -r_{12}{}^{-3}(\boldsymbol{\mu}_1 - 3\boldsymbol{\mu}_1 \cdot \hat{\mathbf{r}}_{12}\,\hat{\mathbf{r}}_{12}) \cdot \hat{\mathbf{z}}$$

$$= -\mu_1 r_{12}{}^{-3}[1 - 3(\hat{\mathbf{r}}_{12} \cdot \hat{\mathbf{z}})^2].$$

The step to the second line follows because the specified alignment of $\boldsymbol{\mu}_1$ means $\boldsymbol{\mu}_1 = \mu_1 \hat{\mathbf{z}}$. We generate precisely the peculiar form of the summand. So Φ does reflect the magnetic field, as seen by a single dipole, that is due to (partial) alignment of all the other magnetic dipoles. Furthermore, we can begin to see why the shape of the sample has such an influence on Φ. The z-component of the field at $\boldsymbol{\mu}_2$ due to $\boldsymbol{\mu}_1$ depends strongly on the position of $\boldsymbol{\mu}_2$ relative to $\boldsymbol{\mu}_1$, as both intuition and the factor $(\hat{\mathbf{r}}_{12} \cdot \hat{\mathbf{z}})^2$ indicate.

Spin-Spin Interactions in Magnetic Cooling

When we analyzed magnetic cooling in section 11.4, we resorted to an effective field H_{eff} to describe the influence of the spin-spin interactions. That is not a bad approach, but one would like to avoid such a crutch. We now have the machinery needed to treat the spin-spin influence in microscopic detail.

To calculate the final temperature achieved by slowly reducing the external field to zero, we need expressions for the information-theory entropy both before and after demagnetization. Since one starts from a large value of H_z and a temperature well above the characteristic spin-spin temperature τ, the spin-spin interactions will have little influence on the initial value of S_I. It is in the final situation, when H_z is zero, that they will be crucial, and so our major concern is with S_I for $H_z = 0$.

We can calculate S_I once we know the partition function. A glance back at equation (12.4.7) shows that in order to include spin-spin effects in the zero-H_z information-theory entropy, we must evaluate $\mathrm{Tr}\ \mathscr{B}^2$, and so that is the next task.

The computation of $\mathrm{Tr}\ \mathscr{B}^2$ is more awkward than that of the other traces, but a little attention to details will yield a quite simple result. We begin by writing the trace as

$$\mathrm{Tr}\ \mathscr{B}^2 = (g\mu_B)^4 \sum_{k<i} \sum_{m<l} r_{ik}^{-3} r_{lm}^{-3}$$
$$\times \mathrm{Tr}(\mathbf{J}_i \cdot \mathbf{J}_k - 3\mathbf{J}_i \cdot \hat{\mathbf{r}}_{ik}\ \mathbf{J}_k \cdot \hat{\mathbf{r}}_{ik})(\mathbf{J}_l \cdot \mathbf{J}_m - 3\mathbf{J}_l \cdot \hat{\mathbf{r}}_{lm}\ \mathbf{J}_m \cdot \hat{\mathbf{r}}_{lm}).$$

The trace on the right is zero unless the dipole pair in the second factor is the same as the pair in the first. Consequently, we need deal only with traces of the form

$$\mathrm{Tr}[\mathbf{J}_i \cdot \mathbf{J}_k\ \mathbf{J}_i \cdot \mathbf{J}_k - 6\mathbf{J}_i \cdot \mathbf{J}_k\ \mathbf{J}_i \cdot \hat{\mathbf{r}}_{ik}\ \mathbf{J}_k \cdot \hat{\mathbf{r}}_{ik} + 9\mathbf{J}_i \cdot \hat{\mathbf{r}}_{ik}\ \mathbf{J}_k \cdot \hat{\mathbf{r}}_{ik} \mathbf{J}_i \cdot \hat{\mathbf{r}}_{ik}\ \mathbf{J}_k \cdot \hat{\mathbf{r}}_{ik}].$$

Let us merely imagine writing this out in component form:

$$\mathrm{Tr}[J_{ix} J_{kx} J_{ix} J_{kx} + J_{ix} J_{kx} J_{iy} J_{ky} + \ldots].$$

A term in this trace will be zero unless the component of the second $\mathbf{J}_i$ is the same as that of the first $\mathbf{J}_i$; an identical relation holds for the two $\mathbf{J}_k$'s. For the nonvanishing terms, previous work indicates

$$\mathrm{Tr}\ J_{ix}^2 J_{ky}^2 = [\tfrac{1}{3}J(J+1)]^2\ \mathrm{Tr}\ 1$$

and likewise for the squares of other components. So the trace we are considering becomes

$$[\tfrac{1}{3}J(J+1)]^2(\text{Tr } 1)[3 - 6\hat{\mathbf{r}}_{ik} \cdot \hat{\mathbf{r}}_{ik} + 9(\hat{\mathbf{r}}_{ik} \cdot \hat{\mathbf{r}}_{ik})^2].$$

Since $\hat{\mathbf{r}}_{ik}$ is a unit vector, the scalar products are unity, and hence the last factor is simply 6. Now we may go back and write

$$\text{Tr } \mathcal{B}^2 = (g\mu_B)^4[J(J+1)]^2(\text{Tr } 1)\frac{6}{9}\sum_{k<i} r_{ik}^{-6}.$$

The remaining sum goes over all distinct pairs of paramagnets and has the dimensions of (volume)$^{-2}$. Let us introduce the conventional abbreviation:

$$Q \equiv \frac{4V^2}{N^3}\sum_{k<i} r_{ik}^{-6}. \tag{12.4.10}$$

The initial factors make Q dimensionless and ensure that its numerical value will be of order ten. As defined, Q is necessarily positive. The argument for the order of magnitude follows now.

To evaluate Q, one must sum over all distinct pairs of dipoles. We may approximate that sum by picking a typical dipole, summing over all others, and then multiplying by $N/2$:

$$\sum_{k<i} r_{ik}^{-6} \simeq \frac{N}{2}\sum_{i=2}^{N} r_{i1}^{-6}.$$

The righthand side has the correct number of terms: $(N/2) \times (N-1)$, the number of distinct pairs. Because of the inverse sixth-power dependence, only dipoles quite near any chosen dipole make a significant contribution. Whether the "typical" dipole (taken here to be dipole 1) is near the surface of the sample or well inside does not appreciably matter, precisely because of the strong distance dependence, and so the approximation for the lefthand side is permissible. Since distant dipoles make a negligible contribution, we may write, in order of magnitude,

$$\sum_{i=2}^{N} r_{i1}^{-6} \simeq \begin{pmatrix}\text{the number of}\\ \text{close neighbors}\\ \text{of dipole } k = 1\end{pmatrix}\begin{pmatrix}\text{the typical}\\ \text{distance to a}\\ \text{neighboring dipole}\end{pmatrix}^{-6}.$$

The quantity in the first parentheses has a value of order 10. The typical distance will be of order $(V/N)^{1/3}$, the cube root of the volume per dipole in the sample. Upon putting together the pieces, we find

$$\sum_{k<i} r_{ik}^{-6} \simeq \left(\frac{N}{2}\right)O(10)\left(\frac{V}{N}\right)^{-2}.$$

A glance back at the relation defining Q indicates that Q will indeed be of order 10, somewhat more informatively, of the order of the number of immediate neighbors of any dipole. Because of the strong inverse sixth-power dependence, Q will not be sensitive to the shape of the sample, the direct opposite of the situation with Φ. For a crystal of simple cubic structure, Van Vleck gives $Q \simeq 16.8$, providing reassurance for the order of magnitude estimate.

With the aid of Q, we may write the contribution to Z of the Tr $\mathcal{B}^2$ term as

$$\frac{1}{2}\beta^2 \text{ Tr } \mathcal{B}^2 = \frac{1}{12} N\left(\frac{\tau}{T}\right)^2 Q \text{ Tr } 1.$$

The factors of $(g\mu_B)^4$ and the like combine to produce τ^2. Since $k\tau$ is the characteristic spin-spin energy, we see that the contribution is quadratic in that energy. Moreover, we find support for the contention that the secondary expansion of Z is in ascending powers of τ/T.

By now we have evaluated Z as far as the following terms:

$$Z = \text{Tr } 1\left\{1 + \frac{1}{12} N\left(\frac{\tau}{T}\right)^2 Q + \cdots\right.$$

$$\left. + \tfrac{1}{6}NH_z^2(g\mu_B)^2 J(J+1)\beta^2\left[1 + \frac{1}{3}\frac{\tau}{T}\Phi + \cdots\right]\right\}. \quad (12.4.11)$$

The dots indicate terms of higher order in τ/T.

Since the entire statistical calculation has been based on the canonical probability distribution, we may calculate the information-theory entropy from the expression

$$\frac{S_I}{k} = \beta\langle E \rangle + \ln Z = -\beta\frac{\partial \ln Z}{\partial \beta} + \ln Z.$$

In computing $\ln Z$ from equation (12.4.11), we will regard the terms following the 1 in the curly bracket as small relative to unity and write

$$\ln Z = \ln(\text{Tr } 1) + \ln\{1 + (\text{terms written out}) + \ldots\}$$
$$\simeq \ln(\text{Tr } 1) + (\text{terms written out}).$$

This may produce some uneasiness, for although the coefficients of N are small relative to unity, in any laboratory sample N itself is quite large. The end result for $\ln Z$ is all right; the intermediate trouble arises because one expands and approximates Z, instead of $\ln Z$ itself. Since $\ln Z$ governs the size of quantities like $\langle M_z \rangle$ and $\langle E \rangle$, it should be of the order of the number of paramagnets in the system, that is, of order N. The expression "terms written out" in the approximation for ln Z is properly of order N. If one could compute Z exactly, one would not run into these mathematical problems. They plague much of statistical mechanics and are to be handled by appeal to physical reasonableness.

> The classical partition function for a gas of interacting molecules provides an analogous situation. A common method of calculation yields, as an approximation for Z_U/V^N, merely the first two terms in the binomial expansion of our expression in equation (8.5.6). Problems arise when one takes the logarithm, for the second of those two terms is large relative to unity. One proceeds to the second virial coefficient with an appeal to physical reasonableness and a belief that a better approximation scheme would remove the embarrassment. That better approximation scheme does exist; our calculation provides it (and there are others, as well).

In the expression for the information-theory entropy, Tr 1 will remain, though it does not contribute to $\langle M_z \rangle$ or $\langle E \rangle$. Earlier we noted that the number of states in the complete orthonormal set is $(2J + 1)^N$. The trace of the unit operator amounts merely to the number of states in the set, and so we have for the needed logarithm

$$\ln(\text{Tr } 1) = \ln(2J + 1)^N = N \ln(2J + 1).$$

Incidentally, this supports the contention that ln Z is of order N.

The information-theory entropy follows quite directly now and is given by

$$\frac{S_I}{Nk} \simeq \ln(2J + 1) - \frac{1}{12}\left(\frac{\tau}{T}\right)^2 Q$$

$$- \tfrac{1}{6}H_z^2(g\mu_B)^2 J(J + 1)\beta^2\left[1 + \frac{2}{3}\frac{\tau}{T}\Phi\right]. \quad (12.4.12)$$

Let us check the physical reasonableness of the two minus signs. In the *absence* of the external field H_z and of the spin-spin interactions, the energies of the energy eigenstates would all be the same (in fact, would all be zero); so the probabilities of the $(2J + 1)^N$ energy eigenstates would all be the same. In that case S_I should be k ln (number of states), precisely what the first term above gives. When an external magnetic field is applied, the energies and associated probabilities differ from state to state. The analysis of the function MI in Section 3.5 indicates that S_I should then be reduced in value; the term in H_z^2 is quite properly negative. Finally we imagine "turning on" the spin-spin interactions. Back in Section 5.2 we noted that, when mutual interactions are neglected, there are typically many distinct energy eigenstates with the same energy. The states differ only in which spins are up and which down, not in how many of each. The spin-spin interactions will alter the situation significantly. For them it does matter whether the up spins are spatially close together or smoothly distributed among the down spins; the sums of the mutual potential energies are different. We can expect two immediate results of "turning on" the spin-spin interactions: (1) the energy eigenstates and eigenvalues will be different from those when the mutual interactions are neglected; and (2) there will be far fewer (if any) sets of distinct energy eigenstates with the same energy. Point (2) suggests a further reduction in the value of S_I; that conjecture is vindicated by the minus sign in the Q term.

At last we get to the heart of this subsection: the influence of the spin-spin interactions on the final temperature in magnetic cooling. Just as we did in Section 11.4, we may drop S_I for the lattice as quantitatively negligible at both initial and final temperatures. If the initial temperature T_i is a few kelvin and the initial external field H_z is of order 10^4 gauss, we may safely drop the spin-spin terms in S_I as small relative to the others. That simplifies matters and leads to

$$\left(\frac{S_I}{Nk}\right)_{\text{initial}} \simeq \ln(2J + 1) - \tfrac{1}{6}H_z^2(g\mu_B)^2 \frac{J(J + 1)}{(kT_i)^2}. \qquad (12.4.13a)$$

Provided the external field is reduced to zero slowly, we may continue to use the canonical probability distribution. In the final situation, at temperature T_f, we certainly may not neglect the spin-spin interactions—nor would we want to, after all the labor. The expression for S_I is then

$$\left(\frac{S_I}{Nk}\right)_{\text{final}} \simeq \ln(2J + 1) - \frac{1}{12}\left(\frac{\tau}{T_f}\right)^2 Q. \qquad (12.4.13b)$$

To determine the final temperature, we invoke the constancy of S_I, equate the two expressions for S_I, and solve for T_f. Judicious rearrangement yields

$$T_f \simeq \left(\frac{Q}{2}\right)^{1/2} \frac{[g\mu_B\sqrt{J(J + 1)(N/V)}]}{H_z} T_i. \qquad (12.4.14)$$

A comparison with equation (11.4.7) indicates that we have found the microscopic counterpart of the effective magnetic field H_{eff}:

$$H_{\text{eff}} \simeq \left(\frac{Q}{2}\right)^{1/2} [g\mu_B\sqrt{J(J + 1)}(N/V)]. \qquad (12.4.15)$$

The order of magnitude is correct, too. The factor $(Q/2)^{1/2}$ is of order unity, and the factor N/V is roughly the inverse cube of the distance between neighboring paramagnets. Therefore the righthand side is roughly the size of the magnetic field produced by a single dipole at the location of its immediate neighbors. For a numerical estimate, let us take $Q = 10$ and use, for the other parameters, the values adopted earlier in estimating τ. This yields

$$H_{eff} \simeq 145 \text{ gauss},$$

in excellent agreement with the general size inferred from experimental data.

The origin of this specific effective magnetic field merits some additional remarks. Aside from factors like μ_B, whose presence is almost automatic, H_{eff} is governed by the quantity Q. That quantity, coming from Tr $\mathcal{B}^2$, arises solely from the mutual interactions of the dipoles. Consequently, we find that this specific effective field does not depend in any way on an external magnetic field. (One can associate another effective field with Φ; that effective field would depend on the external field.) Rather, H_{eff} reflects solely the mutual interactions, as its continued presence (implicit or explicit) in S_I when $H_z = 0$ also makes clear.

Earlier the remark was made that magnetic cooling can produce temperatures as low as order τ. With brute force—huge values of the initial external field, values of order 10^5 gauss—one can in principle get below τ, but the practical difficulties are severe. A rather formal derivation of why order τ is a natural (though penetrable) lower limit to the final temperature is readily given. Low temperature is typically correlated with small values of S_I. Since S_I/Nk is proportional to the function MI, we know that it cannot be negative. The most one can do with strong initial fields is to make it close to zero. Let us take that case. After slow removal of the external field, the final value of S_I/Nk will, of course, also be close to zero. Upon putting the righthand side of equation (12.4.13b) equal to (approximately) zero and solving for the final temperature, we find

$$T_f \simeq \left[\frac{Q}{12 \ln (2J + 1)} \right]^{1/2} \tau.$$

Since the coefficient is of order unity, the final temperature will be of order τ. To be sure, one may regard this as no more than suggestive, for at temperatures of order τ additional terms in the series in powers of τ/T may become important.

A microscopic explanation, albeit only a loose one, is not hard to find. In Section 11.4 we viewed the temperature reduction as occurring because energy is extracted from the lattice vibrations (and the spin-spin interactions) as spins flip from orientations of low to high potential energy in the slowly decreasing external field. When the physical temperature approaches the characteristic temperature τ of the spin-spin interactions, one can expect the mutual interactions to tend to lock groups of spins into fixed orientations. This will hinder the further extraction of energy from the lattice (and spin-spin potential energy), and so will leave the final physical temperature somewhere in the vicinity of τ.

REFERENCES for Chapter 12

D. ter Haar, in a review article in *Reports on Progress in Physics*, **24**, 304 (1961), presents three views of what the density matrix " is " and exhibits numerous applications. There is an extensive list of references. Tolman's book, cited in the references for Chapter 4, gives a clear discussion of the density matrix in statistical mechanics. (The position taken in this book differs somewhat from Tolman's, primarily because of the different meaning ascribed to the notion of " probability.") A paper by U. Fano, *Reviews of Modern Physics*, **29**, 74 (1957), presents a characteristically quantum-mechanical view of the density matrix, a view particularly relevant to discussions of coherence phenomena in optics.

Our development of spin-spin interactions has been based on the paper by J. H. Van Vleck, *Journal of Chemical Physics*, **5**, 320 (1937). The contribution of the spin-spin interactions to the heat capacity was derived by Ivar Waller in *Zeitschrift für Physik*, **104**, 132 (1936). A comparison of theory and experiment at the time of the above papers is given by M. H. Hebb and E. M. Purcell, *Journal of Chemical Physics*, **5**, 338 (1937).

PROBLEMS

12.1. Offer an interpretation of the diagonal elements of the density matrix (considered as probabilities).

12.2. Trace independence. Suppose one changes from a representation with states ψ_j to another representation ψ'_i, with the states linearly related (of course?) as

$$\psi'_i = \sum_j U_{ij}\psi_j.$$

Here U_{lj} is a matrix giving the transformation between the two representations. Remember that "representation" implies that the set of states forms a complete orthonormal set.

a. By considering the integral $\int (\psi'_m)^* \psi'_l$, show that the transformation matrix is *unitary*:

$$\sum_j U_{lj} U_{mj}^* = \delta_{lm}.$$

This equation states that the complex conjugate and transpose of the matrix U_{jm}, that is, U_{mj}^*, provides an inverse matrix for U_{lj}.

b. Use the unitary property to show that the numerical value of the trace of any operator, for example, the operator $\mathcal{M}$, is independent of the representatation used to form the trace:

$$\sum_l \int (\psi'_l)^* \mathcal{M}\psi'_l = \sum_j \int \psi_j^* \mathcal{M}\psi_j.$$

12.3. To check on the absence of a factor of 2 in the mutual potential energy of a pair of magnetic dipoles, one may take a pair of electric dipoles as an example and then argue by analogy. Write out the mutual Coulomb potential energy

Figure P12.3.

of the configuration in figure P12.3 and then determine the lowest nonvanishing term in an expansion in powers of $1/r$. Bear in mind that the dipole moments here are $q_1 s_1$ and $q_2 s_2$. Is the absence of a factor of 2 correct?

12.4. Magnetic susceptibility. The magnetic susceptibility expresses a relation between the estimated magnetic moment per unit volume, $\langle M_z \rangle / V$, and the externally applied magnetic field H_z. Two different, though analogous, definitions are in common use.

$$\begin{pmatrix} \text{magnetic} \\ \text{susceptibility} \end{pmatrix} = \frac{\langle M_z \rangle / V}{H_z} \quad \text{or} \quad \frac{\partial}{\partial H_z}\left(\frac{\langle M_z \rangle}{V}\right).$$

When the external field is not too strong, so that $\langle M_z \rangle$ is approximately linear in H_z, the expressions emerging from the two definitions agree. In this problem we will be dealing with such a situation.

Retain the spin-spin interactions through trace terms linear in the operator $\mathcal{B}$ only and compute the magnetic susceptibility. On the assumption $T \gg \tau$, show that the result may be cast into the conventional form

$$\begin{pmatrix} \text{magnetic} \\ \text{susceptibility} \end{pmatrix} \simeq \frac{\text{constant}}{T - \frac{1}{3}\tau\Phi}.$$

The constant, having the dimensions of a temperature, is commonly called the *Curie constant*. Could one extract the value of the characteristic spin-spin temperature τ from data on the magnetic susceptibility? If so, how?

12.5. Evaluation of Φ. For all sample shapes except a sphere, the explicit evaluation of the function Φ is distinctly nontrivial. Here is an outline of a method by which one can make good numerical estimates, particularly for a long rod. The sum defining Φ, with the notation $k < i$, goes over all distinct pairs of dipoles. Provided we compensate with a factor $\frac{1}{2}$, we may write Φ as a sum over all pairs of indices, without the "distinct" proviso:

$$\Phi = -\frac{1}{2} \frac{V}{N(N/2)} \sum_{i \neq k} r_{ik}^{-3} [1 - 3(\hat{r}_{ik} \cdot \hat{z})^2].$$

The notation $i \neq k$ means a sum with $1 \leq k \leq N, 1 \leq i \leq N$, but $i = k$ excluded. We approximate this sum by a pair of volume integrals, using N/V as the number of dipoles per unit volume:

$$\sum_{i \neq k} r_{ik}^{-3} [1 - 3(\hat{r}_{ik} \cdot \hat{z})^2] \simeq \left(\frac{N}{V}\right)^2 \int_{V_1} \int_{V_2} \left[\frac{\partial}{\partial z_{12}} \left(\frac{z_{12}}{r_{12}^3}\right) - \frac{4\pi}{3} \delta(\mathbf{r}_{12}) \right] d^3x_1 \, d^3x_2.$$

a. Work out the indicated derivative and show that it produces the expression demanded by the lefthand side. The second term, with the Dirac delta function, is a necessary compensation: when $r_{12} = 0$ in the first term, one is, in effect, taking $k = i$, a term excluded on the lefthand side.

b. Integrate with respect to z_1 in the first term and do the integral with the delta function. This will enable you to cast into Φ the form

$$\Phi \simeq \frac{4\pi}{3} - \frac{1}{V} \int_{S_1} \int_{V_2} \frac{\hat{r}_{12} \cdot \hat{z}}{r_{12}^2} \hat{z} \cdot d\mathbf{S}_1 \, d^3x_2.$$

Here $d\mathbf{S}_1$ is an element of surface area with the direction of the outward perpendicular to the surface. This form makes clear the dependence of Φ on the shape of the sample.

c. For reassurance, consider a spherical shape for the sample. With a scheme like that used in the text, show that the remaining integral will cancel with the first term, yielding $\Phi = 0$.

d. Now take the case of a long thin rod oriented along $\hat{z}$, and estimate the order of magnitude of the integral. You should find $\Phi \simeq 4\pi/3 - $ [order of (width/height)^2]. Does this agree with the judgment, made in the text, of the sign of Φ for this case?

12.6. Internal field approximation in paramagnetism. The magnetic dipole-dipole interactions in a paramagnetic system have generated a host of techniques for treating them in an approximate fashion. To the extent that an external magnetic field H_z leads to a net alignment of the dipoles along $\hat{z}$, the dipoles themselves produce an "internal magnetic field" whose spatial average value is proportional to $\langle M \rangle / V$. Let's check the claimed proportionality. Since $\langle M_z \rangle$

is roughly N times μ_B, the quotient $\langle M_z \rangle / V$ is equal, in order of magnitude, to $\mu_B/(V/N)$, and that quantity, we know, is roughly the size of the magnetic field produced by one dipole at its neighbors. One may, therefore, approximate the influence of the dipole-dipole interactions by saying that a typical dipole is subject to a field H'_z given by

$$H'_z = H_z + \varphi \langle M_z \rangle / V,$$

The numerical factor φ is dimensionless and will be of order unity. A calculation of φ, rather tedious, entails two kinds of considerations: (1) the shape of the sample, and (2) compensation for the fact that the dipole is not at an "average" position in the sample but instead is at a lattice site, a special position. A calculation of $\langle M_z \rangle$ with the field H'_z is a kind of self-consistent calculation: one computes $\langle M_z \rangle$ in terms of H_z and $\langle M_z \rangle$ itself.

Use the field H'_z in place of H_z in the analysis in Section 7.3, and then go to the high-temperature limit. This will enable you to solve easily for $\langle M_z \rangle$. You should find

$$\langle M_z \rangle \simeq \frac{\langle M_z \rangle \text{ without internal field}}{1 - \frac{1}{3}\tau\varphi/T} \propto \frac{1}{T - \frac{1}{3}\tau\varphi}.$$

Here τ is the characteristic spin-spin temperature introduced in section 12.4. Comparison with the equations of that section indicates that, to the corresponding orders of approximation, φ and Φ are equivalent.

12.7. **More internal field.** If one takes the internal field approximation of the preceding problem fully seriously, one finds that it predicts a *non-zero* limiting value for $\langle M_z \rangle$ as H_z goes to zero, provided φ is positive and T is below a certain "transition temperature." The prediction is that the paramagnetic system acquires a kind of permanent magnetic moment, becomes somewhat "ferromagnetic."

a. Take the $J = \frac{1}{2}$ case (as the easiest), and derive the self-consistent equation relating $\langle M_z \rangle$ to H_z and itself. You should be able to cast your expression into the following convenient form:

$$\langle M_z \rangle = \frac{1}{2} g\mu_B N \tanh \left[\frac{g\mu_B}{2kT} (H_z + \varphi \langle M_z \rangle / V) \right].$$

b. Show that, as H_z goes to zero, this equation *always* has the solution $\langle M_z \rangle = 0$ and *sometimes* has a second solution with $\langle M_z \rangle \neq 0$. (Hint: sketch each side as a function of $\langle M_z \rangle$ and look for points of intersection; do this for qualitatively different values of the temperature.) Establish a firm criterion for the existence of the second, non-zero solution. Then estimate the order of magnitude of the transition temperature.

c. Expand the hyperbolic tangent in powers of its argument, and determine the temperature dependence of the vanishing of $\langle M_z \rangle$ as the physical temperature approaches the transition temperature from below.

12.8. Spin-spin energy and heat capacity. a. Use the approximate expression for Z developed in Section 12.4 to compute $\langle E \rangle$ for the spin-spin interactions in the absence of an external magnetic field. Note that, to the lowest nonvanishing order of approximation, $\langle E \rangle$ is quadratic in the characteristic spin-spin energy $k\tau$. In a sense, one factor of $k\tau$ represents the energy itself, and the other reflects the weighting with the canonical probability distribution in the formation of the expectation value. For support of this view, expand the exponential implicit in the relation $\langle E \rangle = \mathrm{Tr}\,\mathcal{H}\rho$. Why is a negative value for $\langle E \rangle$ not unreasonable?

b. Determine the temperature dependence of the heat capacity at constant volume and zero external magnetic field (in first approximation). How general a result is this likely to be?

PHYSICAL AND MATHEMATICAL DATA

The ability to make rough estimates " on the back of an envelope " depends, in part, on a physicist's recollection of the exponents in a few physical constants. The arrangement of the decimal points in table A.1 was designed to help the frequent user to remember the correct order of magnitude.

TABLE A.1

Some useful physical constants. The last two entries indicate equivalences.

Constant	Symbol and numerical value	Units
Boltzmann's constant	$k = \begin{cases} 1.381 \times 10^{-16} \\ 0.862 \times 10^{-4} \end{cases}$	ergs/K ev/K
Planck's constant	$h = 6.626 \times 10^{-27}$	erg-secs
$\hbar \equiv h/2\pi$	$\hbar = 1.055 \times 10^{-27}$	erg-secs
Electron rest mass	$m_e = 0.911 \times 10^{-27}$	grams
Proton rest mass	$m_p = 1.673 \times 10^{-24}$	grams
Proton charge	$e = \begin{cases} 4.803 \times 10^{-10} \\ 1.602 \times 10^{-19} \end{cases}$	esu coulomb
Speed of light in vacuum	$c = 2.998 \times 10^{10}$	cm/sec
Bohr magneton $(e\hbar/2m_e c)$	$\mu_B = 0.927 \times 10^{-20}$	ergs/gauss
Bohr radius $(\hbar^2/e^2 m_e)$	$a_0 = \begin{cases} 0.529 \times 10^{-8} \\ 0.529 \end{cases}$	cm Ångström
1 electron volt (ev) is equivalent to	1.602×10^{-12}	ergs
1 Ångström (Å) is equivalent to	$1 \quad \times 10^{-8}$	cm

INTEGRALS OF THE FORM $\int_0^\infty e^{-ax^2} x^n \, dx$

Integrals of the form

$$I(n, a) \equiv \int_0^\infty e^{-ax^2} x^n \, dx, \qquad a > 0,$$

with n being zero or a positive integer, occur frequently in statistical mechanics. Only for two values of n, namely $n = 0$ and $n = 1$, need one work out the integrals in detail, for the others are derivable from a simple recurrence relation. Upon differentiating $I(n, a)$ with respect to the parameter a, one finds

$$\frac{\partial}{\partial a} I(n, a) = \int_0^\infty -x^2 e^{-ax^2} x^n \, dx = -I(n + 2, a).$$

Proceeding from either $I(0, a)$ or $I(1, a)$, as the case may require, one may generate the integral for any desired higher value of n by repeated differentiation.

Case of $n=0$

The evaluation of

$$I(0, a) = \int_0^\infty e^{-x^2} \, dx$$

requires a trick. One takes the square of both sides and then writes the second factor on the right in terms of another dummy variable of integration, y:

$$[I(0, a)]^2 = \int_{x=0}^\infty e^{-ax^2} \, dx \int_{y=0}^\infty e^{-ay^2} \, dy$$

$$= \int_{x=0}^\infty \int_{y=0}^\infty e^{-a(x^2+y^2)} \, dx \, dy.$$

The double integral may be viewed as an integration over the upper right quadrant in the x, y plane, as indicated in figure A.1. The integrand depends

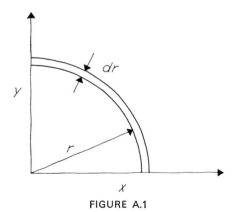

FIGURE A.1

only on $r = \sqrt{(x^2 + y^2)}$, and so is constant on arcs of constant r. The infinitesimal area between an arc of radius r and another arc of radius $r + dr$ is $\frac{1}{4} \times (2\pi r) \times (dr)$. So we may write

$$[I(0, a)]^2 = \int_{r=0}^{\infty} e^{-ar^2} \frac{1}{4} 2\pi r \, dr$$

$$= -\frac{\pi}{4a} \int_{r=0}^{\infty} d(e^{-ar^2}) = -\frac{\pi}{4a} e^{-ar^2} \Big|_{r=0}^{\infty} = +\frac{\pi}{4a}.$$

Upon taking the square root of both sides, we have the desired evaluation:

$$I(0, a) = \frac{1}{2} \sqrt{\frac{\pi}{a}}.$$

Case of $n=1$

The case of $n = 1$ presents no problems:

$$I(1, a) = \int_0^{\infty} e^{-ax^2} x \, dx = \frac{1}{2a}.$$

The values of $I(n, a)$ for higher integral values of n now follow by differentiation with respect to the parameter a. For example, the recurrence relation permits us to calculate $I(2, a)$ as

$$I(2, a) = -\frac{\partial}{\partial a} I(0, a) = -\frac{\partial}{\partial a} \left(\frac{1}{2} \sqrt{\frac{\pi}{a}} \right) = +\frac{\sqrt{\pi}}{4} a^{-3/2}.$$

The explicit results for $n = 0$ to $n = 5$ are as follows.

$$I(a, n) = \int_0^{\infty} e^{-ax^2} x^n \, dx$$

$$I(a, 0) = \frac{1}{2}\sqrt{\pi} a^{-1/2} \qquad I(a, 1) = \frac{1}{2}a^{-1}$$

$$I(a, 2) = \frac{1}{4}\sqrt{\pi} a^{-3/2} \qquad I(a, 3) = \frac{1}{2}a^{-2}$$

$$I(a, 4) = \frac{3}{8}\sqrt{\pi} a^{-5/2} \qquad I(a, 5) = a^{-3}$$

LAGRANGE MULTIPLIERS

The aim here is to show how Lagrange multipliers arise and why they work. The approach is a geometric, "semi-intuitive" one. The results can be established by purely algebraic means, but the algebraic methods fail to give one an idea of why the scheme works at all.

First some preliminaries. Suppose we are set the task of determining the extreme values of some function $f(x, y)$ defined over the entire x, y plane. From the ordinary calculus of minimum-maximum problems, we know that a necessary condition for an extremum at some point (x, y) is that the equations

$$\frac{\partial f}{\partial x} = 0 \quad \text{and} \quad \frac{\partial f}{\partial y} = 0$$

hold at the point in question. When these two equations do hold, the function does not change in value for displacements in either the x or y directions. By introducing unit vectors, $\hat{\mathbf{x}}$ and $\hat{\mathbf{y}}$, along the x-axis and y-axis, respectively, we may write these two equations as a single vector equation. For all points in the plane, let us define the vector $\mathbf{f}'(x, y)$ by

$$\mathbf{f}'(x, y) \equiv \hat{\mathbf{x}} \frac{\partial f}{\partial x} + \hat{\mathbf{y}} \frac{\partial f}{\partial y}.$$

Then a necessary condition for an extremum at a point, when we consider the entire plane, is that the equation

$$\mathbf{f}' = 0$$

hold at that point. In general, the scalar product of $\mathbf{f}'$ with any unit vector gives the rate of change with distance of f in the direction specified by the unit vector. (The vector $\mathbf{f}'$ is usually called the *gradient* of the function f.)

464

With the preceding established as a preliminary result, we may go on to the real problem: the search for an extremum subject to a constraint. Suppose now that we look for the extrema of the function $f(x, y)$ when the values of x and y are constrained to satisfy an equation like

$$g(x, y) = C.$$

Here C is a constant, and $g(x, y)$ is some specified function. Not all values of x, y over the plane are to be considered in looking for the extrema of $f(x, y)$; only those values satisfying the constraint equation are candidates. The constraint equation defines a curve in the x, y plane. So we are, in effect, now looking for the extrema of the function $f(x, y)$ *along* the specified curve in the plane. That $f(x, y)$ may have higher or lower values off the curve than on is now irrelevant. The question is, where does $f(x, y)$ attain its highest or lowest values along the curve?

In terms of the vector $\mathbf{f}'$, the necessary condition for an extremum at a point on the curve—relative to other points on the curve—is that the *component* of $\mathbf{f}'$ along the curve be zero. It is no longer necessary that $\mathbf{f}'$ itself be zero. So long as the component of $\mathbf{f}'$ along the tangent to the curve is zero, the function f will not change (to first order) as one moves to infinitesimally nearby points on the curve.

The tangent to the curve may be found by differentiating the equation for the curve:

$$dg(x, y) = \frac{\partial g}{\partial x}\, dx + \frac{\partial g}{\partial y}\, dy = 0,$$

with the righthand side being zero if the change is from one point on the curve to a nearby point on the curve (dx, dy, further along). This equation relates the increments dx, dy, to the partial derivatives. Then the direction of the tangent $\mathbf{t}$ is determined by the proportionality

$$\mathbf{t} \propto [dx\, \hat{\mathbf{x}} + dy\, \hat{\mathbf{y}}] \propto [\hat{\mathbf{x}} + (dy)/(dx)\hat{\mathbf{y}}]$$

$$\propto \left[\hat{\mathbf{x}} + (-)\left(\frac{\partial g}{\partial x}\right) \middle/ \left(\frac{\partial g}{\partial y}\right)\hat{\mathbf{y}} \right].$$

More relevant for our ultimate scheme of specifying necessary conditions is a vector *perpendicular* to the curve at each point on the curve. The vector $\mathbf{g}'(x, y)$, defined by

$$\mathbf{g}'(x, y) \equiv \hat{\mathbf{x}}\frac{\partial g}{\partial x} + \hat{\mathbf{y}}\frac{\partial g}{\partial y},$$

has this property. Taking the scalar product of the vector $\mathbf{g}'$ with the tangent $\mathbf{t}$ yields zero and thus confirms the claim.

The perpendicular vector **g′** provides another—and more fruitful—way of stating the necessary condition for an extremum along the curve. The first statement was that the component of **f′** along the curve is zero. An *equivalent* statement is that if the "right amount" of a vector perpendicular to the curve is subtracted from **f′**, the result is zero. This also means no component of **f′** *along* the curve.

Turned into equation form, the second statement is this: we look for points (x, y) satisfying

$$g(x, y) = C$$

and such that

$$\mathbf{f}'(x, y) - \lambda \mathbf{g}'(x, y) = 0,$$

where λ is an as-yet undetermined constant that gives the "right amount" of the perpendicular vector **g′** to be subtracted from **f′** at the point. Only if **f′** really does have no component along the curve at a point (x, y) will it be possible for some choice of λ to make the lefthand side of the vector equation zero. This is illustrated in figure B.1.

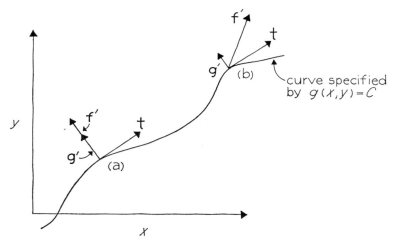

FIGURE B.1

The vector **f′** is defined for all points in the plane, but we care about **f′** only for points along the curve. The vector **g′** is defined for points on the curve and is perpendicular to the tangent **t** to the curve. At point (a) on the curve **f′** has no component along the tangent, indicating that the point (a) satisfies a necessary requirement for f to attain an extreme value at (a) relative to nearby points on the curve. We note that at point (a) it is possible to find a constant λ such that **f′** − λ**g′** is zero at (a), for **f′** and **g′** are parallel. At point (b) the vector **f′** does have a component along the tangent, indicating that point (b) cannot be an extremum. Furthermore, it would be impossible at (b) to find a constant λ such that **f′** − λ**g′** is zero.

When the vector equation is broken up into component form, the preceding means that to find an extremum subject to the constraint we look for values of x, y, *and* λ such that the following three equations are satisfied:

$$\frac{\partial f(x, y)}{\partial x} - \lambda \frac{\partial g(x, y)}{\partial x} = 0;$$

$$\frac{\partial f(x, y)}{\partial y} - \lambda \frac{\partial g(x, y)}{\partial y} = 0;$$

$$g(x, y) - C = 0.$$

With three equations for three unknowns, we can solve the problem if there is a solution at all.

The geometric analysis provides us with the equations that we should write down when we want to find the extrema of a function $f(x, y)$ subject to the constraint equation, $g(x, y) = C$, restricting the values of x and y to be considered. We can now set up a tidy prescription for writing down those equations automatically. Start off by writing a function $\Lambda(x, y, \lambda)$ defined by

$$\Lambda(x, y, \lambda) \equiv f(x, y) + \lambda[C - g(x, y)].$$

The parameter λ is called a *Lagrange multiplier.* Then demand zero values for *all* partial derivatives of $\Lambda(x, y, \lambda)$ as a necessary condition for an extremum subject to the constraint:

$$\frac{\partial \Lambda}{\partial x} = \frac{\partial f}{\partial x} - \lambda \frac{\partial g}{\partial x} = 0,$$

$$\frac{\partial \Lambda}{\partial y} = \frac{\partial f}{\partial y} - \lambda \frac{\partial g}{\partial y} = 0,$$

$$\frac{\partial \Lambda}{\partial \lambda} = C - g = 0.$$

These equations are identical to the geometric ones. The extension to higher dimensions and more constraints taxes one's power of geometric visualization, but this establishes the Lagrange multiplier technique.

It should be borne in mind that the conditions just established are only *necessary* conditions for an extremum. They are not *sufficient* conditions. One must still test a point that meets the necessary conditions to see whether it is indeed an extremum, but usually only a few points require this extra attention.

THE MAGNETIC-MOMENT OPERATOR

In this appendix we will trace in detail the connection between the magnetic moment of a system and the expression $-\partial E_j/\partial H_z$. The connection is established in two stages, by confirming the two following contentions.

1. The z-component of the total magnetic-moment operator may be written as $-\partial \mathcal{JC}/\partial H_z$. As previously, the symbol $\mathcal{JC}$ denotes the energy operator for the system, and H_z denotes the z-component of a uniform external magnetic field.

2. For an energy eigenstate ψ_j the quantum-mechanical expectation value of the above operator is $-\partial E_j/\partial H_z$.

We begin by examining a system consisting of a single particle of charge q and mass m. The particle is in a uniform external magnetic field, and is acted on also by electric forces derivable from an electric scalar potential φ. The first task is to construct the energy operator $\mathcal{JC}$ for the system. It should consist of a kinetic energy part and a part representing the electric potential energy. The *classical* (nonrelativistic) energy expression would be

$$E = \tfrac{1}{2}m\,\mathbf{v}\cdot\mathbf{v} + q\varphi. \tag{C.1}$$

The transcription to a quantum-mechanical operator requires that we express a velocity $\mathbf{v}$ in terms of a momentum $\mathbf{p}$. A generalized momentum, different in some physical situations from merely mass times velocity, must be used. Remarkably, it is through the connection between $\mathbf{v}$ and $\mathbf{p}$ that the magnetic field enters the energy expression.

A magnetic field can always be calculated as the curl of a vector potential $\mathbf{A}$:

$$\mathbf{H} = \text{curl } \mathbf{A}.$$

The vector potential plays an essential role in the development. Indeed, the proper (nonrelativistic) connection between the generalized momentum **p** and the velocity **v** for the single particle is

$$\mathbf{p} = m\mathbf{v} + \frac{q}{c}\,\mathbf{A}. \tag{C.2}$$

Though seemingly a strange result, this is derivable in a standard way from a generalized definition of momentum.* Special relativity theory provides a rather direct argument for the relation; although one must concede that it is only a plausibility argument, it is worth pursuing.

In the absence of all forces, the full relativistic energy of a particle is $mc^2(1 - v^2/c^2)^{-1/2}$. This contains both the contribution from the "rest energy" mc^2 and the relativistic kinetic energy:

$$\frac{mc^2}{\sqrt{1 - v^2/c^2}} = mc^2 + (\tfrac{1}{2}mv^2 + \ldots) \quad \text{when} \quad \frac{v^2}{c^2} \ll 1.$$

When electric and magnetic fields are present, the energy expression for the particle must be augmented by the addition of the electric potential energy:

$$E_{\text{rel}} = \frac{mc^2}{\sqrt{1 - v^2/c^2}} + q\varphi.$$

We may look on this as a generalized energy.

Now we proceed in a parallel fashion with momentum. In the absence of forces, the relativistic momentum is merely $m\mathbf{v}(1 - v^2/c^2)^{-1/2}$. Should something be added to produce a generalized momentum when electromagnetic fields are present? The quantities

$$\left\{ \frac{mc}{\sqrt{1 - v^2/c^2}}, \frac{m\mathbf{v}}{\sqrt{1 - v^2/c^2}} \right\}$$

transform as components of a four-dimensional vector under Lorentz transformations. So do the quantities $\{\varphi, \mathbf{A}\}$, the scalar and vector potentials of electromagnetism. (The last claim can be made plausible by noting that φ depends on external charge density and **A** on external current density. Since current density is rather like charge density times velocity, it is plausible that those densities form the components of a four-dimensional vector, just

---

* Herbert Goldstein, *Classical Mechanics* (Reading, Mass.: Addison-Wesley, 1956), Sections 1.5 and 7.3.

as do energy and momentum. Indeed, they do, and so it is reasonable that φ and $\mathbf{A}$ do likewise. Now back to the main argument.) If we add $q\varphi$ to $mc^2(1 - v^2/c^2)^{-1/2}$ in order to get a generalized relativistic energy, then symmetry indicates that we should add $(q/c)\mathbf{A}$ to $m\mathbf{v}(1 - v^2/c^2)^{-1/2}$ to get a generalized relativistic momentum:

$$\mathbf{P}_{\text{rel}} = \frac{m\mathbf{v}}{\sqrt{1 - v^2/c^2}} + \frac{q}{c}\mathbf{A}.$$

When the ratio v^2/c^2 is small relative to one, this reduces to the connection stated in equation (C.2). The expression is known as the *generalized* or *canonical momentum*. Just as the nonrelativistic form of the generalized energy is used as "the energy" in the transcription to quantum-mechanical operator form, so should the nonrelativistic form of the generalized momentum be used as "the momentum."

Now we are in a position to construct the energy operator for the system. Upon taking over the classical expression in equation (C.1) and substituting for the velocity in terms of the momentum and vector potential, we arrive at the energy operator:

$$\mathcal{H} = \frac{1}{2m}\left(\mathbf{p} - \frac{q}{c}\mathbf{A}\right) \cdot \left(\mathbf{p} - \frac{q}{c}\mathbf{A}\right) + q\varphi. \tag{C.3}$$

The magnetic field enters through the vector potential.

Thus far there has been no restriction on $\mathbf{A}$. It could be the vector potential for any magnetic field. We specialize now to an expression for $\mathbf{A}$ that represents a constant uniform magnetic field $\mathbf{H} = H_z\hat{\mathbf{z}}$. A calculation of the curl indicates that the expression

$$\mathbf{A} = -\tfrac{1}{2}yH_z\hat{\mathbf{x}} + \tfrac{1}{2}x H_z\hat{\mathbf{y}}$$

will do. When we substitute this into equation (C.3) and then collect terms with like powers of H_z, we find

$$\mathcal{H} = \frac{1}{2m}\mathbf{p} \cdot \mathbf{p} - \frac{q}{2mc}(xp_y - yp_x)H_z$$

$$+ \frac{q^2}{8mc^2}(x^2 + y^2)H_z{}^2 + q\varphi.$$

Finally, upon taking the partial derivative with respect to H_z and applying a minus sign, we arrive at

$$-\frac{\partial\mathcal{H}}{\partial H_z} = \frac{q}{2mc}(xp_y - yp_x) - \frac{q^2}{4mc^2}(x^2 + y^2)H_z. \tag{C.4}$$

The first term is proportional to the angular momentum along the z-axis, and gives the paramagnetic contribution to the magnetic moment along that axis. The second term is independent of the sign of the charge q and has a sign opposite to that of H_z. It gives the diamagnetic contribution. We do indeed have the z-component of the magnetic moment operator for a spinless charged particle.

> If more were needed for the identification, one could " undo " the quantum-mechanical expression by writing the momentum components in terms of velocity and vector potential components. The ensuing (effectively classical) expression would be $(q/2c)(xv_y - yv_x)$. If a classical charged particle were moving in a circular orbit in the x, y plane, that expression would equal the electric current times the orbit area over c. That product is perhaps a more familiar expression for a magnetic moment and should confirm the identification made above.

The permanent magnetic moment associated with a spin, such as for an electron, presents no problem. Such a moment $\boldsymbol{\mu}$ will enter the energy operator in the form $-\boldsymbol{\mu} \cdot \mathbf{H}$. The differentiation operation will yield the appropriate component of that moment, too. The extension to many particles entails nothing more than a summing of many individual contributions.

The first of the two contentions is confirmed. We may summarize and write

$$\begin{pmatrix} \text{the } z\text{-component of} \\ \text{the total magnetic} \\ \text{moment operator} \end{pmatrix} = -\frac{\partial \mathcal{H}}{\partial H_z}. \tag{C.5}$$

The energy operator $\mathcal{H}$ refers to the entire system. (Analogous statements can be made for the x-component and y-component operators.)

With the magnetic-moment operator in hand, we can compute the quantum-mechanical expectation value of the moment in any state. Of primary concern are the energy eigenstates. So let us start by writing

$$\begin{pmatrix} \text{the quantum-mechanical expecta-} \\ \text{tion value of the } z\text{-component} \\ \text{of the magnetic moment in an} \\ \text{energy eigenstate } \psi_j \end{pmatrix} = \int \psi_j{}^* \left(-\frac{\partial \mathcal{H}}{\partial H_z} \right) \psi_j. \tag{C.6}$$

For the sake of brevity, no indication, other than that provided by the integral sign, is given of the integration over the variables on which the wave function depends. As the first move toward confirming the second contention, we pull the partial differentiation with respect to H_z out in front.

To compensate for the differentiations of ψ_j and of ψ_j^* implied by this, we must add two counterterms. As a whole, the step yields a more useful version of the righthand side of equation (C.6):

$$-\frac{\partial}{\partial H_z}\int \psi_j^* \mathcal{H}\psi_j + \int \frac{\partial \psi_j^*}{\partial H_z}\mathcal{H}\psi_j + \int \psi_j^* \mathcal{H} \frac{\partial \psi_j}{\partial H_z}.$$

The first integral is the expectation value of the energy in an energy eigenstate; hence that integral yields precisely E_j. The wave function ψ_j will almost certainly depend upon H_z, and therefore the two counterterms will not individually be zero. We can, however, show that their sum must be zero. Let $\mathcal{H}$ operate to the right in the first counterterm and to the left in the second; in each case the operator yields E_j times the original wave function. Thus the sum of the two counterterms may be written as

$$E_j \int \frac{\partial \psi_j^*}{\partial H_z}\psi_j + E_j \int \psi_j^* \frac{\partial \psi_j}{\partial H_z} = E_j \frac{\partial}{\partial H_z}\int \psi_j^* \psi_j = 0.$$

The last integral is merely unity, for the wave function is properly normalized to unity. From this follows the zero value of the derivative. The upshot is the equation

$$\int \psi_j^*\left(-\frac{\partial \mathcal{H}}{\partial H_z}\right)\psi_j = -\frac{\partial E_j}{\partial H_z}. \tag{C.7}$$

Upon tracing matters back from equation (C.7) to equation (C.6), we reach the conclusion of this appendix:

The expression $-\partial E_j / \partial H_z$ gives the quantum-mechanical expectation value of the z-component of the total magnetic moment for the energy eigenstate ψ_j.

THE CONNECTION WITH THERMODYNAMICS

This appendix is written for the reader who is already familiar with thermodynamics and who would like to see the connection between the theoretical entities of that discipline and those of statistical mechanics.

The conceptual bases of the two disciplines are sufficiently different that one can expect only a close *correspondence* between the theoretical quantities, not a true equality or identity. To clarify this, let us look at the energy of an isolated macroscopic system in thermal equilibrium at a known temperature. Thermodynamics ascribes to the system a definite energy. Let us write that as E_{th}. (It is often called the "internal energy.") In statistical mechanics one would concede that one doesn't know the precise energy of the system. The data given are inadequate. One can, however, assign probabilities for various values of the energy and then form an estimate of the system's energy. One would certainly employ the canonical probability distribution, and would estimate the energy by computing the expectation value $\langle E \rangle$. At the same time one could calculate a root mean square estimate of the anticipated deviations from $\langle E \rangle$. If the system is macroscopic, as is typical of the systems to which thermodynamics is applied, one would find that the anticipated deviations are exceedingly small relative to $\langle E \rangle$. Indeed, a probability distribution for various values of the energy (formed from the canonical probability distribution by grouping states with nearly the same energy) would show a fantastically sharp peak. One could be quite confident that the system has an energy very close to $\langle E \rangle$. The sharpness of such a peak provides the justification for asserting a close *correspondence* between $\langle E \rangle$ and E_{th}. Let us write this as

$$\langle E \rangle \cong E_{th}, \tag{D.1}$$

473

with $\triangleq$ indicating "corresponds to." The two energies are thought of quite differently; they are *not identical,* but for practical calculations they are *numerically equivalent.* Thus the symbol $\triangleq$ denotes "not conceptually, but numerically, equivalent."

An analogous correspondence holds between pressure in thermodynamics and estimated pressure in statistical mechanics:

$$\langle p \rangle \triangleq p_{\text{th}}. \tag{D.2}$$

The list can be extended to total magnetic moment and to other variables having a "mechanical" definition.

The connection between thermodynamic entropy and information-theory entropy is more subtle. Indeed, the failure to preserve a *distinction* is often a stumbling block on the path to an appreciation of both. Let us suppose that the canonical distribution is appropriate for describing the system statistically, and let us write the information-theory entropy as

$$(S_I)_{\text{c.p.d.}} = \left(-k \sum_j P_j \ln P_j \right)_{\text{c.p.d.}}$$
$$= k \Big(\beta \langle E \rangle + \ln Z \Big). \tag{D.3}$$

A subscript c.p.d. is appended so we will not forget that we are evaluating the general expression for S_I with the canonical probability distribution. The last form is taken from equation (7.6.4). Now suppose we slowly change the volume of the system by a small amount dV. Also, we transfer a small amount of energy to the system as heat (from a body of nearly the same temperature). In Section 7.4 we noted that it is at least reasonable to continue to use the canonical distribution to describe the system statistically. (The volume change will induce a slight change in the energy eigenstates and eigenvalues, and a slight change in assigned temperature may be required.) On the assumption that a continued use of the canonical distribution is appropriate, let us calculate the slight change in the information-theory entropy. We form the differential of equation (D.3):

$$d(S_I)_{\text{c.p.d.}} = k \left(d\beta \langle E \rangle + \beta \, d\langle E \rangle + \frac{\partial \ln Z}{\partial \beta} \, d\beta + \frac{\partial \ln Z}{\partial V} \, dV \right)$$
$$= \frac{1}{T} \Big(d\langle E \rangle + \langle p \rangle \, dV \Big). \tag{D.4}$$

The step to the second line follows when (1) we note that the first and third terms cancel, as follows from equation (7.1.4), and (2) we recognize the derivative in the fourth term as $\beta \langle p \rangle$, as follows from equation (7.2.4).

Now we prepare for a comparison of equation (D.4) with the differential relation for the thermodynamic entropy S_{th}. The latter would be

$$dS_{th} = \frac{\text{(energy transferred as heat)}}{T_{th}}$$

$$= \frac{1}{T_{th}}\left(dE_{th} + p_{th}\,dV\right). \qquad \text{(D.5)}$$

The temperature in thermodynamics has been written as T_{th}. The first line indicates that T_{th} is the integrating factor for the small (infinitesimal, really) amount of energy transferred as heat. The step to the second line follows from the first law of thermodynamics, which amounts to energy conservation.

The quantities in parentheses in equations (D.4, D.5) are numerically equivalent. This follows from the correspondences established earlier in this appendix. So it is both appealing and reasonable to infer two additional correspondences:

$$T \cong T_{th} \qquad \text{(D.6a)}$$

and

$$d(S_I)_{c.p.d.} \cong dS_{th}. \qquad \text{(D.6b)}$$

Little need be said about the temperature correspondence. We introduced temperature into statistical mechanics with the aid of a dilute-gas thermometer, and then examined the deeper meaning in Section 7.5. The scale was extended in Section 11.5 with the explicit aid of the canonical distribution. Now we find that temperature so specified is numerically equivalent to temperature as defined in thermodynamics. Fine.

The second correspondence is more provocative. Since it asserts a correspondence between differentials of the thermodynamic entropy and the information-theory entropy when the latter refers to the canonical probability distribution, one is tempted to look for a correspondence between the quantities themselves. There is a minor problem, for the second law of thermodynamics deals only with entropy differences. The zero of thermodynamic entropy, or the thermodynamic entropy for any one condition of the physical system, is left free and unspecified. There is, however, a convention for the thermodynamic entropy of a pure substance in the limit of zero temperature: it is to be taken to be zero. Let us adopt this convention. Then we must ask about the numerical value of $(S_I)_{c.p.d.}$ at absolute zero. If the ground state of the system is nondegenerate, that will be zero, as the analysis in Section 7.6 showed. Except in possible anomalous cases of high degeneracy for the lowest energy level, $(S_I)_{\text{c.p.d.}}$ at absolute zero will be small enough (if not precisely zero) so that we may assert a numerical equivalence between it and the conventional value of S_{th} at absolute zero. On the basis of the convention and this analysis, we may extend the correspondence between differentials in equation (D.6b) to a correspondence between the quantities themselves:

$$(S_I)_{\text{c.p.d.}} \cong S_{\text{th}}. \tag{D.7}$$

One point is essential: it is S_I *when evaluated with the canonical probability distribution*, not S_I in general, that corresponds to S_{th}. The overwhelming importance of this point will become clear in a moment.

THE INCREASE OF THERMODYNAMIC ENTROPY

In Section 11.6 we examined the evolution of a system when an external parameter is changed rapidly. The correspondences which we have now established enable us to construct a demonstration, based on statistical mechanics, that the thermodynamic entropy will increase in such a process. Let us recall the process: at time t_1 the system is in equilibrium and is described with the canonical distribution; between times t_1 and t_2 an external parameter is changed and thus the energy operator is time-dependent; for times later than t_2 the energy operator is independent of time, and the system settles down to a new equilibrium (at least effectively). The demonstration is based on four elements, all already established in one place or another. They will be listed and briefly commented on.

$$S_I(t_1) \cong S_{\text{th}} \text{ (when } t = t_1). \tag{D.8}$$

The correspondence at time t_1 holds because at that time we describe the system with the canonical probability distribution. The quantity $S_I(t_1)$ is calculated with a canonical distribution, and so equation (D.7) implies the correspondence asserted here.

$$S_I(t) = S_I(t_1) \qquad \text{for } t \geqslant t_1. \tag{D.9}$$

This is simply equation (11.1.6). The intuitive meaning is this: since we are able to follow the evolution of the physical system with the Schrödinger equation, we neither gain nor lose information about it.

$$\left(\begin{array}{l} (S_I)_{\text{c.p.d.}} \text{ of the c.p.d.} \\ \text{adopted when } t \gg t_2 \end{array} \right) \geqslant S_I(t). \tag{D.10}$$

After the system has settled down to equilibrium at times $t \gg t_2$, we may adopt a new canonical probability distribution for estimating macroscopic properties and many microscopic properties. The analysis in Section 11.6 led to the above relation, with the equality sign holding only if the change in external parameter was exceedingly slow.

$$\begin{pmatrix} (S_I)_{\text{c.p.d.}} \text{ of the c.p.d.} \\ \text{adopted when } t \gg t_2 \end{pmatrix} \hat{=} S_{\text{th}} \text{ (when } t \gg t_2). \tag{D.11}$$

This states a correspondence, based on equation (D.7), after the system has settled down (at least effectively) to its new equilibrium. Only then is one again permitted in thermodynamics to speak of the thermodynamic entropy.

Now it is merely a matter of running one's eye over the relations, (D.8) to (D.11), in order to establish the claimed inequality:

$$S_{\text{th}} \text{ (when } t \gg t_2) \geqslant S_{\text{th}} \text{ (when } t = t_1). \tag{D.12}$$

Only if the change in external parameter was made exceedingly slowly will the equality sign hold. For a fast change we are assured that the inequality will prevail. Note well that we have used the *constancy* of S_I for the evolved probability distribution, in element (D.9), to demonstrate an *increase* in S_{th}. This bears out the claim made in Section 7.6 that wholesale identification of S_I with S_{th} is unjustifiable and would lead to grief. The limited correspondence is that given in equation (D.7).

> The immediately preceding statement is a little too restrictive. Sometimes probability distributions different from the canonical distribution are used in statistical mechanics to describe systems in equilibrium. They are based on different hypotheses, on different data given about the system. (Problem 4.5 developed one, the microcanonical distribution.) One can generally establish a correspondence between S_I for them and S_{th}. A more precise statement would be that only for probability distributions designed to describe equilibrium can a correspondence between S_I and S_{th} be established.

What we have here is something akin to a proof, based on statistical mechanics, of the second law of thermodynamics. The demonstration is simple—and somewhat deceptively so. When one starts with statistical mechanics, the property truly difficult to establish theoretically is the (effective) attainment of equilibrium at some reasonable time after the external parameter has ceased changing. Somewhat more technically: one needs to show that, for purposes of estimating macroscopic properties, one may eventually replace the probability distribution that evolves rigorously according to the Schrödinger equation with a canonical distribution (giving the same total energy estimate). This is indeed a difficult theoretical task, but the work of Prigogine and others (cited in Section 11.6) strongly indicates that the replacement is permissible. The replacement is crucial, for it is in this step that one throws away information about the system. The inequality stated as element (D.10) makes this clear mathematically. We can understand this intuitively as well: the evolved probability distribution can and does contain information about the system's past history that the newly adopted canonical distribution cannot possibly contain. In the present context the increase in thermodynamic entropy arises because information is wilfully thrown away.

PARTITION FUNCTION AND FREE ENERGY

Once the correspondence between $(S_I)_{c.p.d.}$ and S_{th} has been established, the correspondences with the host of other thermodynamic functions are readily established. Only that with the Helmholtz free energy need concern us here; the others may be derived in an analogous fashion. We begin with the correspondence

$$\langle E \rangle - T(S_I)_{c.p.d.} \triangleq E_{th} - T_{th} S_{th} \equiv F_{th},$$

in which the Helmholtz free energy appears on the right, first in terms of the more basic thermodynamic functions and then abbreviated as F_{th}. The left-hand side can be written in terms of the partition function. If one glances back at equation (D.3) and mentally rearranges that equation, one can confirm the correspondence

$$-kT \ln Z \triangleq F_{th}. \qquad (D.13)$$

The logarithm of the partition function and F_{th} play analogous roles, their partial derivatives providing a host of useful physical quantities.

ANSWERS TO SELECTED PROBLEMS

CHAPTER 2

2.4. $(\frac{1}{2})^{20}$; $(20!)(\frac{1}{2})^{20}/(10!)^2$; the second case includes the first.

2.5. (b) $\langle n \rangle = 7$; (c) $\Delta n = 2.4$.

2.6. (a) yes; yes; 1; (c) as in (a); $P(A_1 \mid H^*) = 2 - C$ is reasonable.

2.7. (b) $(\frac{1}{2})^N$; same; (c) zero; symmetry; (d) $lN^{1/2}$.

CHAPTER 3

3.3. $\pm(2)^{-1/2}$.

3.4. Set (1).

CHAPTER 4

4.1. 362 K.

4.3. For macroscopic volume and reasonable temperature, yes.

4.4. $P_j \propto \exp(-\beta E_j)$ for $E_j \leqslant 1,000\tilde{E}$; no; length of tail is largely irrelevant.

CHAPTER 5

5.1. $N = 25,000$; volume $= O(10^{-17})$ cm^3.

5.2. (a) $e^{-\beta\varepsilon}/(2e^{-\beta\varepsilon} + 1)$; $T \to \infty$;

(b) $2\varepsilon e^{-\beta\varepsilon}/(2e^{-\beta\varepsilon} + 1)$; (c) zero.

CHAPTER 6

6.1. (a) $A = s_0^{-2}$; (b) $2s_0$; (c) s_0; (d) $7e^{-6} = 0.017$.

6.2. (a) $\langle v \rangle = (8kT/\pi m)^{1/2}$; (b) I think not.

6.3. 4×10^4 cm/sec; no, for collisions lead to zigzags.

6.4. $(\beta m/2\pi)^{1/2} \exp[-\frac{1}{2}\beta m v_x^2]dv_x$; Gaussian profile; zero; kT/m; symmetry.

6.5. (a) $\langle v \rangle = v_0$; (b) $v_0(kT/m)^{1/2}/c$; (c) $I(v) = \text{const} \exp[-(mc^2/2kTv_0^2)$ $(v - v_0)^2]$; reasonable but not accurate because of nonrelativistic treatment.

6.6. $2^{1/2} \times (3kT/m)^{1/2}$.

6.7. (a) $\mathcal{P}(\varepsilon) = 2\pi^{-1/2}\beta^{3/2}e^{-\beta\varepsilon}\varepsilon^{1/2}$; (c) $\varepsilon_{mp} = \frac{1}{2}kT$; no; no.

6.8. $2kT$; energetic particles are more likely to escape the oven.

6.9. (c) $\mathcal{P}(z)\,dz = \beta mg(1 - e^{-\beta mgL})^{-1}e^{-\beta mgz}dz$;
   (e) $\langle mgz \rangle = [1 - (1 + \beta mgL)e^{-\beta mgL}]/\beta(1 - e^{-\beta mgL})$; no; equipartition theorem does not apply; roughly $(mg) \times (\frac{1}{2}L)$; you should.

6.10. (a) $\mathcal{P}(\theta) = \frac{1}{2}\beta\mu H(\sinh \beta\mu H)^{-1}e^{\beta\mu H \cos \theta} \sin \theta$;
   (b) $\langle \mu_z \rangle = \mu[\coth \beta\mu H - (\beta\mu H)^{-1}]$;
   (c) $\frac{1}{3}\mu(\mu H/kT)$ when $\mu H/kT \ll 1$.

6.15. $\Delta n/\langle n \rangle = N^{-1/2}[(1 - \tilde{p})/\tilde{p}]^{1/2}$, where $\tilde{p} = \mathcal{P}_b(v)dv$; roughly, $10^{-2} \leqslant \tilde{p} \leqslant 10^{-1}$, and so the ratio is, roughly, 10^{-7} or smaller.

CHAPTER 7

7.1. No; E^*; no; no; factor of $\exp(-\beta E^*)$; $-\beta E^*$.

7.2. $\langle E \rangle = \frac{3}{2}NkT - aN^2/V$; for $\langle p \rangle$ see equation (8.3.1).

7.3. (a) $Z = e^{-\beta h\nu/2}/(1 - e^{-\beta h\nu})$; (b) $T \geqslant h\nu/k \ln 2$; 4,800 K; negligible vibrational excitation under room conditions.

7.4. $\langle KE \rangle = \frac{1}{2}kT$ by equipartition theorem; $\langle U \rangle = kT/2n$.

7.8. Yes.

CHAPTER 8

8.1. (a) $\Delta E/\langle E \rangle = (\frac{2}{3})^{1/2} N^{-1/2}$; (b) set $N = 1$ above, (c) reinforces conclusions of section 5.2.

8.2. Order 0.01 per cent.

8.3. $\tilde{u} \simeq 0.006$ electron volts; $r' \simeq 3$, $r'' \simeq 6$ Ångströms.

8.4. $a \simeq 5 \times 10^{-36}$ erg-cm^3; $b \simeq 9 \times 10^{-23}$ cm^3; poor at low T.

8.5. (a) $\frac{3}{2}NkT + (N^2/2V)\int_0^\infty ue^{-\beta u}4\pi r^2 dr$; (b) no; interactions influence $\langle E \rangle$ and $\langle p \rangle$ differently; (c) infinite repulsive core will appear only through density dependence of attractive contribution to $\langle E \rangle$ in higher order in density expansion.

8.6. $\langle E \rangle = \frac{3}{2}NkT - (2\pi N^2/3V)(r''^3 - r'^3)\tilde{u}\exp(\tilde{u}/kT)$; disaster at low temperature; step on p. 246 may not be taken.

8.7. (a) $C = \frac{3}{2}Nk + \langle(U - \langle U \rangle)^2\rangle/kT^2$; (b) *classical* statistical mechanics fails at low temperatures.

CHAPTER 9

9.4. (c) $\Delta p/\langle p\rangle = (\tfrac{2}{3}N)^{1/2}[1 \mp 5N\lambda^3/16(2)^{1/2}(2S + 1)V]$.

9.6. See equations (5.1.9, 11.4.4).

CHAPTER 10

10.1. $(3/\pi)^{2/3}(3h^2/40m)N(N/V)^{2/3}$; through the required antisymmetry of the wave function for the entire system, the fermions "know" about one another (despite the neglect of all real forces between them).

10.2. $E_1 = h^2N^2/8\pi mA$; $C \simeq 4\pi^3 mAk^2 T/3h^2$.

10.3. $T^{1/2} \exp[-\varepsilon^*/kT]$.

10.6. 3.3×10^{-6}; 2.5×10^7.

10.7. (b) $h\nu/kT = 2.8$ at peak.

10.8. (d) $v_D = (\tfrac{3}{4}\pi)^{1/3}(N/V)^{1/3}c_s$; it should; (f) order 200 to 400 K; (h) $C = (12\pi^4/5)Nk(T/\Theta_D)^3$; see equation (11.4.5).

10.9. $T^{3/2}$; no; failure to include interparticle forces.

10.10. $T_F(0) = N\tilde\varepsilon/k = 7.2 \times 10^4$ K implies that 300 K is low temperature; $\langle E\rangle \simeq \tfrac{1}{2}N^2\tilde\varepsilon + (\pi kT)^2/6\tilde\varepsilon$.

CHAPTER 11

11.1. $S_I/k = \tfrac{3}{2}N \ln (TV^{2/3}) - (2\pi r_0{}^3 N^2/3V)[1 + u_0{}^2/6k^2 T^2] + \text{const}$; the repulsive forces lead to more temperature drop; the attractive forces, to less.

11.2. (b) $\langle N\rangle = 8\pi V(kT/ch)^3 \int_0^\infty x^2(e^x - 1)^{-1}dx$; (c) $\langle E\rangle/\langle N\rangle = 2.7kT$; 0.1 ev, 1.4 ev.

11.4. A dilute gas is certainly less difficult to analyze.

CHAPTER 12

12.1. The probability, given our information, that measurement will yield the eigenvalues characterizing that state used to form the diagonal matrix element.

12.7. (b) $T < \tfrac{1}{3}\tau\varphi$; (c) $\langle M_z\rangle \propto T(\tfrac{1}{3}\tau\varphi - T)^{1/2}$; vertical slope.

12.8. (a) $-Nk\tau^2 Q/6T$; (b) T^{-2} behavior will emerge in high-temperature limit if system has a finite number of energy eigenstates and the latter have finite energies.

INDEX